T0271921

Membrane and Membrane-Based Processes for Wastewater Treatment

The proposed book mainly sorts out emerging and burning issues faced day to day by municipal and industrial wastewater treatments. It also provides a comprehensive view of recent advances in hybrid treatment technologies for wastewater treatment and addresses the current limitations and challenges of applying these tools to wastewater treatment systems. This book gives insight into recent developments in membrane technology for wastewater treatment. Industrial wastewater contains a large variety of compounds, such as heavy metals, salts and nutrients, which makes its treatment challenging. Thus, the use of conventional water treatment methods is not always effective. In this sense, membrane-based hybrid processes have emerged as a promising technology to treat complex industrial wastewater. The present book analyses and discusses the potential of membrane-based hybrid processes for the treatment of complex industrial wastewater along with the recovery of valuable compounds and water reutilization. In addition, recent and future trends in membrane technology are highlighted.

FEATURES

1. The properties, mechanisms, advantages, limitations and promising solutions of different types of membrane technologies are discussed.
2. The optimization of process parameters is addressed.
3. The performance of different membranes is described.
4. The potential of nanotechnology to improve the treatment efficiency of wastewater treatment plants is presented.
5. The application of membrane and membrane-based hybrid treatment technologies for wastewater treatment is covered.

Wastewater Treatment and Research

Series Editor: Maulin P. Shah

Wastewater Treatment: Molecular Tools, Techniques, and Applications
Maulin P. Shah

Advanced Oxidation Processes for Wastewater Treatment: An Innovative Approach
Maulin P. Shah, Sweta Parimita Bera, Hiral Borasiya and Gunay Yildiz Tore

Emerging Technologies in Wastewater Treatment
Maulin P. Shah

Bio-Nano Filtration in Industrial Effluent Treatment: Advanced and Innovative Approaches
Maulin P. Shah

Membrane and Membrane-Based Processes for Wastewater Treatment
Maulin P. Shah

Phycoremediation Processes in Industrial Wastewater Treatment
Maulin P. Shah

For more information, please visit: www.routledge.com/Wastewater-Treatment-and-Research/book-series/WASTEWATER

Membrane and Membrane-Based Processes for Wastewater Treatment

Edited by Maulin P. Shah

CRC Press
Taylor & Francis Group
Boca Raton London New York

CRC Press is an imprint of the
Taylor & Francis Group, an **informa** business

First edition published 2023
by CRC Press
6000 Broken Sound Parkway NW, Suite 300, Boca Raton, FL 33487–2742

and by CRC Press
4 Park Square, Milton Park, Abingdon, Oxon, OX14 4RN

CRC Press is an imprint of Taylor & Francis Group, LLC

Library of Congress Cataloging-in-Publication Data
Names: Shah, Maulin P., editor.
Title: Membrane and membrane-based processes for wastewater treatment / edited by Maulin P. Shah.
Description: Boca Raton : CRC Press, 2023. | Series: Wastewater treatment and research |
    Includes bibliographical references.
Identifiers: LCCN 2022048872 (print) | LCCN 2022048873 (ebook) | ISBN 9780367759841 (hardback) |
    ISBN 9780367759858 (paperback) | ISBN 9781003165019 (ebook)
Subjects: LCSH: Water—Purification—Membrane filtration. | Membrane separation.
Classification: LCC TD442.5 .M425 2023 (print) | LCC TD442.5 (ebook) | DDC 628.1/64—dc23/eng/20221205
LC record available at https://lccn.loc.gov/2022048872
LC ebook record available at https://lccn.loc.gov/2022048873

ISBN: 978-0-367-75984-1 (hbk)
ISBN: 978-0-367-75985-8 (pbk)
ISBN: 978-1-003-16501-9 (ebk)

DOI: 10.1201/9781003165019

Typeset in Times
by Apex CoVantage, LLC

Contents

Editor Biography

Maulin P. Shah is very interested in genetic adaptation processes in bacteria, the mechanisms by which they deal with toxic substances, how they react to pollution in general and how we can apply microbial processes in a useful way (like bacterial bioreporters). One of his major interests is studying how bacteria evolve and adapt to use organic pollutants as novel growth substrates. Bacteria with new degradation capabilities are often selected in polluted environments and have accumulated small (mutations) and large genetic changes (transpositions, recombination, horizontally transferred elements). His work has been focused to assess the impact of industrial pollution on microbial diversity of wastewater following cultivation-dependent and cultivation-independent analysis. He has more than 280 research publications in highly reputed national and international journals. He is an editorial board member of *CLEAN-Soil, Air, Water* (Wiley), *Current Pollution Reports* (Springer Nature), *Environmental Technology & Innovation* (Elsevier), the *Journal of Biotechnology & Biotechnological Equipment* (Taylor & Francis), *Current Microbiology* (Springer Nature), *Ecotoxicology (Microbial Ecotoxicology)* (Springer Nature), *Geomicrobiology* (Taylor & Francis), *Applied Water Science* (Springer Nature), *Archives of Microbiology* (Springer), the *Journal of Applied Microbiology* (Wiley), *Letters in Applied Microbiology* (Wiley), *Green Technology, Resilience and Sustainability* (Springer), *Biomass Conversion and Biorefinery* (Springer), the *Journal of Basic Microbiology* (Wiley), *Energy Nexus* (Elsevier), *e Prime* (Elsevier), *IET Nanobiotechnology* (Wiley), *Cleaner and Circular Bioeconomy* (Elsevier) and *International Microbiology* (Springer) and has edited 75 books in wastewater microbiology and industrial wastewater treatment. He has edited 20 special issues on the theme of industrial wastewater treatment and research in high-impact-factor journals with Elsevier, Springer, Wiley and Taylor & Francis.

Contributors

J. Anandkumar
National Institute of Technology Raipur
Chhattisgarh, India

Ritwik Banerjee
University of Engineering & Management
Kolkata, India

Anjali Bishnoi
LD College of Engineering
Gujarat, India

Sumalatha Boddu
Vignan's Foundation for Science
Andhra Pradesh, India

Mamta Chahar
Nalanda College of Engineering
Bihar, India

Anusha Chandra
Vignan's Foundation for Science
Andhra Pradesh, India

Parmesh Kumar Chaudhari
National Institute of Technology
Raipur, India

Atif Aziz Chowdhury
University of Kalyani
Kalyani, West Bengal, India

Pranjal P. Das
Indian Institute of Technology Guwahati
Assam, India

Ankita Dey
Maulana Abul Kalam Azad University of
   Technology
West Bengal, India

Sandeep Dharmadhikari
Guru Ghasidas Vishwavidyalaya (A Central
   University)
Bilaspur, India

Prangan Duarah
Indian Institute of Technology Guwahati
Assam, India

Sivakumar Durairaj
Guru Ghasidas Vishwavidyalaya (A Central
   University)
Bilaspur, India

Sayantani Garai
University of Engineering & Management
Kolkata, India

Sougata Ghosh
RK University
Rajkot, India

Vandana Gupta
National Institute of Technology Raipur
Chhattisgarh, India

Ekramul Islam
University of Kalyani, Kalyani
West Bengal, India

Ghoshna Jyoti
Guru Ghasidas Vishwavidyalaya (A Central
   University)
Bilaspur, India

Anoar Ali Khan
Vignan's Foundation for Science
Andhra Pradesh, India

Akhilesh Khapre
National Institute of Technology Raipur
Chhattisgarh, India

Sarita Khaturia
Mody University
Rajasthan, India

Vijyendra Kumar
National Institute of Technology Raipur
Chhattisgarh, India

Dibyajit Lahiri
University of Engineering & Management
West Bengal, India

Jaya Lakkakula
Amity University
Mumbai, India

Jyoti S Mahale
DBT-ICT Centre for Energy Biosciences,
    Institute of Chemical Technology
Mumbai, India

Jagamohan Meher
Kalasalingam Academy of Research and
    Education
Tamil Nadu, India

Rajanandini Meher
Kalasalingam Academy of Research and
    Education
Tamil Nadu, India

Piyal Mondal
Indian Institute of Technology Guwahati
Assam, India

Dipro Mukherjee
University of Engineering & Management
Kolkata, India

Moupriya Nag
University of Engineering & Management
West Bengal, India

Sachin Palekar
Ramnarain Ruia Autonomous College
Mumbai, India

Preeti H. Pandey
DBT-ICT Centre for Energy Biosciences,
    Institute of Chemical Technology
Mumbai, India

Smaranika Pattanaik
Sambalpur University
Odisha, India

Sanchita Patwardhan
Amity University
Mumbai, India

Hitesh S. Pawar
DBT-ICT Centre for Energy Biosciences,
    Institute of Chemical Technology
Mumbai, India

Nitin Pawar
National Institute of Technology Raipur
Chhattisgarh, India

Sinu Poolachira
National Institute of Technology
Calicut, India

Abhinesh Prajapati
IPS Academy
Indore, India

Mihir K. Purkait
Indian Institute of Technology Guwahati
Assam, India

Raja. C
National Institute of Technology Raipur
Chhattisgarh, India

Rina Rani Ray
Maulana Abul Kalam Azad University of
    Technology
West Bengal, India

B.P. Sahariah
Chhattisgarh Swami Vivekanand Technical
    University
Bhilai, India

Bishwarup Sarkar
University of Baroda
Gujarat, India

Deepak Sharma
National Institute of Technology
Raipur, India

Pankaj Shinde
DBT-ICT Centre for Energy Biosciences,
    Institute of Chemical Technology
Mumbai, India

Har Lal Singh
Mody University
Rajasthan, India

Raghwendra Singh Thakur
Guru Ghasidas Vishwavidyalaya (A Central
    University)
Bilaspur, India

Tejas M. Ukarde
DBT-ICT Centre for Energy Biosciences,
    Institute of Chemical Technology
Mumbai, India

Ayush Vasishta
DBT-ICT Centre for Energy Biosciences,
    Institute of Chemical Technology
Mumbai, India

Sivasubramanian Velmurugan
National Institute of Technology
Calicut, India

Nilesh S. Wagh
Amity University
Mumbai, India

1 Potential of NF Membranes for the Removal of Aquatic Pollutants from Industrial Wastewater
A Review

*Prangan Duarah, Piyal Mondal,
Pranjal P. Das, and Mihir K. Purkait*

CONTENTS

1.1 INTRODUCTION

The globe is facing new challenges as the human population continues to expand and natural wealth supplies get depleted. Many countries continue to place a high demand on the planet's natural resources, while residents in other countries face food and water shortages. Furthermore, the spread of manufacturing and urbanization has resulted in considerable aquatic contamination. Large polluting sectors, for instance, the textile, paper, leather, pharmaceuticals, fertilizer, and dyeing sectors, generate massive volumes of wastewater containing a wide range of contaminants, including hazardous heavy metals, dyes, phenolic organic compounds, saline effluents, and other persistent organic pollutants [1]. These pollutants have wreaked havoc on the environment and human health, jeopardizing the achievement of Target 6.3 of Sustainable Development Goal 6, which has hastened the adoption of more stringent emission and recovery regulations in developing nations [2]. The increasing presence of variables including total

DOI: 10.1201/9781003165019-1

suspended solids (TSS), total dissolved solids (TDS), chemical oxygen demand (COD), and biochemical oxygen demand (BOD), contributes to toxicity in wastewater produced by industrial activities [3].

Water treatment techniques appear to be the most important approach for reducing pollution effects in the aqueous phase and aquatic systems. The aforementioned environmental issues are addressed by all wastewater and water treatment plants. As a result, various technologies for treating industrial waste have been developed throughout the years, including precipitation, electrocoagulation, ion-exchange, membrane filtering, adsorption, and so on [4]. Among these approaches, membrane filtration offers several benefits over other traditional methods, including high separation selectivity, minimal energy requirements, and extremely quick reaction kinetics. Based on the pressure gradient across the membrane, membrane filtering techniques are categorized as microfiltration (MF), ultrafiltration (UF), nanofiltration (NF), and reverse osmosis (RO). Membrane separation, which involves pressurizing and forcing treated water through a semipermeable membrane, applies to both RO and NF. Among all these membranes, NF has a unique appeal in a variety of applications such as metal recovery from effluents, water reuse, the treatment of industrial effluents, the demineralization of water, and the drinking water sector. The distinctive features of NF membranes, such as complete water softening without significant change in water salinity, lower investment costs, increased removal of dissolved and uncharged organic compounds, and high specificity of water flow achieved at moderately low operating pressures, have made them a popular choice.

The current chapter presented the recent advancements in NF membrane–based processes for the treatment of industrial wastewater with a brief discussion on conventional membrane–based approaches utilized to treat the impurities. In addition, the basic structure, different kinds, and characteristics of NF membranes have been highlighted. More particular, the chapter includes the current developments in several NF membrane types for the removal from industrial effluent of heavy metals, dyes, pesticides, herbicides, and medicines. In order to illustrate the feasibility of the procedure, commercial considerations of accessible NF membranes are discussed. In addition, many perspectives and restrictions related to scaling up the process are extensively described, with an aspiration toward future advancements.

1.2 OVERVIEW OF NF MEMBRANES

1.2.1 PROPERTIES OF NF MEMBRANES AND THEIR SYNTHESIS

Initially, the NF membranes were created as a low-pressure alternative for water-softening applications to reverse osmosis membranes. The nominal molecular weight, which has been taken off from an NF membrane, ranges from 100–1000 Da and shows the active layer of the NF membrane to be approximately 1 nm in thickness. NF membranes are usually polymeric and asymmetrical and feature a low resistance layer with an active porous top layer and a macroporous structure underneath the membranes. These three layers determine the membrane's functionality. Permeability, resistance to fouling, ionic selectivity, hydrophilicity, and roughness are all determined by the active layer quality. The supporting layer features the mechanical strength of the membrane. The last layer underlying the medium layer is macroporous. In addition, it is clear that the overall performance of NF membranes is dictated by a thin "active polymer layer" that comes from interfacial polymerization, which is helpful for their scalability in commercial production and capability of producing NF membranes less than 250 nm in thickness [5]. Three distinct separating potentials are caused by ion refusal across this active layer: steric impairment (porous effects), exclusion of Donnan (fixed surfaced load), and dielectric exclusion (by Born effect and image forces).

Many scientists reported that the NF membrane's rejection capability is affected mostly by its pore dimension and its charge density. In addition, the thickness of the membrane determines the resistance to hydrodynamics and subsequently the membrane flux. The membrane's performance can be quantified if these certain membrane properties are familiar to characterize the influence on solution transport. Two prime aspects are broadly covered while modeling NF membrane, that is, predicting flux and the rejection for large-scale installation. In view of that, researchers have constructed several models in the past. Initially, the separation model of NF membranes was described with the Donnan Steric Splitting Poric (DSPM) model, in which the membrane is assumed to be porous, taking steric barriers into account and the size effects into consideration [6]. However, due to DSPM's failure to forecast the rejection of divalent cations, a later steric, electric, and dielectric (SEDE) model was devised. SEDE is the four-parameter model, that is, effective membrane porosity, volume charging density, thickness-to-porosity ratio, and the dielectric solution constant of the membrane pores, which relatively well forecast NF membranes' rejection performance [7].

Over the past decades, numerous approaches such as interfacial polymerization, phase inversion, nanomaterial deposition, mussel-inspired deposition, deep coating, self-assembly, metal polyphenol complexion, and graft polymerization, are studied to construct NF membranes [8]. Among these methods interfacial polymerization (IP) technique is commonly applied. The polymerization of two monomeric reactors, one dissolved in an aqueous phase and one in an organic phase, is the fundamental premise of interface polymerization. A polymer network forms on the supporting surface when the two phases are put in contact with a porous membrane medium. The expansion of the polymeric network limits additional contact at a later stage between the two reactants, which creates an ultra-thin selective layer. In IP, different types of monomers are employed such as bisphenol A (BPA), tannic acid, m-phenylenediamine (MPD), trimesoyl-chloride (TMC) polyvinylamine, and isophthaloyl chloride, which produce the thin active film layer. An IP-produced NF membrane can be controlled by the type of polymer, solvent, additive type, and casting conditions. In view of that, different approaches to improve membrane efficiency have been developed during IP, including polyamide membrane surface fluorination, facile zwitterionization, membrane thickness restriction, hyperbranching of polyesters, and on-site Mg/Al hydrotalen exfoliation [9]. Most NF membranes prepared via the IP method are charged negatively, because of the carboxylic acid hydrating of unreacted TMC acyl groups. The synthesis of positively charged NF membranes was of great interest because of the potential of the Donnan exclusion mechanism to increase the selectivity of multivalent cations. In view of that, NF membranes with a positive charge that are not interfacially polymerized were produced with increased selectivity. Various techniques have been used, such as crosslinking and quaternization of p-xylylene dichloride (XDC) amine groups or UV grafting of quaternary amine groups [10].

1.2.2 Various Types of NF Membranes

NF membranes are categorized into nonporous, isotropic microporous, electrically charged, dense, asymmetric, ceramic, and liquid membranes depending on their structure and pore shape. Based on the surface charge characteristics, the NF membrane can be broadly categorized into two categories, that is, negatively charged and positively charged NF membrane. The majority of the NF membranes produced by the IP method are inherently negatively charged due to the hydration of unreacted acyl groups from TMC into carboxylic acids. In the case of multivalent anions (e.g., SO_4^{2-}, PO_4^{3-}), the negatively charged TFC-NF membranes often have a greater rejection rate than multivalent cations regarding the influence of charging-repulsive action (e.g. Fe^{2+}, Mn^{2+}). Positively charged NF membranes are more appropriate for eliminating multivalent cations (e.g. Fe^{2+}, Ca^{2+}), as well as for the rejection of amino acids below isoelectric points, cationic dyes purification, or the rehabilitation of cathode–electrophorus lacquers. Chloromethylation, IP of TMC with aliphatic amines or triethanolamine, and quaternization can be done in preparing NF membranes with a positive charge [10].

Based on the flow path, NF membranes can be categorized into two types: crossflow with concentrate recycle (CFCR) and flow system with a dead-end. The utilization of a high-pressure water source allows for crossflow filtering via the membrane. *Permeate* refers to the membrane or filtering component, whereas *concentrate* or *reject* refers to the surface that remains just minimally flowing with the membrane without separation or filtration. The concentrates are mainly constituted by all rejected salts and are typically concentrated with all unwanted elements. The flow system with a dead-end unit involves the process of collecting rejects until backwashing is required. During the backwashing operation, a washing volume of 2% to 5% of the total entry solution is utilized to flush and dispose of the collected concentrates.

To achieve the necessary membrane surface working area per unit of membrane element volume, the NF membrane can be arranged in various configurations. Different NF membrane configurations include plate and frame module ($60-300$ m^2/m^3), tubular membrane module ($60-200$ m^2/m^3), spiral-wound module ($300-800$ m^2/m^3), and hollow-fiber membrane module ($20,000-30,000$ m^2/m^3) [11].

1.3   VARIOUS FORMS OF INDUSTRIAL AQUATIC POLLUTANTS

Water scarcity and pollution are regarded as unresolvable global issues and a huge effort to be accomplished. Water pollution is commonly associated with a high volume of wastewater released into the environment from numerous industrial sources. Furthermore, the composition and classification of effluents are drastically different and exceedingly complicated due to the wide range of pollution sources, such as dwellings, hospitals, industries, veterinary services, and agriculture, as well as their variable application processes. Table 1.1 summarized the numerous types of contaminants emitted by various sectors and their impact on human health. Among the substances that may occur in wastewaters, heavy metals are one of the principal contaminants that can be found in large quantities in the aquatic environment. Heavy metals are defined as metals with atomic weights ranging from 63.5 to 200.6 and densities of more than 5 g/m^3. Copper (Cu), zinc (Zn), iron (Fe), manganese (Mn), nickel (Ni), and cobalt (Co) are heavy metals that play an essential part in biochemical activities in the human body. Excessive exposure to these metal ions, on the other hand, might be harmful. Other toxic elements from the same group, such as arsenic (As), lead (Pb), cadmium (Cd), chromium (Cr), and mercury (Hg), are harmful even at low concentrations (parts per billion, ppb), since they are nonbiodegradable and can bioaccumulate in the primary systems of the human body [12].

Industrial dyes are now extensively recognized for water contamination. Dyes are essentially coloring pigments that add color to the substrate and the dye intermediates are complexes formed during the dye production process. These are mostly consumed by sectors such as leather, tannery, textile, and paper and pulp. When the dye has served its purpose, the majority of the dye components are dumped into surrounding bodies of water. Effluent released from dye-intermediate industrial discharge has a high COD concentration, a dark color, a high content of organic compounds, and a high acidity. In general, the BOD/COD ratio is relatively low. The raw ingredients used for producing dyes, such as benzene, toluene, naphthalene, phenol, anthracene, pyridine, and others, enhance the contamination concentration of carbon-based pollutants in dye-intermediate effluent [13, 14].

The so-called emerging organic pollutants are one of the major contaminants that pose risks to humans and ecosystems today. This phrase covers newly found substances such as drugs and personal care items in the environment (PPCPs). Other organic pollutants commonly identified and monitored in aquatic systems include pesticides, hexachlorobenzenes (HCBs), polychlorinated polychlorine-pulmonary furans (PCDFs), dibenzo-p-dioxins (PCDDs), organochlorine pesticides (OCPs), polycyclic aromatic hydrocarbons, and polychlorinated biphenyls [15].

TABLE 1.1

Contaminants Found in Various Industrial Wastewater and Their Effects on the Environment and Human Health

Type of contaminants	Major contaminants	Major industries	Major impact	Reference
Heavy metals	Ni, Fe, Zn, Cr, Cd, As, Cu, Mn, and Mg	Pulp and paper	• Cancers of various sorts, myocardial infections, hypertension, and diabetes • Joint pain, Knee pain, fatigue, and other eye disorders	[12]
Dye	Methyl orange, sunset yellow, Direct Red 16, Acid Red 18, Acridine Orange, Amaranth, bright blue, bromothymol blue, cationic Red X-GTL, Congo red, Lanasol Blue 3R, malachite green, methyl blue, methyl orange, reactive blue 21, rose bengal	Textile, dyeing, printing	• Decrease in water bodies' solar infiltration and changes in photosynthetic activity and the demand for biological oxygen • Cause serious health issues such as renal malfunction, reproductive system dysfunction, brain and liver malfunction, and central nervous system difficulties	[46–48]
Personal care products	Diethyltoluamide, 4-benzophenone, galaxolide, tonalide	WWTP effluent	• Adverse effect on aquatic ecosystem • Acute toxicity to algal, invertebrate, and fish • Interferes with the activities of both animal and human hormone systems • Increase the risk of cancer	[49]
Pharmaceutically active complexes	Phenol, diazepam, ciprofloxacin, metoprolol, diclofenac, carbamazepine, clorfibric acid, testosterone.	Pharmaceutical industries, hospitals	• Changes in the reproductive health of humans • Increase the risk of cancer (especially breast and prostate cancers)	[49]
Pesticides and herbicides	Chlorpyrifos, phenanthrene, metaldehyde, butachlor, epoxiconazole, trazine, and prometryn	Agricultural industries, fertilizer industries	• Induces reproductive hormonal changes • effects on immune, central nervous, endocrine, and reproductive systems	[15]
Saline effluents	Brine solutions, dry salt (NaCl)	Agrofood, petroleum, and leather industries	• Generate anoxic condition on seabed • Change the lighting condition of aquatic environment	[50, 51]

1.4 CONVENTIONAL MEMBRANE–BASED APPROACHES FOR INDUSTRIAL WASTEWATER TREATMENT

A number of standard strategies for contaminated water treatment have been widely researched throughout the last few decades. But millions of people in today's world are subjected to exponential growth in contamination of drinking water. In that context, the most popular treatment methods used to remediate such contaminants from drinking water include electrocoagulation, membrane filtration, photochemical oxidation, ion exchange, and adsorption. Among the various other technologies, membrane filtration has been extensively utilized for the treatment of industrial wastewater [16–18]. Membrane filtration processes are classified into four types based on pore size: MF, UF, NF, and RO. The largest pores in MF membranes range from 0.1 microns to 10 microns, which provide high water permeability. The MF membrane can eliminate large amounts of contaminants, such as suspended particles, germs, and colloids. An RO membrane is a thick membrane with no detectable pores. When compared to other filtration systems, it filters polluted water utilizing a solution diffusion mechanism with very low water permeability and the greatest pressures. Its high energy consumption cost prevents it from being extensively used. UF membranes, on the other hand, feature tiny pores (2–100 nm) to filter out particles on the sub-micro to nano dimensions. Their permeability is substantially lower than that of MF; therefore, the needed water pressure is significantly large. Despite the numerous benefits of UF membranes, their application in diverse industrial sectors is limited by their large pore size. For example, UF membranes are ineffective for dye recovery because a large amount of dye can flow through porous UF membranes, resulting in poor dye recovery during a UF procedure [19]. A similar issue is associated with microfiltration membranes, which limit their applicability in the removal of small particle-sized contaminants such as heavy metals, dyes, and other organic substances. In this context, RO and NF membranes have been extensively investigated for the elimination of various contaminants from industrial effluents. NF membranes with higher permeation flux and adequate rejection operate at lower pressures than RO membranes with severe fouling, poor permeation flux, and high operating pressure and energy demands. Donnan electrostatic repulsion and size exclusion processes allow NF membranes to successfully reject heavy metal ions. NF membranes feature smaller pores than UF, which is approximately 1.0 nm. This indicates that the NF membrane can eliminate comparatively small organic compounds, such as aromatic compounds, and dyes, which are dissolved with roughly 300 molecular weights. Different researchers have undertaken different studies to investigate NF membrane feasibility in industrial wastewater treatment, which are briefly discussed in the next section.

1.5 APPLICATION OF NF MEMBRANES FOR AQUATIC POLLUTANT REMOVAL

NF applications in drinking water and industrial wastewater treatment are increasing exponentially, with the gradual growth in RO and UF applications in drinking water and wastewater treatment. NF membranes are also currently employed in numerous applications to replace ROs such as water, as well as to extract fine and costly components to generate profits and minimize industrial energy costs [11]. In this section, we describe the use of NF membranes to remove aquatic contaminants from wastewater.

1.5.1 APPLICATION OF NF MEMBRANE FOR HEAVY METAL REMOVAL

Due to its pores being smaller than those in UF and MF membranes, the elimination of heavy metals with NF membranes has received considerable interest. This approach is effective for removing greater metal concentrations of up to 12,000 ppm. NF membranes can handle wastewater containing more than one heavy metal with an 80% efficiency for elimination. Table 1.2 summarized the use of various types of NF membranes for removing heavy metal from contaminated water. The number and kinds of accessible functional groups employed for physical or chemical interaction impact an adsorbent's capacity to capture diverse pollutants. Graphene oxide (GO) has the ability to bind heavy metal ions and aromatic pollutants into its layers and edges and is composed of functional groups, for instance, hydroxyl (OH), epoxy (C–O–C), and carboxyl (COOH).

TABLE 1.2

List of Multiple NF Membranes Used to Remove Heavy Metals from Industrial Wastewater

Type membrane	Substrate	Type heavy metal	Initial concentration (ppm)	pH	Pressure (bar)	Zeta potential (mV)	Rejection	References
Negatively charged	PES	Cu^{2+}	20	5.0	5.0	−39.8	92	[21]
	PAN/TiO$_2$	Ni^{2+}	50	–	10	–	88.1	[52]
		Cr^{6+}					80.3	
	PVDF/APTES	Cu^{2+}	5	5.5	5	–	47.9	[53]
		Cd^{2+}					44.2	
		Cr^{6+}					52.3	
Positively charged	PAI	Pb^{2+}	1000	5.34	3.0	–	95.88±0.93	[25]
		Ni^{2+}					99.74±0.18	
		Zn^{2+}					98.07±0.27	
	PES	$Cu^{2+}, Pb^{2+}, Cd^{2+} As^{5+}$	200	5.0–11	10	–	99	[23]
		Ni^{2+}, Zn^{2+}					98	
		As^{3+}					97.6	
	PEI	Mg^{2+}	500	10	10	–	96	[24]
		Ca^{2+}					96	
		Li^+					32	
	HFC	$Zn^{2+},$	49.63	4.5	–	–	93.33	[22]
		Cu^{2+}	75.51				92.73,	
		Ni^{2+}	40.05				90.45	
		Pb^{2+}	49.3				88.35	

Zhang et al. (2015) described a new twofold strategy for manufacturing structurally stable, GO framework–based membranes for heavy metal elimination. The fabricated GO-based NF membrane showed a high affinity for Mg^{2+}, Pb^{2+}, Ni^{2+}, Cd^{2+}, and Zn^{2+} removal [20]. Abdi et al. (2018) prepared a negatively charged magnetic graphene-based composite (MMGO) embedded polyethersulfone (PES) polymer via phase inversion induced by immersion precipitation technique. Constructed 0.5% MMGO hybrid NF membrane had of best affinity for the elimination of Cu^{2+} (92%) [21]. Li et al. (2021) prepared a TFC-NF membrane macroporous hollow-fiber ceramic (HFC) that was favorable for mining wastewater treatment. The prepared membrane exhibited remarkable rejection efficiency for Zn^{2+} (93.33%), Cu^{2+} (92.73%), Ni^{2+} (90.45%), and Pb^{2+} (88.35%) [22].

As already stated, most of the prepared NF membrane is negatively charged, which is unfavorable for heavy metal removal. Thus, weakening or even turning the membrane positive becomes an appropriate technique to enhance the ability of the NF membrane to remove heavy metals. In view of that, Zhu et al. (2015) constructed a modified hollow fiber TFC-NF membranes by grafting poly(amidoamine) (PAMAM) dendrimer on the interfacially polymerized layer of PES membranes for heavy metal elimination. The rejections of various heavy metals, including Cu^{2+}, Pb^{2+}, Cd^{2+}, Ni^{2+}, Zn^{2+}, and As^{5+}, were documented at about 99.0% at a pressure of 10 bars. Additionally, the prepared membrane possess rejection over 97% for As^{3+} [23]. Qi et al. (2019) adopted grafting to develop a hydrophilic positively charged membrane with exceptional anti-fouling capabilities for cationic active agents. For the purposes of this investigation, the comparison proposal consisted of three distinct TFC membranes: TFC, DP-TFC, and DPC-TFC. TFC samples were created using the original interface polymerization techniques of PIP and DDM with TMC. The direct utilization of a large polyethylene molecule (PEI) is used in DP-TFC. DPC-TFC denotes membrane fabrication through grafting in the presence of CMP. The removal efficacy of the divalent cation Mg^{2+} was improved after the grafted and fixed membrane of the PEI, showing positive charges ($MgCl_2 > Na_2SO_4$). Sulfate rejection decreased from 94% to 84%, and the magnesium chloride retrieval efficiency was raised to 96.6% because of its abnormal charging impact. In comparison to the other two TFC membranes, the higher rejection of monovalent Na^+ and Li^+ ions may be attributed mostly to the primary function of the effect of the membrane's pores [24]. Zhang et.al (2019) fabricated a hollow-fiber NF membrane (HFNF-GO) composed of graphene oxide (GO) layers. Figure 1.1 describes the layer-by-layer deposition of GO for the preparation of the NFNF-GO membrane. During the preparation process, free amine groups are grafted on the surface of Torlon® 4000T-MV polyamide-imide (PAI) substrate by crosslinking it with hyperbranched polyethyleneimine (HPEI), which provided a bonding site for GO sheets. The prepared membrane demonstrated excellent rejection capability for Pb^{2+}, Zn^{2+} and Ni^{2+} ions [25].

Many studies have investigated the performance of NF membranes under different operating conditions. For instance, Meher et al. (2014) investigated impacts on the removal of Pb^{2+} and Ni^{2+} from drinking water by utilizing a commercial NF membrane under a number of operating conditions, such as pH value, pressure, and feed flow. Based on the outcomes of these efforts, the rate of Pb^{2+} (86.0%) and Ni^{2+} (93.0%) rejections was enhanced. Despite the greater rejection rate, greater feed concentration and pH resulted in increased scaling accumulation on the surface of the membranes. This does not benefit the long-term application because it would adversely affect the overall performance of NF membranes [26]. Regarding flux permeability, an increase in feed solution concentration has been documented as a result of decreasing flux of water. NF membranes can

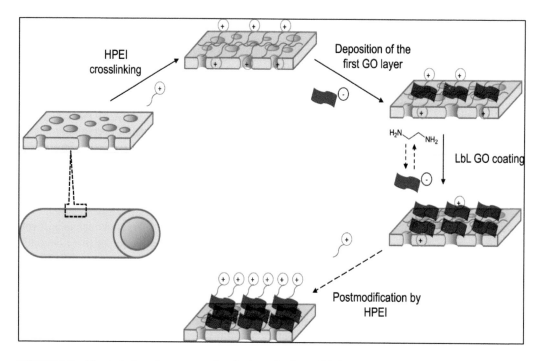

FIGURE 1.1   The technique for constructing the layer-by-layer GO framework membrane.

Source: Reproduced with permission from [25]

more efficiently reject a broad spectrum of heavy metal ions in comparison with UF mixed matrix membranes (MMMs). However, the water flux of the NF membrane is significantly less than the UF membrane. MMMs give better water permeabilities; however, they often have a low rejection of high concentrations of metal. This makes the NF membrane better suited for industrial wastewater with a high metal ion content and UF MMMs are more suitable for treating wastewater with a low metal ion content [27].

1.5.2 APPLICATION FOR TEXTILE INDUSTRY WASTEWATER TREATMENT

Dyes are the principal contaminants released by the textile industry that might affect the receiving water bodies, for example, a decrease in solar infiltration, photosynthesis, and biochemical oxygen requirements [28, 29]. The removal of dye molecules in water depends mostly on electrostatic forces, and thus, a number of studies from the literature on positive charges NF membranes are summarized in Table 1.3.

TABLE 1.3
List of NF Membranes Used for Various Contaminant Removal from Textile Industry Wastewater

Type membrane	Substrate	Type of contaminant	Initial concentration (mg/L)	pH	Pressure (bar)	Zeta potential (mV)	Rejection (%)	References
Negatively charged	PES	Direct red 16	30	5.0	5.0	−39.8	99	[21]
	PVDF/APTES	Direct red 16	100	5.5	5	–	94.9	[53]
	PSf	Congo red	100	3–10	4	–	99.0	[31]
		Coomassie brilliant blue	100	3–10	4	–	95	
		Evans blue	100	3–10	4		95	
	PSf	Methylene blue	7.5	–	3.44	–	46–66	[32]
		Rodamine-WT	7.5	–	3.44	–	93–95	
		NaCl	–	–	3.44	–	6–19	
		Na$_2$SO$_4$	–	–	3.44	–	26–46	
	PES	Congo Red	100	–	6.8	–	99.8	[33]
		NaCl	100	–	6.8	–	6.1	
		Na$_2$SO$_4$	100	–	6.8		2.2	
	PES/SWCNT	Direct Red 80	50	2–9	–	−18.5	99.5	[34]
		Direct Red 23	50	2–9	–	−18.5	99.3	
		Congo red	50	2–9	–	−18.5	99.5	
		NaCl	1000	2–9	–	−18.5	10	
	TFC-NF	Cibacron black B	25	–	2.76	–	94.6	[54]
		Cibacron red RB	15	–	2.76	–	93	
Positively charged	PEI	Victoria blue B	200	10	10	–	99.2	[24]
		Semixylenol orange	200	10	10	–	99	
		Tropaeolin O	200	10	10	–	98.3	
		Neutral	200	10	10	–	98.2	
	sPPSU	Safranin O	–	–	–	+35.28	99.60	[30]
			–	–	–	+2.02	99.98	
		Orange II	–	–	–	+35.28	60	
			–	–	–	+2.02	87	

Zhong et al. (2012) fabricated a positively charged membrane by using direct sulfonated poly-phenylenesulfone with 2.5 mol% 3,3'-di-sodiumdisulfate-4,4'-dichlorodiphenylsulfone as substrate. Two grafting monomers with hydrophilic properties, diallyldimethylammonium chloride and [2-(methacryloyloxy)ethyl]trimethyl ammonium chloride, are used to obtain positively charged NF1 and NF2, respectively. In order to investigate the effect of solute charge on NF membrane, two dye molecules of different charge properties are chosen, namely, Safranin O (+ve charge) and Orange II sodium salt (−ve charge). Both membranes exhibited excellent rejection for Safranin O, 99.60% for NF1 and 99.98% NF2. NF1 is reported to be 18 times more positively charged than that of NF2 due to which it attracted negatively charged Orange II more strongly and displayed low rejection efficiency (60%) for Orange II, whereas NF2 exhibited a rejection efficiency of 87%. It is evident that zeta potential has a significant impact on the selectivity of NF membranes in rejecting cationic macromolecules and this effect can be efficiently managed by choosing an appropriate monomer [30]. Yang et al. (2020) prepared a negatively charged NF membrane using 4,4'-diaminodiphenylmethane (DADPM) and PSf as a substrate for the rejection of negatively charged dyes. The rejection rate of on neutral dye, Rhodamine B (RB), and three negatively charges dyes, that is, Congo red (CR), Coomassie brilliant blue (BBR), and Evans blue (EB), were examined. The membrane displayed an exceptional rejection rate for negatively charged dyes and the lowest rejection rate for RB [31].

Apart from the dyes, the textile effluents also contain various salts, including sodium chloride (NaCl) and sodium sulfate (Na_2SO_4). Therefore, it is of considerable importance to synthesize an efficient NF membrane to separate the salt and dyes from the effluents of textile industries. In view of that, Hu et al. (2013) prepared loose NF membrane (LNF) by cross-linking GO on a DA-coated PSf substrate. The resultant negatively charged membrane displayed significant rejection (93–95%) of Rhodamine-WT dye (negatively charged) and moderate rejection rate for positively charged methylene blue (46–66%), while relatively low rates to NaCl (6–19%) and Na_2SO_4 (26–46%) with a permeability of 8.0–27.6 LMH·bar^{-1} [32]. Similarly, Li et al. (2019) fabricated an LNF membrane by crosslinking between polyethylenimine (PEI) and tannic acid (TA) on PES substrate via a green rapid coating (GRC) process, which is solvent-free. The prepared LNF membrane displayed remarkable permeability (40.6 LMH·bar^{-1}) and dye rejection for CR (99.8%), while considerably low rejection rate to NaCl (6.1%) and Na_2SO_4 (2.2%) [33]. Lu et al. (2020) fabricated an acid-tolerant polyarylate NF membrane (PAR-NF) on the top of a single-walled carbon nanotube (SWCNT) for the elimination of dyes from acidic saline solutions of textile effluent. The SWCNT supports the ultra-thin thickness of the active layer PAR and a high-permeability membrane fluid of approximately 210 L m^{-2}h^{-1}bar^{-1}, while the inclusion of 5,5',6,6'-tetrahydroxy-3,3,3',3'-tetramethyl-1,1'-spirobisindane (TTSBI) monomer provides a negative charge PAR active layer surface which permits a high dye rejection (direct red 80, direct red 23, CR) of more than 99% for a wide pH and salinity range of feed solutions. In addition, while the membrane is being treated in a pH range of 2 to 9 and NaCl feed levels between 1000 and 5000 ppm, a high selective dye, and NaCl was seen with steady NaCl retention of approximately 10% [34].

1.5.3 Application of NF Membrane for Pollutant Removal from Pharmaceutical Industry Wastewater

Wastewater from pharmaceutical production often comprises large amounts of organic and inorganic hazardous, biodegradable/nonbiodegradable chemicals along with pharmaceutical residues. NF membrane–based treatment processes have shown encouraging results on pollutants rejection of wastewater and other emerging micropollutants from the pharmaceutical industry wastewater. Table 1.4 summarizes the use of various types of NF membranes for the removal of pharmaceuticals from the aquatic environment.

Yoon et al. (2007) examined the elimination of endocrine disrupting compounds, pharmaceuticals, and personal care products (EDC/PPCPs) of 27 chemicals by NF membranes and compared its rejection rate with UF membranes. The analysis indicated that both hydrophobic adsorption and size exclusion mechanisms are important in retaining EDC/PPCP for the NF membrane, but the

TABLE 1.4
List of NF Membranes Used for Various Contaminant Removal from Pharmaceutical Industries

Type of membrane	Type of contaminant	Initial concentration (ppm)	pH	Pressure (bar)	Zeta potential (mV)	Rejection (%)	References
NF-90	sulfamethoxazole	0.5	8	–	–7.9	100	[36]
	ibuprofen	0.5	8	–	–7.9	100	
HF-NF	carbamazepine	1	7	4	~5	91.1	[37]
	carbamazepine-10,11-epoxide					81.2	
	primidone					86.8	
	2-ethyl-2-phenylmalonamide					85.6	
	enrofloxacin					86.1	
	ciprofloxacin					87.3	
	irinotecan					90.1	
	7-ethyl-10-hydroxy-camptothecin					91.3	
	estradiol benzoate					86.5	
	medroxyprogesterone 17-acetate					90.74	
	b-estradiol 17-enanthate					88.68	
	estradiol cypionate					91.26	
Nanomax–50	amoxicillin	100	2.5	1–5	–	98.2	[55]
		100	10	1–5	–	97.5	

UF membrane retained mostly hydrophobic EDC/PPCPs owing to hydrophobic adsorption [35]. For the retention of three pharmaceutical compounds, sulfamethoxazole, carbamazepine (CBZ), and ibuprofen, Nghiem et al. (2005) investigated the performance of NF-270 and NF-90. Results demonstrate that pharmaceuticals retained by a tight NF membrane (NF-90) are dominated by steric exclusion while the retention of ionizable pharmaceutical with an LNF membrane (NF270) is governed by electrostatic repulse and steric exclusion [36].

It is evident that most commercial NF membranes, such as NF-270, NF-90, DS-5DK, and NF-200, are negatively charged, which restricts the performance of NF membranes for various pharmaceutical-based contaminants removal. In view of that, there has been increasing interest in the use of positively charged NF membranes. To date, several scientists have tried to develop and employ positively charged NF membranes on a number of different contaminants. Wei et al. (2021) fabricated positively charged hollow fiber NF membranes to trap pharmaceutical contaminants, including eight PPCPs and four environmental estrogenic hormones, in the aquatic environment. During the investigation, excellent rejection rates were observed for neutral CBZ (91.1%), carbamazepine-10,11-epoxide (81.2%), primidone (86.8%), and 2-ethyl-2-phenylmalonamide (85.6%). For the positively charged CPT-11 and 7-ethyl-10-hydroxy-camptothecin, the rejection rate was rather significant due to their high molecular weights and the electrostatic repulsion of the positive PEI–NF membrane surface. Despite their larger molecular weight than CBZ, the negatively charged ENR and CIP have been displayed a low rejection rate that is, 86.1% and 87.3%, respectively [37].

1.5.4 APPLICATION FOR PESTICIDE AND HERBICIDE POLLUTANT REMOVAL

Various studies have demonstrated that the rejection rate of NF membrane for pesticides is regulated by a number of critical parameters including molecular size and molecular weight of the compound, geometry, charge, polarity, and hydrophobicity of the membrane [38]. Table 1.5 summarizes the use of several

TABLE 1.5

List of NF Membranes for Pesticide and Herbicide Pollutant Removal

Type of membrane	Type of contaminant	Contaminant	Initial concentration (ppm)	pH	Pressure (bar)	Rejection (%)	References
HF-NF	Pesticide	Propiconazole	0.120	6	3	82.3	[39]
		Carbaryl (NAC)	0.111			70.9	
		Chlorothaloni	0.045			69.5	
		Propyzamide	0.114			64.9	
		Chloroneb	0.102			88.4	
		Methyl dymron	0.152			64.4	
		Fenobucarb	0.137			64.1	
		Tricyclazole	0.194			58.1	
		Esprocarb	0.134			55.4	
		Mefenace	0.173			41.0	
TFC-NF	Pesticide	diazinon	300	–	–	98.8	[40]
ZrO_2 NF	Pesticide	carbofuran	200		5	89	[41]
			40		5	82	
NF200	Herbicides	Atrazine				~80	[15]
		Isoproturon				~80	
		Prometryn				97	

types of NF membranes for pesticide and herbicide removal. The table shows that, compared to other types of contaminants, the research on NF membranes for pesticide and herbicide removal is limited.

In 2004, Chan et al. examined the influences of various parameters such as molecular size, molecular weight, flux, and recovery on pesticide removal by NF membranes. Eleven aromatic pesticides were studied by using the NF70 membrane. It has been shown that, based on its molecular weight, length, fluxes, and recoveries, the NF membrane can eliminate pesticides from 46% to 100%. A rapid increase to a completed refusal (100%) was noted to about MW 200 Da as the molecular weight increased [38]. Later, Jung et al. (2005) investigated the rejection of 10 aromatic pesticides using a hollow-fiber NF membrane (HF-NF). The rejections for pesticides were in the range of 41.0% to 88.4% [39]. Palaks et al. (2006) investigated the retention of three particular herbicides (atrazine, isoproturon, prometryn) in single-solute or dissolute solutions by two commercially available hydrophilic NF membranes (NF270, NF200) and one hydrophobic NF membrane (ESNA, an aromatic polyamide membrane). Herbicide retention in single-solute aqueous solution was shown to rely on each molecule's specific features, including molecular solubility, size, polarity, and the membrane's physical characteristics. The herbicide prometryn, the largest of the three investigated chemicals, had the highest retention, followed by atrazine and isoproturon [15].

Karimi et al. (2016) fabricated a TFC-NF membrane for two pesticide removal (atrazine and diazinon) and examined the influence of the addition of triethylamine (TEA) as an accelerator in the aqueous phase for the prepared membrane. The permeability of water and the rejection of diazinon rose from 22 L/m²/h and 95.2% to roughly 41.56 L/m²/h and 98.8% for the unmodified membrane and for the TEA-modified membrane, respectively [40]. Qui et al. (2020) fabricated an yttria-stabilized ZrO_2-NF membrane and compared its carbofuran rejection capacity with γ-Al_2O_3 UF membrane. The maximum rejection rate of γ-Al_2O_3 UF membrane for carbofuran is less than half of the removal rate ZrO_2-NF membrane; indicating the poor rejection of carbofuran, yet the corresponding flux (130 L m^{-2} h^{-1}) of γ-Al_2O_3 UF membrane is seven times higher than that of ZrO_2-NF membrane (17.75 L m^{-2} h^{-1}). ZrO_2-NF membrane displayed a rejection capacity of 82% when the initial concentration is 40 ppm and 89% when the initial concentration is 200 ppm [41].

1.6 COMMERCIAL ASPECTS OF THE UTILIZATION OF NF MEMBRANE FOR INDUSTRIAL AQUATIC POLLUTANT REMOVAL

Due to their great performance and inexpensive cost, NF membranes have dominated the world market since 1980. American and Japanese industries are currently the major manufacturer of NF membranes. Dow Filmtec (USA), Hydranautics, Toray (Japan), Synder (USA), Toyobo (Japan), Nitto (Japan), Trisep (USA), and GE-Osmonics (USA) are some of the top NF-membrane manufacturers. Currently, Trisep offers two types of NF membranes: XN45, TS80, TS82, and TS83. The TS80 is mostly utilized in municipal water softening, but the XN45 is capable of removing monovalent ions and low-molecular-weight organic materials. FilmTec manufactures composite polyamide membranes NF70, NF270, and NF90. Nitto primarily manufactures NTR-7400 series membranes, while TORAY predominantly manufactures UTC series membranes [42]. From the literature available, it can be seen that TFC membranes dominate the current market due to their excellent performance. Polyamides, cellulose diacetate, cellulose acetate, cellulose triacetate, piperazine, and others are some of the important polymers for creating commercial RO and NF membranes [43].

NF's potential in treating waste and water reuse is remarkable, but commercial implementation is prevented by high operating costs. Because of the high expense of NF operation, its use commercially in water treatment plants is very limited as compared to the other types of the membranes [11]. Additionally, most NF membranes such as NF-700 (Dow FilmTech), UTC-60 (Toray), and DK NF membrane (GE) have relatively poor water permeability and require high pressures of operation in order to produce moderate flow. This leads to higher energy usage that further hinders their potential uses. The commercial viability of the NF process is decided by two primary factors: reusing and/or reducing water loss in the NF rejected section using appropriate technology and recovering and utilizing the inorganic and organic ions contained in NF/RO/UF reject as value-added products. Many studies have demonstrated that combining two or more methods can lead to an efficient and favorable elimination of contaminants from wastewater in a short time, which is advantageous for industrial applications [44]. However, research into such hybrid technologies for the NF membrane is quite limited to the best of our knowledge.

1.7 CHALLENGES AND FUTURE PERSPECTIVES

Several attempts at the fabrication and applications of the NF membrane have received massive significance of research interests among global researchers over the last few decades; however, a mechanistic understanding of the rejection rate and functionality of the NF membrane under various conditions is still required. According to the literature, NF has limited industrial uses owing to the membrane's pore size, which is confined to nano-pore size. UF and RO are selected because, they can span the UF range effectively without the cost constraint of NF (high initial, operating, maintenance costs). Because NF membrane replacement is a function of TDS, NF membranes are replaced in a shorter time than the actual filter life span, increasing the cost of the overall filtration process. Various fabrication methods for NF membranes have been developed over the years, including interfacial polymerization, electron beam irradiation, UV or photografting, plasma grafting, and the layer-by-layer technique. However, when large-scale membrane manufacture is necessary, these processes will have limitations and issues [45].

The main challenges of NF membranes for which solutions are still being developed are prevention and mitigation of fouling, improved solute separation, further concentrate treatment, improved chemical resistance, limited membrane lifetime, insufficient rejection of contaminants in water treatment, and the need for modeling and simulation tools. The information collected demonstrates the development and utilization of many NF membranes in diverse applications including metal recovery, dye removal, heavy metals removal, and other organic contamination removals from industrial wastewater. However, a very handful quantity of NF membranes is utilized efficiently and commercialized. Therefore, researchers should also focus on bringing down the cost of the overall NF membrane filtration process for its practical feasibility from an economical context.

1.8 SUMMARY

NF membranes offer appropriate selectivity to offer the best separation depending on the application of interest. Several mechanisms govern the NF process, including electrostatic repulsion, steric effect, hydrophobic interactions, and diffusion. NF membrane–based treatment processes have shown encouraging results on pollutants rejection of wastewater and other emerging micropollutants from the textile, dyeing, printing, leather, pharmaceutical and fertilizer industries and industrial wastewater. Many methodologies are explored for the fabrication of the NF membrane, including interfacial polymerization, phase inversion, deposition of nanomaterials, inspired molding, deep coating, self-installation, polymetal integration, and graft polymerization. The most frequent technique includes interfacial polymerization. Most NF membranes prepared via the IP method are negatively charged. It is observed from various studies that turning the NF membrane positive become an appropriate technique to enhance the capability of the NF membrane to eliminate heavy metals and dyes. Various techniques have been taken, such as crosslinking and quaternization of p-xylylene dichloride amine groups or UV grafting of quaternary amine groups to fabricate positively charged NF membranes. A promising application of NF membrane is also evident for pharmaceuticals and personal care products. However, a very limited number of reports are available for pesticide and herbicide removal by using NF membrane.

Despite the promising characteristics, applications of NF membrane confront hurdles such as stabilities under varying conditions, commercialization, and overall process cost-effectiveness. Overall, the process of NF is a powerful tool to treat industrial effluents. To overcome the existing bottlenecks, substantial progress is already made including membrane performance and efficient substrate. As a matter of fact, the exploitation of NF membrane for various industrial wastewater treatment is still in its early stage, and furthermore, there is a long way to go for its large-scale industrial applications.

BIBLIOGRAPHY

1. Ghaedi, M., et al., *Activated carbon and multiwalled carbon nanotubes as efficient adsorbents for removal of arsenazo(III) and methyl red from waste water*. Toxicological & Environmental Chemistry, 2011. **93**(3): p. 438–449.
2. Mao, G., et al., *A bibliometric analysis of industrial wastewater treatments from 1998 to 2019*. Environmental Pollution, 2021. **275**: p. 115785.
3. Sharma, P., et al., *Role of microbial community and metal-binding proteins in phytoremediation of heavy metals from industrial wastewater*. Bioresource Technology, 2021. **326**: p. 124750.
4. Jana, S., M.K. Purkait, and K. Mohanty, *Removal of crystal violet by advanced oxidation and microfiltration*. Applied Clay Science, 2010. **50**(3): p. 337–341.
5. Valentino, L., et al., *Development and performance characterization of a polyimine covalent organic framework thin-film composite nanofiltration membrane*. Environmental Science & Technology, 2017. **51**(24): p. 14352–14359.
6. Schaep, J., et al., *Analysis of the salt retention of nanofiltration membranes using the donnan: Steric partitioning pore model*. Separation Science and Technology, 1999. **34**(15): p. 3009–3030.
7. Lanteri, Y., A. Szymczyk, and P. Fievet, *Influence of steric, electric, and dielectric effects on membrane potential*. Langmuir, 2008. **24**(15): p. 7955–7962.
8. Guo, S., et al., *Loose nanofiltration membrane custom-tailored for resource recovery*. Chemical Engineering Journal, 2021. **409**: p. 127376.
9. Mohammad, A.W., et al., *Nanofiltration membranes review: Recent advances and future prospects*. Desalination, 2015. **356**: p. 226–254.
10. Léniz-Pizarro, F., et al., *Positively charged nanofiltration membrane synthesis, transport models, and lanthanides separation*. Journal of Membrane Science, 2021. **620**: p. 118973.

11. Abdel-Fatah, M.A., *Nanofiltration systems and applications in wastewater treatment: Review article.* Ain Shams Engineering Journal, 2018. **9**(4): p. 3077–3092.

12. Haldar, D., P. Duarah, and M.K. Purkait, *MOFs for the treatment of arsenic, fluoride and iron contaminated drinking water: A review.* Chemosphere, 2020. **251**: p. 126388.

13. Purkait, M.K., et al., *Cloud point extraction of toxic eosin dye using Triton X-100 as nonionic surfactant.* Water Research, 2005. **39**(16): p. 3885–3890.

14. Purkait, M.K., S. DasGupta, and S. De, *Resistance in series model for micellar enhanced ultrafiltration of eosin dye.* Journal of Colloid and Interface Science, 2004. **270**(2): p. 496–506.

15. Plakas, K.V., et al., *A study of selected herbicides retention by nanofiltration membranes: The role of organic fouling.* Journal of Membrane Science, 2006. **284**(1): p. 291–300.

16. Bhattacharyya, D., M. Moffitt, and R. Grieves, *Charged membrane ultrafiltration of toxic metal oxyanions and cations from single-and multisalt aqueous solutions.* Separation Science and Technology, 1978. **13**(5): p. 449–463.

17. Jana, S., M.K. Purkait, and K. Mohanty, *Preparation and characterizations of ceramic microfiltration membrane: Effect of inorganic precursors on membrane morphology.* Separation Science and Technology, 2010. **46**(1): p. 33–45.

18. Das, P.P., et al., *Integrated ozonation assisted electrocoagulation process for the removal of cyanide from steel industry wastewater.* Chemosphere, 2021. **263**: p. 128370.

19. Al Aani, S., T.N. Mustafa, and N. Hilal, *Ultrafiltration membranes for wastewater and water process engineering: A comprehensive statistical review over the past decade.* Journal of Water Process Engineering, 2020. **35**: p. 101241.

20. Zhang, Y., S. Zhang, and T.-S. Chung, *Nanometric graphene oxide framework membranes with enhanced heavy metal removal via nanofiltration.* Environmental Science & Technology, 2015. **49**(16): p. 10235–10242.

21. Abdi, G., et al., *Removal of dye and heavy metal ion using a novel synthetic polyethersulfone nanofiltration membrane modified by magnetic graphene oxide/metformin hybrid.* Journal of Membrane Science, 2018. **552**: p. 326–335.

22. Li, P., et al., *Thin-film nanocomposite NF membrane with GO on macroporous hollow fiber ceramic substrate for efficient heavy metals removal.* Environmental Research, 2021. **197**: p. 111040.

23. Zhu, W.-P., et al., *Poly(amidoamine) dendrimer (PAMAM) grafted on thin film composite (TFC) nanofiltration (NF) hollow fiber membranes for heavy metal removal.* Journal of Membrane Science, 2015. **487**: p. 117–126.

24. Qi, Y., et al., *Polyethyleneimine-modified original positive charged nanofiltration membrane: Removal of heavy metal ions and dyes.* Separation and Purification Technology, 2019. **222**: p. 117–124.

25. Zhang, Y., et al., *Layer-by-layer construction of graphene oxide (GO) framework composite membranes for highly efficient heavy metal removal.* Journal of Membrane Science, 2016. **515**: p. 230–237.

26. Maher, A., M. Sadeghi, and A. Moheb, *Heavy metal elimination from drinking water using nanofiltration membrane technology and process optimization using response surface methodology.* Desalination, 2014. **352**: p. 166–173.

27. Abdullah, N., et al., *Recent trends of heavy metal removal from water/wastewater by membrane technologies.* Journal of Industrial and Engineering Chemistry, 2019. **76**: p. 17–38.

28. Taghizadeh, F., et al., *Comparison of nickel and/or zinc selenide nanoparticle loaded on activated carbon as efficient adsorbents for kinetic and equilibrium study of removal of Arsenazo (III) dye.* Powder Technology, 2013. **245**: p. 217–226.

29. Mondal, S., M.K. Purkait, and S. De, *Advances in dye removal technologies.* 2018: Springer.

30. Zhong, P.S., et al., *Positively charged nanofiltration (NF) membranes via UV grafting on sulfonated polyphenylenesulfone (sPPSU) for effective removal of textile dyes from wastewater.* Journal of Membrane Science, 2012. **417–418**: p. 52–60.

31. Yang, C., et al., *Novel negatively charged nanofiltration membrane based on 4,4′-diaminodiphenylmethane for dye removal.* Separation and Purification Technology, 2020. **248**: p. 117089.

32. Hu, M. and B. Mi, *Enabling graphene oxide nanosheets as water separation membranes.* Environmental Science & Technology, 2013. **47**(8): p. 3715–3723.

33. Li, Q., et al., *Tannic acid-polyethyleneimine crosslinked loose nanofiltration membrane for dye/salt mixture separation.* Journal of Membrane Science, 2019. **584**: p. 324–332.

34. Lu, Y., et al., *A microporous polymer ultrathin membrane for the highly efficient removal of dyes from acidic saline solutions.* Journal of Membrane Science, 2020. **603**: p. 118027.

35. Yoon, Y., et al., *Removal of endocrine disrupting compounds and pharmaceuticals by nanofiltration and ultrafiltration membranes.* Desalination, 2007. **202**(1): p. 16–23.

36. Nghiem, L.D., A.I. Schäfer, and M. Elimelech, *Pharmaceutical retention mechanisms by nanofiltration membranes.* Environmental Science & Technology, 2005. **39**(19): p. 7698–7705.

37. Wei, X., et al., *Removal of pharmaceuticals and personal care products (PPCPs) and environmental estrogens (EEs) from water using positively charged hollow fiber nanofiltration membrane.* Environmental Science and Pollution Research, 2021. **28**(7): p. 8486–8497.

38. Chen, S.-S., et al., *Influences of molecular weight, molecular size, flux, and recovery for aromatic pesticide removal by nanofiltration membranes.* Desalination, 2004. **160**(2): p. 103–111.

39. Jung, Y.-J., et al., *Rejection properties of aromatic pesticides with a hollow-fiber NF membrane.* Desalination, 2005. **180**(1): p. 63–71.

40. Karimi, H., A. Rahimpour, and M.R. Shirzad Kebria, *Pesticides removal from water using modified piperazine-based nanofiltration (NF) membranes.* Desalination and Water Treatment, 2016. **57**(52): p. 24844–24854.

41. Qin, H., et al., *Preparation of yttria-stabilized ZrO2 nanofiltration membrane by reverse micelles-mediated sol-gel process and its application in pesticide wastewater treatment.* Journal of the European Ceramic Society, 2020. **40**(1): p. 145–154.

42. Wang, S., et al., *A review of advances in EDCs and PhACs removal by nanofiltration: Mechanisms, impact factors and the influence of organic matter.* Chemical Engineering Journal, 2021. **406**: p. 126722.

43. Yang, Z., et al., *A review on reverse osmosis and nanofiltration membranes for water purification.* Polymers, 2019. **11**(8): p. 1252.

44. Changmai, M., et al., *Hybrid electrocoagulation: Microfiltration technique for treatment of nanofiltration rejected steel industry effluent.* International Journal of Environmental Analytical Chemistry, 2020: p. 1–22.

45. Van der Bruggen, B., M. Mänttäri, and M. Nyström, *Drawbacks of applying nanofiltration and how to avoid them: A review.* Separation and Purification Technology, 2008. **63**(2): p. 251–263.

46. Ghaedi, M., et al., *Removal of methyl orange by multiwall carbon nanotube accelerated by ultrasound devise: Optimized experimental design.* Advanced Powder Technology, 2015. **26**(4): p. 1087–1093.

47. Roosta, M., et al., *Ultrasonic assisted removal of sunset yellow from aqueous solution by zinc hydroxide nanoparticle loaded activated carbon: Optimized experimental design.* Materials Science and Engineering: C, 2015. **52**: p. 82–89.

48. Januário, E.F.D., et al., *Advanced graphene oxide-based membranes as a potential alternative for dyes removal: A review.* Science of the Total Environment, 2021. **789**: p. 147957.

49. Cizmas, L., et al., *Pharmaceuticals and personal care products in waters: Occurrence, toxicity, and risk.* Environmental Chemistry Letters, 2015. **13**(4): p. 381–394.

50. Lefebvre, O. and R. Moletta, *Treatment of organic pollution in industrial saline wastewater: A literature review.* Water Research, 2006. **40**(20): p. 3671–3682.

51. Sahu, P., *A comprehensive review of saline effluent disposal and treatment: Conventional practices, emerging technologies, and future potential.* Journal of Water Reuse and Desalination, 2020. **11**(1): p. 33–65.

52. Hosseini, S.S., et al., *Fabrication, tuning and optimization of poly (acrilonitryle) nanofiltration membranes for effective nickel and chromium removal from electroplating wastewater.* Separation and Purification Technology, 2017. **187**: p. 46–59.

53. Zeng, G., et al., *Novel polyvinylidene fluoride nanofiltration membrane blended with functionalized halloysite nanotubes for dye and heavy metal ions removal.* Journal of Hazardous Materials, 2016. **317**: p. 60–72.

54. Chakraborty, S., et al., *Nanofiltration of textile plant effluent for color removal and reduction in COD.* Separation and Purification Technology, 2003. **31**(2): p. 141–151.

55. Oulebsir, A., et al., *Treatment of artificial pharmaceutical wastewater containing amoxicillin by a sequential electrocoagulation with calcium salt followed by nanofiltration.* Journal of Environmental Chemical Engineering, 2020. **8**(6): p. 104597.

2 Membrane Separation

An Advanced Tool for the Development of a Wastewater Treatment Process

*Ayush Vasishta, Jyoti S Mahale, Preeti H. Pandey,
Tejas M. Ukarde, Pankaj Shinde, and Hitesh S. Pawar*

CONTENTS

DOI: 10.1201/9781003165019-2

LIST OF ABBREVIATIONS

BLM	bulk liquid membrane
CP	concentration polarization
CWMI	Composite Water Management Index
ED	electrodialysis
EDR	electrodialysis reversal
ELM	emulsion liquid membrane
FO	forward osmosis
GDP	gross domestic product
MBR	membrane bioreactor
MD	membrane distillation
MF	microfiltration
MIEX	magnetic ion exchange
NF	nanofiltration
NOM	natural organic matter
PV	pervaporation
RO	reverse osmosis
SLM	solid liquid membrane
TDS	total dissolved solids
TMP	transmembrane pressure
UF	ultrafiltration

2.1 INTRODUCTION

The availability of clean water is a major concern for day-to-day life activities in densely populated countries. India is the second-largest population among the globe and thus is facing the world's worst water crisis for accessible, cleaner and safer drinking water. The survey by NITI Aayog Government of India depicted that more than 50% country population does not have access to safe drinking while about 2 lakh people were died every year due to the absence of access to clean and safe drinking water. The CWMI 2018 indicated that 6% of economic GDP will be lost by 2050, while water demand will exceed the available supply by 2030 [1]. However, water pollution has adversely affected the health of India's population (due to outbreaks of several diseases), economy and environment. The UNICEF data reported that the economic burden of waterborne diseases is approximately US$600 million due to chemical contamination in potable water [2]. Therefore, growing water demands is a major concern due to pollution, which is ultimately hampering the availability of drinkable water. In order to combat the crisis and provide the solution, many types of wastewater treatment methods were applied for purification of wastewater such as anaerobic digestion, photocatalytic reaction, oxidation process, membrane separation and electrodialysis, among others. Among these different wastewater treatment processes available in the market, membrane technology is most widely employed for reclaiming water from different wastewater process streams due to its economical and industrial advantages over the other technological options.

A membrane is basically a selective barrier between water and dissolved organic and inorganic molecules that allows some organic and inorganic molecules to pass through and resists others to pass through the membrane layer. Membrane technology is a process in which molecules get separated in liquid solutions or gas mixtures through semipermeable membranes, thereby retaining large molecules and allowing smaller molecules to pass through in permeate from a region of high concentration to low concentration [3]. Briefly, membrane technologies are differentiated based on the driving force, permeate streams and retentate streams. The semipermeable membrane acts as a barrier that retains larger molecules while allowing smaller molecules to pass through the membrane into the permeate. A resulting flux is calculated based on the volumetric flow rate per unit area of the membrane used [4]. The transmembrane pressure is generated due to the force barrier, and it can be defined as the pressure difference between the feed and the permeate stream that needs to be maintained properly.

$$J = TMP / \mu * Rt \tag{2.1}$$

where
J = flux through membrane,
TMP = transmembrane pressure (bar),
μ = is the viscosity (kg/ms), and
R_t = the total resistance (membrane flow and cake resistance).

Basically, membrane technologies are divided into main categories such as (a) pressure-driven, which is subdivided into MF, UF, NF, and RO; (b) pressure thermally driven (membrane distillation); (c) non-pressure-driven (FO and liquid membrane); and (d) non-pressure electrically driven (electrodialysis). Different types of membrane-based processes are discussed in detail in the following sections. Figure 2.1 shows different categories of membrane-based technologies.

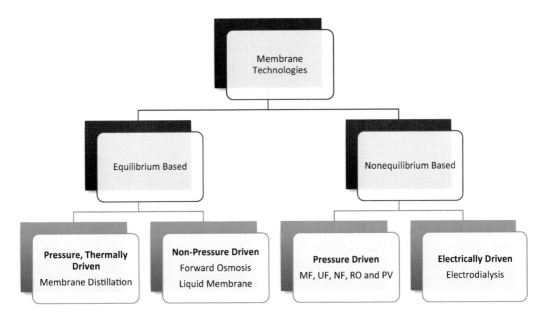

FIGURE 2.1 Different categories of membrane-based technologies.

2.2 PRESSURE-DRIVEN MEMBRANE PROCESSES

Pressure-driven membrane processes are the most commonly used technology for wastewater treatment. This technology is used for re-concentrating the dilute solution based on the application of pressure to separate the permeate and retention phases. The permeate phase has low solute content compared to retention and feed solution. The applied pressure determines the total operational cost of the system. Based on the molecular size of solute molecules retained by the membrane, it can be further classified into different types.

2.2.1 MF

MF is a pressure-driven physical process of filtering the contaminated fluid through membranes having pore sizes ranging from 0.1 to 10 μm [5]. MF membranes can separate particles of molecular weights less than 100,000 g/mol [6]. It is mostly used for removing suspended large particulates, colloids and microorganisms from process streams. MF has a wide range of applications, mostly employed in water, food, beverage and bioprocessing industries [7]. MF membranes are very vital for primary disinfection of water containing pathogens, which are responsible for outbreak of several diseases. MF membranes can also be used for treating secondary wastewater effluents to remove turbidity particularly. MF membranes are also employed in the cold sterilization of beverages and pharmaceuticals to remove bacteria and other undesired suspended particles from liquids. In petroleum refining, where removal of particulates from flue gases is a major concern, today, MF membranes are used. MF membranes are also applied in the dairy industry, particularly for milk and whey processing and the removal of bacteria and harmful species from milk. MF membrane are of two types: (1) crossflow filtration, where the fluid is passed through tangentially concerning the membrane [8], and (2) dead-end filtration, where the process fluid and particles larger than the pore size of the membrane are stopped at its surface and treated at once, subject to cake formation [9].

2.2.2 UF

UF is a similar pressure-driven process like MF but with smaller pore sizes ranging from 0.01 to 0.1 μm, leading to a separation through a semipermeable membrane [10]. UF membranes are helpful in the removal of viruses and polypeptides and are widely used in protein concentration. UF membranes have a wide range of applications in industries such as chemical and pharmaceutical manufacturing, food and beverage processing, and wastewater treatment [11]. To produce potable water, UF is used to remove particulates and macromolecules. A UF system is used as a substitute for secondary and tertiary filtration, which involves coagulation, flocculation, sedimentation, chlorination, and so on. UF is advantageous since no additional chemical is required, it requires a small plant size and more than 90% removal of pathogens can be achieved [12].

2.2.3 NF

NF is a membrane filtration process with pore sizes ranging from 1 to 10 nm [13]. The process stream consists of less TDS and is mostly used for softening water and removal of organic matter [14]. Nanofilters soften water by retaining scale-forming, hydrated divalent calcium, and magnesium ions while passing smaller hydrated monovalent ions [15]. During this process, additional sodium ions are not required when filtration is carried out [16]. NF membranes can process a large volume of the feed stream and continuously yield use products.

2.2.4   RO

RO is a process of separating ions and other unwanted inorganic molecules and particles through a semipermeable membrane with pore sizes ranging from 0.0001 to 0.001 μm [17]. The applied pressure is greater than the osmotic pressure in the high-concentration region. RO membranes are useful in removing dissolved and suspended chemical and biological species from the feed stream to produce drinking water. As a result, solute is retained on the pressurized side of the membrane, and pure solvent is permitted to pass to the other side. RO is mostly used for seawater purification to remove salt and other contaminates from the seawater. In industries, RO is used to remove minerals from boiler water [18]. It is also used to clean brackish groundwater and is used in the production of deionized water. The reverse osmosis process does not require heat energy and flow can be regulated by high-pressure pumps. RO is now commonly employed in the desalination process because of its low energy consumption [19]. Figure 2.2 depicts the pressure-driven membrane processes in decreasing order of the particle size retained by the membrane.

2.3   FO

FO has attracted the attention of worldwide researchers as it is a potential membrane process for the treatment of wastewater and an alternative to RO for producing high-quality water. Natural osmotic pressure difference acts as a driving force in FO [20].

That FO has no requirement for external hydraulic pressure is one of the major advantages of FO. In addition, FO requires low capital cost. The advantages associated with FO are (a) it is capable of removing almost all solute particles; (b) it has excellent durability, reliability and water quality; (c) high salt rejections; and (d) no need for feed pretreatment and the like. FO can treat many complex feeds like industrial wastewater, landfill leachate, nutrient-rich liquid streams, activated sludge, municipal sewage water, and nuclear wastewater, among others. However, FO has some disadvantages, such as partial elimination of trace organic pollutants, recovery step in closed, little flux and solute escape and more [21]. Growing interest in FO for wastewater treatment tends toward the

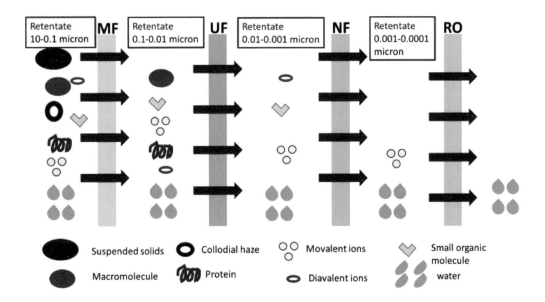

FIGURE 2.2   Membrane processes in terms of particle size distribution.

commercialization of FO. For the applicability of FO at a commercial scale, the fouling tendency of FO membranes needs to be resolved. The characteristics of an ideal FO membrane for wastewater treatment are (a) a dense, ultra-thin and active-separating layer; (b) an open, thin hydrophobic support layer with high mechanical stability; and (c) a high hydrophilicity to enhance the flux and reduce fouling tendency of the membrane. Recently, the first FO desalination facility with a capacity of 200 m³/day has been established in Oman, which makes FO technology more trustworthy [22].

2.4 MD

MD is a potential technology that can help reduce global water-energy stress sustainably. MD works based on the equilibrium of vapor and liquid, which requires heat energy to achieve the feed solution's latent heat of vaporization. In the case of MD, one side of the membrane must be in contact with the feed solution, which allows only the vapors of volatile compounds to pass through the membrane and retains the nonvolatile solutes and liquid molecules due to hydrophobicity. Based on the difference in the partial vapor pressure, the vapors of the volatile compounds pass through the membrane and get liquified on other side. MD has been used for applications such as desalination, the removal of small contaminants and the recovery of other components. Various MD configurations have been utilized to retain driving force on both sides of membrane [23]. MD has been categorized into configurations that are differentiated only by their course of condensation, the recovery of vapor and the technique by which the driving force is applied: (a) direct contact membrane distillation, (b) vacuum membrane distillation, (c) air gap membrane distillation and (d) sweeping gas membrane distillation. MD is a potential alternative technology for wastewater treatment as compared to present desalination practices owing generation of high-purity distillate and the probability of working at lower temperatures. In addition, MD has several associated advantages, such as (a) reduced vapor space over conventional distillation, (b) the flexible mechanical properties of the membrane, (c) no need for pretreatment, (d) negligible organic fouling, (e) the compactness of system, (f) easy installation, (g) low energy cost and more [24]. In spite of these advantages, MD is associated with several drawbacks, such as (a) little flux in the produced vapor, (b) huge sensitivity of the produced vapor flux, (c) a chance of membrane wetting, (d) the extreme cost of the membrane, (e) little thermal energy recovery and (f) huge energy utilization, among others. However, the use of other energy sources like solar thermal energy, waste heat and others can make MD energy-efficient, cost-effective and environmentally pleasant process for wastewater treatment [25]. Figures 2.3a and 2.3b show the schematic of the FO and MD processes, respectively.

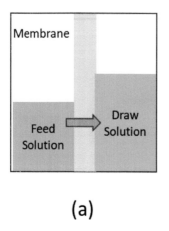

(a)

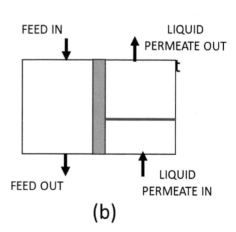

(b)

FIGURE 2.3   (a) FO. (b) MD.

2.5 LIQUID MEMBRANES

A liquid membrane (LM) system has been utilized in multiple sectors such as chemical, biotechnological, biomedical engineering and wastewater treatment for a spectrum of applications like gas partitions, the retrieval of prized or lethal metals, the exclusion of organic compounds, the development of sensing equipment and more. LMs are water-supported or unsupported immiscible suspensions consisting of surfactants and various other reagents in a hydrocarbon solvent, which seizes globules of an aqueous solution of suitable reagent for eliminating wastewater contaminants [26] LMs have been categorized into the following types: (a) ELM, (b) SLM and (c) BLM. SLMs are composed of an inorganic film impregnated with an organic solvent and required complexing agent while BLM consists of U-tube with an immiscible liquid phase. Of all liquid membranes, emulsion liquid membranes are preferred due to its cost-effectiveness, energy-saving nature, elevated interfacial area and provide very fast mass transfer [27]. In addition, a high solute transfer rate, fast extraction, little energy utilization, easy and simple installation and low capital, as well as operational cost, high selectivity and more, are some additional advantages associated with ELM.

An ELM is composed of a uniform slim layer of organic liquid that is present between two aqueous phases with distinctive compositions. It can be explained as a bubble within a bubble in which the outer one carries an extractant or complexing agent while the inner one carries an internal phase reagent. The efficiency of ELM mainly depends on the compositions of the organic phase, the emulsifier, the internal phase, the diluents and the carriers/extractants. Kumar at al. reviewed the characteristics of ELM, the stability of EML and the parameters that influence ELM's efficiency, among others, in detail [28]. However, the requirement of control on the stability of the emulsion and the requirement of emulsion breakdown to recover the receiving and carrier phases are the major drawbacks of ELM.

2.6 ELECTRODIALYSIS AND ELECTRODIALYSIS REVERSAL

ED is an electrically driven membrane separation that is commonly used for wastewater treatment. The ED process is driven by the difference in electric potential by placing an ion-exchange membrane between the anode and the cathode. Due to the difference in electric potential between the ions, the ions travel through a selective membrane barrier toward the respective electrodes. The efficiency of ED process is dependent on factors like current density, pH, flow rate, ED cell structure, feed-water ionic concentration, and the properties of the ion-exchange membrane, among others [29].

ED is a proficient process and favors a feed solution with a small salt concentration. And the electric current required for treating wastewater can be derived from the rate of ions being transported through the ion-exchange membrane. The ED process has been used extensively on a commercial scale for water desalination and salt preconcentration. Using ED, a 50–99% exclusion of pollutants, contaminants or salts can be accomplished [30]. Besides, several of the advantages of ED include no requirement of additional chemicals, high selectivity, small electrical resistance, high mechanical and chemical stability and more. However, fouling, which increases membrane resistance and reduces selectivity, is one of the main disadvantages of ED.

Several methods have been suggested to reduce the ED fouling such as pretreatment of feed, zeta-potential control, membrane characteristics adjustment, flow rate intensification and so on Of all suggested methods, EDR is recommended for reducing fouling of membrane because there is no requirement for extra chemicals and the improved membrane life span.

EDR can also be used as an independent wastewater treatment separately. EDR identifies as a highly potential wastewater treatment owing to its reasonable energy utilization, anti-scaling, and antifouling characteristics. Besides, EDR has some important advantages such as being capable of generating highly concentrated brine and being an easy process [31]

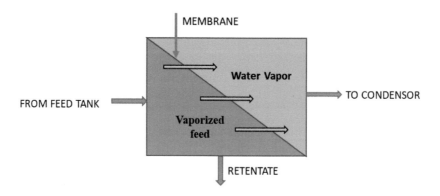

FIGURE 2.4   Schematic of PV membrane process.

2.7   PV

PV is a pressure-driven membrane practice utilized for separating liquid mixtures. In PV, the membrane plays is the selective obstacle between two phases, that is, the feed liquid phase and the permeate vapor phase. The feed liquid phase contacts one side of the membrane, which selectively allows only certain components to pass through. The permeate vapor phase is enriched in the selective components. The driving force for PV is the difference in pressure generated by cooling and condensing during the permeate vapor phase [32] According to the solution–diffusion model, PV the process can be split into three stages: (a) adsorption of permeate from the feed on the membrane, (b) diffusion of permeate in vapor phase through the membrane and (c) desorption of permeate in the vapor phase from the membrane. The separation achieved by PV depends on the permeation rate of specific constituents of the feed across the membrane. Higher efficiency, low power consumption, cost-effectiveness, no pollution and easy operation and installation are the associated advantages of PV [33].

PV is primarily utilized for the elimination of water from organic compounds, the exclusion of low-concentration organics from aqueous mixtures and organic–organic separations. PV is an advantageous separation process compared to conventional distillation due to its low energy requirement and its capability of separating the azeotropic mixtures. PV can be classified into two types depending on the permeating components: hydrophilic and hydrophobic. Water/alcohol separation is a renowned illustration of PV practice in the chemical industry. The foremost industrial plant for removing water from alcohols using PV was established in the late 1980s. PV with a hydrophobic membrane has the highest potential to eliminate a low concentration of organic components from the wastewater stream [34]. Figure 2.4 depicts the PV membrane process.

2.8   HYBRID MEMBRANE PROCESS

A hybrid process is a combination of two processes one is a conventional membrane process and another conventional process. The hybrid process can be categorized into two groups: a combination of two or more different membrane processes and a combination of a membrane process and another process. Membrane processes include MF, UF, NF, RO, PV and MD. In the first group, two processes are combined in different permutations and combinations depending on the requirements like MF–RO, UF–RO, NF–RO, NF–MD, PV–RO, UF–MD and UF–NF–RO–MD. In the second group, MF, NF, UF RO and so on are combined with coagulations, adsorption, ion-exchange membranes, reactors and the like. The advantage of the hybrid process is that it can be a unique combination for a specific application [35]. The hybrid process has several advantages that single-membrane technologies cannot offer. It offers high-purity products [36]. A hybrid process can overcome physical and chemical restrictions over any single process [36].

2.9 MBRs

An MBR is also one of the emerging technologies for wastewater treatment. MBR is a combination of membrane processes like ultrafiltration or microfiltration with biological treatment like conventional activated sludge [37]. An MBR is an economical, environmentally friendly way to separate solid and liquid and is independent of sludge concentration and quality, compared to other conventional processes. An MBR is the most innovative in wastewater treatment, as it overcomes the drawback of the conventional activated sludge process (ASP) [38]. The use of MBR technology replaces the requirement of the secondary and tertiary clarifiers [38]. The microorganism is trapped in the bioreactor because of the membrane, and this gives better control over the biological reactions and modifying the conditions of the microorganisms in the aerated tank [39].

2.9.1 CONFIGURATION OF MBRS [41]

2.9.1.1 Submerged/Immersed MBRs

In a submerged/immersed MBR (iMBR), the membrane module is submerged in a bioreactor directly. A suction pump is attached to draw effluent and sludge will be trapped in the membrane.

2.9.1.2 Crossflow MBRs

A crossflow membrane bioreactor (cMBR) is a combination of membrane filtration and a traditional bioreactor in a single process unit. An MBR is an alternative for the second clarifier, a conventional biological treatment system. A cMBR allows the membrane to be easily cleaned in situ and can be easily cleaned in situ and operated with high sludge concentration in the MBR reactor.

2.9.1.3 Hybrid MBRs

A hybrid MBR (hMBR) is similar to iMBR. It consists submerged membrane module with some carriers in the bioreactor. The carrier is used to stabilize the treatment process.

2.9.1.4 Biocatalytic MBR [40]

In a biocatalytic MBR, the catalyst is embedded in the membrane; thus, the membrane plays a role not only in the separation but in the reaction as well. The biological catalyst is preferably used for biocatalytic membranes. The biocatalyst is embedded in a membrane, which allows the continuous processing, higher efficiency and low fouling of the membrane. Figure 2.5 shows a schematic of the membrane bioreactor.

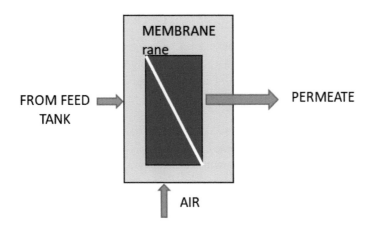

FIGURE 2.5 Schematic of an MBR.

2.9.2   Advantages of MBRs over Conventional Methods [41]

a. Very stable process
The conventional process is dependent on wastewater composition. Variation in the composition of wastewater like the presence of toxic chemicals, high salt concentrations, and oxygen content affects the effluent.

b. Compact design
Due to the membrane separation, the concentration of microorganism can be maintained 4–5 times of conventional systems. It eliminates the large space requirement for secondary and tertiary treatment.

c. High effluent quality
Membranes withhold microorganisms and suspended solids and provide clear effluent compared to conventional treatment.

d. Low sludge production
MBRs can operate at a low F/M ratio, being the feed of organic substance per number of microorganisms per time unit.

e. Treatment wastewater up to 60°C

f. Shorter hydraulic retention times

g. MBRs have a smaller footprint
The secondary and tertiary clarifiers are removed, which reduces the plant's overall footprint.

2.9.3   Disadvantages of MBR [42]

a. High operational and capital cost

b. Membrane complexity and fouling

c. Energy cost

d. MBRs have a membrane with a pore size of less than 0.1 μm, making them resistant to certain chlorine-resistant pathogenic bacteria and viruses in sludge.

2.10   ION-EXCHANGE MEMBRANES

Ion-exchange membranes are semipermeable membranes in which ionic groups are attached with a polymeric backbone. The concentration and ionic groups have helped in different applications [43]. Ion-exchange membranes can be classified by the ions' functionality and the polymer backbone. The foremost driving force for an ion-exchange membrane is the electrochemical interaction between the molecules.

2.10.1   Classification of Ion-Exchange Membranes Based on Functionality [44]

2.10.1.1   Cation-Exchange Membranes

Cation-exchange membranes consist of an anion in the polymeric backbone, which has a selective permeability for cations. The ions used for cation exchange used are $-SO_3^-$, $-COO^-$, $-PO_3^{2-}$, $-PHO_2^-$ and others.

2.10.1.2   Anion-Exchange Membranes

An anion-exchange membrane consists of the cation in the polymeric backbone, which has a selective permeability for anions. The ions used for anion exchange used are $-NH_2R^+$, $-NHR_2^+$, $-NR_3^+$ and $-SR_2^+$.

2.10.1.3   Amphoteric Ion-Exchange Membranes

Amphoteric ion-exchange membranes consist of both cations and anions that are equally and randomly distributed over the backbone.

2.10.1.4   Bipolar Membranes
A bipolar membrane consists of an anion and a cation membrane layered together.

2.10.1.5   Mosaic Ion-Exchange Membranes
Mosaic ion-exchange membranes are composed of pores that consist of cation and anion ions. Mosaic membranes consist of two-layered membranes fixed with cations and ions in parallel resins separated by a neutral polymer.

2.10.2   DESIRED PROPERTIES FOR ION-EXCHANGE MEMBRANES [45]

a. High permeable selectivity: An ion-exchange membrane should be highly permeable for counterions but should be impermeable to co-ions.
b. Low electrical resistance: The permeability of an ion-exchange membrane for the counterions under the driving force of an electrical potential gradient should be as high as possible.
c. Good mechanical and form stability: The membrane should be mechanically strong and should have a low degree of swelling or shrinking in the transition from dilute to concentrated ion solutions.
d. High chemical stability: The membrane should be stable over the entire pH range from 1 to 14 and in the presence of oxidizing agents.

2.11   CP

CP is defined as a phenomenon in which the particle concentration near the membrane is greater than the bulk. CP is commonly observed in all membrane processes. Most of the membrane system works with pressure applied on one side of the membrane; due to the applied pressure, the solvent molecules cross the membrane barrier, and the solute molecules will be retained at the rejection side. The retention of the solute molecule is responsible for its accumulation on the membrane surface. Due to the membrane's solute retention, the concentration of the solute in the permeate is lower than the concentration in the bulk. CP affects the permeate flux because of the increase in the concentration of solute around the membrane surface. Methods to reduce CP have been achieved by pretreating the feed, membrane modification, flow rate and effective cleaning.

2.12   MEMBRANE FOULING AND PRETREATMENT STRATEGIES

All the membrane filtration systems, such as MF, UF, RO and NF, use semipermeable membranes to remove the particles from liquids. During this process of capturing contaminant particles, many of such particles get adsorbed by the surface of the filtration membrane or get deposited within the membrane's pores [46]. As a result, membrane fouling occurs, which will restrict the flow of liquids through the membrane's pores. There are many contributing factors to fouling, such as the presence of an excess amount of organic, biological and colloidal particles in the source water and the choice of unsuitable processing parameters like temperature, pressure, pH and flow rate, along with the inappropriate choice of membrane material.

Membrane fouling is classified based on the type or origin of foulants and on the fouling reversibility. Based on the type of the origin from which these foulants are derived, membrane fouling can be called inorganic fouling, biofouling, organic fouling and particulate fouling. Biofouling generally occurs due to the formation of biofilms of colonies formed by aquatic organisms such as algae [47]. Thus, to prevent this, chemical cleaning is usually employed in the case of low-pressure membranes. Organic fouling is more serious than biofouling, and thus, researchers are more worried about this [48, 49]. Natural organic matter (NOM) is the main source of concern because it is ubiquitous in natural water and heterogenic in nature, and it was found to be one of main reasons for organic fouling [50]. The other type of fouling is inorganic, which occurs due to the precipitation of metal oxide and hydroxide particles and results in a gel or solid cake-type layer formation over the membrane surface [51]. Last but the least, particulate fouling takes place due to the accumulations of inert and colloids particles like silt and clay materials inside as well as on the surface of the membrane [52]. All these types of membrane fouling cause serious issues that disturb the quality and flux of permeate and water recovery and increase the operational cost. They also shorten the life span of the membrane [53, 54]. Thus, to prevent membrane fouling, little maintenance is required in membrane separation technology as compared to other separating methods.

Thus, there are some fouling remediations that involve the pretreatment of the feed to prevent the fouling of membranes and improve the antifouling properties of the membrane by proper cleaning and backwash process.

2.12.1 FEED PRETREATMENT

The pretreatment of the feed is very important to prevent all types of membrane fouling as well as scaling. Such methods should be able to control the fouling to such an extent so that it can be achieved at the practical scale also. Different pretreatment methods are used in the case of low-pressure membranes.

2.12.1.1 Coagulation

In coagulation, different chemical coagulant such as ferrous sulfate, ferric chloride, poly aluminum chloride, and alum, among others, are added to make the size of particles of the feed stream large before subjecting it to a filtration unit. This method has been proved to be useful for reducing reversible fouling while irreversible fouling remains the same. This is because small particles do not get coagulated and thus the chances of irreversible fouling still exist. The factors that affect the coagulation process are coagulant dosage, pH, Ca^{2+} content present in the feed and the nature of the dissolved organic matter [55]. This helps with reducing the membrane fouling by minimizing the pore plugging and increases the membrane backwashing's efficiency. Coagulation can be done by an inline coagulation process in which coagulants are added into the feed stream, which forms flocs, and then it is allowed to enter the filtration unit. This changes the mechanism from pore blocking to cake formation, which can be removed easily by backwashing.

The sedimentation process is followed by the addition of a coagulant that forms the flocs, which are allowed to settle by a sedimentation process before removing. Then the supernatant is then fed to the filtration unit [56]. In addition to this, an alternative process known as coagulation-adsorption, in which foulants are adsorbed by using powder activated carbon as adsorbent between the coagulation step and ultrafiltration. Also, flocculation is another pretreatment method used to remove the particles so to improve the permeate flux. The flocculation is done before membrane filtration to reduce clogging of the membrane. It is used in combination with coagulation in which the large flocs formed from particles aggregated by coagulation.

2.12.1.2 MIEX

This chemical process involves the adsorption of charged species and ions to polymer beads. These beads are recoverable once it becomes saturated and can also be regenerated by preparing a brine solution that allows the desorption of the species and ions. This method is very useful for removing the molecules with high charge density from low- to medium medium-molecular-weight compounds. In many cases, it performs in synergy with the coagulation method to remove dissolved organic carbons [57]. It was found that a combination of both methods is more effective for removing dissolved organic carbon. About 90% trihalomethane and halo acetic acids were removed from water using a combined method of coagulation and MIEX [58]. Also, MIEX alone could be able to remove more than 80% of NOM before subjecting it to the filtration unit.

2.12.1.3   Micellar-Enhanced UF

This method involves the removal of foulants in the form of micelles by adding surfactants to the feed. Today, different types of surfactants are used depending on the charge of the hydrophilic part of the molecule, which is known as anionic, cationic, non-anionic and zwitterionic. These surfactants increase the particle size and thus they can be removed using membranes with large pore sizes. These surfactants can also disrupt the bacterial cell wall. The choice of surfactants is dependent on the compatibility of the solids which has to be removed. Cationic surfactants like dodecyl amine and anionic surfactant like dodecylbenzenesulfonic acid were used for removing Pb and as content from municipal wastewater [59].

2.12.1.4   Modification of Membrane Properties

The properties of membrane play an important role since it affects the solute and membrane interactions and hence affect the adsorption and fouling processes also. For protein filtration, the hydrophilic membranes (such as cellulose esters and aliphatic polyamides) help in reducing membrane fouling. Similarly, chemical medications were also performed like the sulfonation of polysulfone and the blending of a hydrophobic polymer (made up of polyetherimide and polyvinylidene fluoride) with a hydrophilic one (polyvinylpyrrolidone) to enhance the antifouling property of membranes. Another way of improving the solute–membrane interaction is by pretreating the membrane with hydrophilic surfactants or enzymes. The methods like plasma treatment, polymerization or grafting of the membrane by ultraviolet light, heat, or chemicals; interfacial polymerization; and the introduction of polar and ionic groups on the membrane surface by performing their reaction with bromine, fluorine, strong bases and strong acids were used to modify the conventional ultrafiltration membranes, such as polysulfone, polyether sulfone or polyvinylidene fluoride [60, 61].

2.13   ADVANTAGES OF MEMBRANE SEPARATION TECHNIQUES

The membrane separation techniques have offered many advantages as compared to other methods [62, 63]:

a. Membrane separation methods are applicable at both the molecular and the scale-up level, and thus, many separations need to be met by the membrane process.
b. There is no need to change the phase to make membrane separation processes. So the energy requirement is lower unless it needs to be increased to increase the pressure of the feed stream to drive the permeate stream across the membrane.
c. The membrane techniques offer a simple, low economic-based, and easy operational service to separate unwanted components from wastewater. Also, there is no need for complex control systems.
d. Membranes are manufactured with a high selectivity according to the components that need to be separated. The selectivity values are generally higher for membrane separation than the common values for volatility for distillation operations.
e. As many polymers and inorganic compounds can be used to make membranes, there are more chances of having control over the separation selectivity.
f. Membrane techniques are also able to recover the minor components from the feed stream without increasing the energy cost value.
g. Membrane techniques are economical and environmentally friendly ones because they are simple, efficient and based on nonharmful materials.

2.14   LIMITATIONS OF MEMBRANE TECHNIQUES

The drawbacks associated with membrane separation techniques are as follows [64]:

a. The membrane process cannot be staged easily compared to the distillation process because sometimes membrane separation methods consist of two or three stages. And therefore, the membrane must have a higher selectivity for a given separation rather than relative volatilities in distillation. Thus, it should be performed with high selectivity and involving few stages for membrane processes rather than with low selectivity including many stages or steps for other processes.

b. It was observed that the membrane has chemical incompatibilities, especially for typical chemical industry-based feed solutions that have a high content of organic compounds. And for that type of feed solution, polymer-based membranes are usually used, but they suffer from low selectivity and low lifetime too.

c. The membrane modules cannot be operated at much higher than room temperature because the polymer-based membranes do not maintain their physical reliability at a temperature above 100°C. Thus, this temperature limit restricts the membrane to some chemical-based separations.

d. The membrane process has many membrane modules arranged in a parallel fashion that need to be replicated over and over so as to scale up the process. But this task is not easy with a large volume of the feed stream.

e. The membrane process is burdened with membrane fouling, which sometimes is difficult to manage, affects the separation process and makes it unsuitable for such applications.

2.15   APPLICATIONS OF MEMBRANE TECHNOLOGY

There is an infinite number of industries in which membrane separations are used for different purposes [65]. In every sector, such as food (dairy and sugars), pharma, biotechnology, and chemical, this method is applied as follows:

a. Food industry: In this sector, it is mostly used for concentrating egg white; clarification; preconcentrating fruit juices, bovine or bone gelatin; extracting and concentrating ashes of porcine; clarifying meat brine for removing bacteria and its reuse; concentrating vegetables like soy, oats and canola; and removing alcohol from wine and beer.

b. Starch and sweetener industry: In this industry, it is mainly used for clarifying corn syrups like dextrose and fructose, concentrating the rinse water from starch, for depyrogenating dextrose syrup and its enrichment and concentrating maceration water.

c. Sugar industry: Membrane separation is mainly used for clarifying unprocessed juice by using different clarifiers. It is also used for concentrating/dividing various sugar solutions in the production process.

d. Chemical industry: Membrane separation is used for desalination and diafiltration and for purifying dyes, pigments and optical brighteners to clean wastewater. It is also used for concentrating and dehydrating the minerals like kaolin clay, TiO_2 and $CaCO_3$; clarifying caustic agents; producing polymers; and recuperating metals.

e. Pharmaceutical industry: This method is used when harvesting cells and recuperating biomass, which are important steps during the fermentation process for manufacturing antibiotics. Membrane filtration helps with reducing labor and maintenance costs. These membrane separation techniques are very important for producing enzymes and concentrating the enzymatic solutions before subjecting them to other processes.

2.16 TECHNICAL BARRIERS AND CHALLENGES

Although membranes and membrane-based technologies provide us effective solutions for wastewater treatment, they certainly have some technical barriers and challenges to overcome. The first and foremost barrier limiting the performance of the membrane is fouling. The accumulation of feed material, such as impermeable dissolved solutes, suspended particles and the like, on the membrane's surface and within the pores of the membrane reduces flux and affects the efficiency of the process. Another challenge is the life-cycle assessment analysis that is the environmental impact of membrane filtration technologies. Global warming, ecotoxicity and eutrophication are possible environmental impacts. The maintenance costs of pressure-driven filtration processes are high when accounting for fouling and replacement. RO, as installed, in our homes consumes more water due to low back pressure, resulting in recovering of a lower percentage of water entering the system and the remaining water is discharged as waste.

2.17 CONCLUSION AND FUTURE PERSPECTIVES

Regardless, the significant enhancement standards in potable water, many other water resources are contaminated with chemical and bio-pollutants. This leads to an outbreak of many diseases. Soon India might not be able to provide replenishable water sources. Therefore, there is an urgent need to provide a solution for treating wastewater. This chapter provided an overview of membrane and membrane-based strategies for wastewater treatment. Pressure-driven membrane processes such MF, UF, NF and RO are useful in separating contaminants and purifying wastewater based on respective pore sizes. This chapter also explained the non-pressure-driven (FO, liquid membrane), pressure thermally driven (membrane distillation), and non-pressure electrically driven (electrodialysis) processes in detail. But there is a scope for new developments to counter technical barriers, such as membrane fouling, and increase the life cycle of membrane filtration systems. To promote coagulation, there is a need for a modified construction of membranes and additives that can result in a reduction in fouling membranes. More energy-efficient membranes can be incorporated into MBR systems. Reducing energy consumption is one of the major aspects that can be accounted for to minimize the scaling of membranes. To increase the performance of membranes, permeate flux can be enhanced. Recent development involves integrating reverse osmosis with electrodialysis which results in good recovery of products in permeate and reduces the amount of retentate for discharge. Therefore, such developments should require lower energy consumption, and reduced cost should be introduced for membranes systems to treat wastewater effectively.

BIBLIOGRAPHY

1. Petersen-Perlman, J.D., Veilleux, J.C.; Wolf, A.T. (2017). "International water conflict and cooperation: challenges and opportunities". *Water International*. **42**(2): 105–120.
2. Environmental Outlook to 2050: The Consequences of Inaction. (2012). OECD.
3. Obotey Ezugbe, E.; Rathilal, S. (2020). "Membrane technologies in wastewater treatment: A review". *Membranes*. **10**: 89. https://doi.org/10.3390/membranes10050089
4. Ghosh, R. (2006). *Principles of Bioseparations Engineering*. Word Scientific Publishing Co. Pte. Ltd, Toh Tuck Link, p. 233.
5. Baker, R. (2012). *Microfiltration, in Membrane Technology and Applications*, 3rd edn. John Wiley & Sons Ltd, San Fransisco, CA, p. 303.
6. Microfiltration/Ultrafiltration (2008). Hyflux Membranes, accessed 27 September 2013.
7. Perry, R.H.; Green, D.W. (2007). *Perry's Chemical Engineers' Handbook*, 8th edn. McGraw-Hill Professional, New York, p. 2072–2100.
8. Seadler, J.; Henley, E. (2006). *Separation Process Principles*, 2nd edn. John Wiley & Sons Inc, Hoboken, NJ, p. 501.

9. Laîné, J.-M.; Vial, D.; Moulart, P. (1 December 2000). "Status after 10 years of operation: Overview of UF technology today". *Desalination*. **131** (1–3): 17–25.

10. American Water Works Association Research Foundation . . . Ed. group Joël Mallevialle (1996). *Water Treatment Membrane Processes*. McGraw Hill, New York [u.a.]. ISBN 9780070015593.

11. American Water Works Association Research Foundation . . . Ed. group Joël Mallevialle (1996). *Water Treatment Membrane Processes*. McGraw Hill, New York [u.a.]. ISBN 0070015597.

12. Cheryan, M. (1998). *Ultrafiltration and Microfiltration Handbook*. CRC Press, Boca Raton, FL. ISBN 1420069020.

13. Roy, Y.; Warsinger, D.M.; Lienhard, J.H. (2017). "Effect of temperature on ion transport in nanofiltration membranes: Diffusion, convection and electromigration". *Desalination*. **420**: 241–257.

14. Rahimpour, A.; et al. (2010). "Preparation and characterisation of asymmetric polyethersulfone and thin-film composite polyamide nanofiltration membranes for water softening". *Applied Surface Science*. **256** (6): 1657–1663.

15. Labban, O.; Liu, C.; Chong, T.H.; Lienhard V, J.H. (2017). "Fundamentals of low-pressure nanofiltration: Membrane characterization, modeling, and understanding the multi-ionic interactions in water softening" (PDF). *Journal of Membrane Science*. **521**: 18–32.

16. Baker, L.A.; Martin, C. (2006). "Current Nanoscience". *Nanomedicine: Nanotechnology, Biology and Medicine*. **2** (3): 243–255.

17. Crittenden, J.; Trussell, R.; Hand, D.; Howe, K.; Tchobanoglous, G. (2005). *Water Treatment Principles and Design*, 2nd edn. John Wiley and Sons, Hoboken, NJ.

18. Shah, V. (ed.). (2008). *Emerging Environmental Technologies*. Springer Science, Dordrecht, p. 108. ISBN 978-1402087868.

19. Warsinger, D.M.; Mistry, K.H.; Nayar, K.G.; Chung, H.W.; Lienhard V, J.H. (2015). "Entropy generation of desalination powered by variable temperature waste heat". *Entropy*. **17** (11): 7530–7566.

20. Coday, B.D.; Xu, P.; Beaudry, E.G.; Herron, J.; Lampi, K.; Hancock, N.T.; Cath, T.Y. (2014). "The sweet spot of forward osmosis: Treatment of produced water, drilling wastewater, and other complex and difficult liquid streams". *Desalination*. https://doi.org/10.1016/j.desal.2013.11.014.

21. Lutchmiah, K.; Verliefde, A.R.D.; Roest, K.; Rietveld, L.C.; Cornelissen, E.R. (2014). "Forward osmosis for application in wastewater treatment: A review". *Water Research*. https://doi.org/10.1016/j.watres.2014.03.045.

22. "Modern water commissions Al Najdah FO plant". (2012). *Membr. Technol.* https://doi.org/10.1016/s0958-2118(12)70202-1.

23. Jafari, A.; Kebria, M.R.S.; Rahimpour, A.; Bakeri, G. (2018). "Graphene quantum dots modified polyvinylidenefluoride (PVDF) nanofibrous membranes with enhanced performance for air gap membrane distillation". *Chem. Eng. Process. – Process Intensif.* https://doi.org/10.1016/j.cep.2018.03.010.

24. Belessiotis, V.; Kalogirou, S.; Delyannis, E. (2016). *Thermal Solar Desalination: Methods and Systems*. Academic Press, New York. https://doi.org/10.1016/C2015-0-05735-5. ISBN 978-0-12-809656-7.

25. Ding, Z.; Liu, L.; El-Bourawi, M.S.; Ma, R. (2005). "Analysis of a solar-powered membrane distillation system". *Desalination*. https://doi.org/10.1016/j.desal.2004.06.195.

26. Kitagawa, T.; Nishikawa, Y.; Frankenfeld, J.; Li, N. (1977). "Wastewater treatment by liquid membrane process". *Environ. Sci. Technol.* **11** (6): 602–605.

27. Kumbasar, R.A. (2010). "Selective extraction of chromium (VI) from multicomponent acidic solutions by emulsion liquid membranes using tributhylphosphate as carrier. *J. Hazard. Mater.* https://doi.org/10.1016/j.jhazmat.2010.02.019.

28. Kumar, A.; Thakur, A.; Panesar, P.S. (2019). "A review on emulsion liquid membrane (ELM) for the treatment of various industrial effluent streams". *Reviews in Environmental Science and Biotechnology*. https://doi.org/10.1007/s11157-019-09492-2.

29. Akhter, M.; Habib, G.; Qamar, S.U. (2018). "Application of electrodialysis in waste water treatment and impact of fouling on process performance". *J. Membr. Sci. Technol.* https://doi.org/10.4172/2155-9589.1000182.

30. Oztekin, E.; Altin, S. (2016). "Wastewater treatment by electrodialysis system and fouling problems". *The Online Journal of Science and Technology*. **6** (1): 91–99.

31. Zhao, D.; Lee, L.Y.; Ong, S.L.; Chowdhury, P.; Siah, K.B.; Ng, H.Y. (2019). "Electrodialysis reversal for industrial reverse osmosis brine treatment". *Sep. Purif. Technol.* https://doi.org/10.1016/j.seppur.2018.12.056.

32. Favre, E. (2003). "Temperature polarization in pervaporation". *Desalination*. https://doi.org/10.1016/S0011-9164(03)80013-9.

33. Toth, A.J.; Haaz, E.; Nagy, T.; Tarjani, A.J.; Fozer, D.; Andre, A.; Valentinyi, N.; Solti, S.; Mizsey, P. (2018). "Treatment of pharmaceutical process wastewater with hybrid separation method: Distillation and hydrophilic pervaporation". *Waste Treat. Recover.* https://doi.org/10.1515/wtr-2018-0002.

34. Lipnizki, F.; Hausmanns, S.; Ten, P.K.; Field, R.W.; Laufenberg, G. (1999). "Organophilic pervaporation: Prospects and performance". *Chemical Engineering Journal.* **73**: 113–129.

35. www.mdpi.com/journal/membranes/special_issues/hybrid-processes.

36. Ezugbe, E.O.; Rathilal, S. (2020). "Membrane technologies in wastewater treatment: A review". *Membranes.* **10** (5): 89.

37. Bagheri, R.; Pearson, R.A. (1996). "Role of particle cavitation in rubber-toughened epoxies: 1. Micro void toughening". *Polymer.* **37** (20): September, 4529–4538.

38. www.amtaorg.com/Membrane_Bioreactors_for_Wastewater_Treatment.html.

39. Mazzei, R.; Drioli, E.; Giorno, L. (2010). "Biocatalytic membranes and membrane bioreactors". *Catalytic Membranes and Membrane Reactors.* 195–213.

40. Iorhemen, O.T.; Hamza, R.A.; Tay, J.H. (2016). "Membrane bioreactor (MBR) technology for wastewater treatment and reclaimation: Membrane fouling". *Membrane.* **33**: 1–29.

41. Sari Erkan, H.; Bakaraki Turan, N.; Önkal Engin, G. (2018). "Chapter five – membrane bioreactors for wastewater treatment", in *Fundamentals of Quorum Sensing, Analytical Methods and Applications in Membrane Bioreactors*, vol. 81. Elsevier, Amsterdam, pp. 151–200.

42. Hassanvand, A.; Wei, K.; ChenI, S.T.G.Q.; Kentish, S.E. (2017). "The role of ion exchange membranes in membrane capacitive deionisation". *Membrane.* **7** (54): 1–23.

43. Preparation and Characterization of Ion-Exchange Membranes. (2004). *Ion-Exchange Membrane Separation Processes, 89–146.*

44. Strathmann, H.; Grabowski, A.; Eigenberger, G. (2013). "Ion-exchange membranes in the chemical process industry". *Ind. Eng. Chem. Res.* **52**: 10364–10379.

45. Li, N.N.; Fane, A.G.; Ho, W.S.W.; Matsuura, T. (2008). *Advanced Membrane Technology and Applications.* John Wiley & Sons, Hoboken, NJ, p. 989.

46. Rosas, I.; Collado, S.; Gutiérrez, A.; Díaz, M. (2014). "Fouling mechanisms of pseudomonas putida on PES microfiltration membranes". *J Memb Sci.* **465**: 27–33.

47. Zularisam, A.W.; Ismail, A.F.; Salim, M.R.; Sakinah, M.; Ozaki, H. (2007). "The effects of natural organic matter (NOM) fractions on fouling characteristics and flux recovery of ultrafiltration membranes". *Desalination.* **212**: 191–208.

48. Lee, E.K.; Chen, V.; Fane, G. (2008). "Natural organic matter (NOM) fouling in low pressure membrane filtration: Effect of membranes and operation modes". *Desalination.* **218**: 257–270.

49. Gao, W.; Liang, H.; Ma, J.; Han, M.; Chen, Z.; Han, Z.; et al. (2011). "Membrane fouling control in ultrafiltration technology for drinking water production: A review". *Desalination.* **272**: 1–8.

50. Stoquart, C.; Servais, P.; Bérubé, P.R.; Barbeau. B. (2012). "Hybrid membrane processes using activated carbon treatment for drinking water: A review". *J Memb Sci.* **411–412**: 1–12.

51. Koo, C.H.; Mohammad, A.W.; Suja', F.; Meor Talib, M.Z. (2012). "Review of the effect of selected physicochemical factors on membrane fouling propensity based on fouling indices". *Desalination.* **287**: 167–177.

52. Munla, L.; Peldszus, S.; Huck, P.M. (2012). "Reversible and irreversible fouling of ultrafiltration ceramic membranes by model solutions". *J Am Water Works Assoc.* **104**.

53. Chon, K.; Kim, S.J.; Moon, J.; Cho, J. (2012). "Combined coagulation-disk filtration process as a pretreatment of ultrafiltration and reverse osmosis membrane for wastewater reclamation: An autopsy study of a pilot plant". *Water Res.* **46**: 1803–1816.

54. Li, N.N.; Fane, A.G.; Winston, W.S.H.; Matsuura, T. (2008). *Advanced Membrane Technology and Applications.* John Wiley& sons Inc., Hoboken, NJ. ISBN:9780471731672.

55. Kruithof, J.C.; Nederlof, M.M.; Hoffman, J.A.M.H.; Taylor, J.S. (2004). *Integrated Membrane Systems.* Research Foundation and American Water Works Association, Elbert, CO.

56. Drikas, M.; Christopher, W.; Chow, K.; Cook, D. (2003). "The impact of recalcitrant on disinfection stability, trihalomethane formation and bacterial regrowth: A magnetic ion exchange resin (MIEX) and alum coagulation". *Journal Water Supply Research.* **52** (7): 475–487.

57. Singer, P.C.; Bilyk, K. (2002). "Enhanced coagulation using a magnetic ion exchange resin". *Water Research.* **36**: 4009–4022.

58. Ferella, F.; Prisciandaro, M.; Michelis, I.; Veglio, F. (2007). "Removal of heavy metals by surfactant-enhanced ultrafiltration from wastewaters". *Desilination.* **207**: 125–133.

59. Park, Y.G. (2002). "Effect of ozonation for reducing membrane fouling in the UF membrane". *Desalination.* **147**: 43–48.

60. Mulder, M.H.V. (1993). *Membranes in Bioprocessing, Theory and Application.* Chapman and Hall, London, p. 13.

61. Strathmann, H. (2004). "Membrane separation processes: Current relevance and future opportunities". *AIChE Journal.* **47** (5): 1077–1087.

62. He, Y.; Bagley, D.M.; Leung, K.T.; Liss, S.N.; Liao, B.-Q. (2012). "Recent advances in membrane technologies for biorefining and bioenergy production". *Biotechnology Advances.* **30** (4): 817–858. doi:10.1016/j.biotechadv.2012.01.015.

63. Ezugbe, E.O.; Rathilal, S. (2020). "Membrane technologies in wastewater treatment: A review". *Membranes.* **10**: 89.

64. Marella, C.; Muthukumarappan, K.; Metzger, L.E. (2013). "Application of membrane separation technology for developing novel dairy food ingredients, Marella et al.". *J Food Process Technol.* **4**: 9.

65. Membrane Separation Principles and Applications: From Material Selection to Mechanisms and Industrial Uses, 978-0-12-812815-2, 2018. https://doi.org/10.1016/C2016-0-04031-7

3 Current Advances in Bio-Membrane Technology for Wastewater Treatment

Atif Aziz Chowdhury and Ekramul Islam

CONTENTS

DOI: 10.1201/9781003165019-3

3.1   INTRODUCTION

Water is by far the most accessible natural resource on our planet, but only a limited amount is usable and suitable for human activity. As the human population grows, tons of wastewater are released daily through the household, industries and agricultural sectors. However, freshwater reservoirs are not being replenished to satisfy an ever-increasing population and the demands for water consumption. Many parts of the world do not have access to safe and clean drinking water and desperately require economical, efficient and successful treatment methods for local water supplies.

To date, various approaches, including physicochemical (e.g., adsorption, coagulation and precipitation, oxidation, membrane isolation, etc.) and biological methods (e.g., aerobic therapy, anaerobic digestion, etc.) or a combination thereof have been used for the treatment of effluents (Kamali & Khodaparast, 2015). Membrane bioreactors (MBRs) are an updated and sophisticated weapon against wastewater. Wastewater treatment in MBR systems requires two processes, namely, biological processing in a suspended growth bioreactor for biochemical reactions (e.g., bio-oxidation, nitrification and denitrification) and a physical membrane filtration method (Hamedi et al., 2019). Globally, MBR is being used in mitigating both industrial and municipal wastewater. It has been reported that the annual growth rate of MBRs in the global market is about 15% (Judd, 2016).

This chapter is an effort to summarize the use of membrane technology, focusing on membrane biotechnology in wastewater mitigation. It also looks at MBR types, uses and setbacks; fouling and their potential antifouling method; and a potential future framework.

3.2   MEMBRANE PROCESSES

A membrane functions as a selective barrier among two homogeneous phases. Moreover, membranes can conduct much of the separation process and can supplement or pose as an alternative to chemical processes, namely, distillation, extraction, fractionation and adsorption. The benefits of membrane processes include low energy consumption, continuous separation and easy scaling-up. Membranes can be organic or inorganic depending on the constituent material. Synthetic organic polymers compose organic membranes for pressure-driven separation processes (microfiltration [MF], ultrafiltration [UF], nanofiltration [NF] and reverse osmosis [RO]) are mostly made of synthetic organic polymers (summarized in Table 3.1). This includes, among others, polyethylene (PE), polytetrafluorethylene (PTFE), polypropylene and cellulose acetate (Aliyu et al., 2018).

3.2.1   MF

MF is a low-pressure-driven technique in which the substances segregated are 0.1–0.2 μm in diameter (Werber et al., 2016). It is used as a first pretreatment of NF and RO membrane processes, thereby reducing the risk of fouling of the NF or RO membrane. It is suitable for isolating suspensions and emulsions and can retain up to approximately 40% organic pollutants (Kumar et al., 2019).

3.2.2   UF

UF membranes are extremely popular low-energy water filters and serve to eliminate pathogenic microorganisms, macromolecules and suspended matter (Krüger et al., 2016). These membranes have pore sizes up to about 0.1 μm in dimension. However, its drawbacks include an inability to remove some dissolved inorganic contaminants from water and frequent cleaning to ensure the proper pressure stream of water (Zhang et al., 2016).

TABLE 3.1

General Characteristics and Application of Pressure-Driven Membranes (adapted from Obotey Ezugbe & Rathilal, 2020; Zirehpour & Rahimpour, 2016)

Properties	Microfiltration	Ultrafiltration	Nanofiltration	Reverse osmosis
		Membranes		
Membrane type	Symmetric polymer or ceramic membranes	Asymmetric polymer composite or ceramic membrane	Asymmetric polymer composite or ceramic membrane	Thin-film composite membrane
Pore size (nm)	100–10,000	2–100	0.5–2	<0.5
Average Permeability (L/m^2 h bar)	500	150	10–20	5–10
Retained diameters (μm)	10^{-1}–10	10^{-3}–1	10^{-3}–10^{-2}	10^{-4}–10^{-3}
Molecular weight cut off (kilo Dalton)	100–500	20–150	2–20	0.2–2
Pressure (bar)	0.1–2	0.1–5	3–20	5–120
Separation mechanism	Molecular sieve	Molecular sieve	Solution diffusion	Solution diffusion
Solutes Retained	Bacteria, fat, oil, grease, colloids, organics, micro-particles	Proteins, pigments, oils, sugar, organics, microplastics	Pigments, sulfates, divalent cations, divalent anions, lactose, sucrose, sodium chloride	All contaminants including monovalent ions
Material passed	Water, dissolved solutes	Water, dissolved salts	Water, monovalent salts	Water
Applications	Urban and municipal wastewater, Synthetic emulsified oily wastewater	Vegetable oil factory & poultry slaughterhouse wastewater, Metal finishing industry, Oily wastewater, phenolic wastewater	Dumpsite leachate, textile, phenolic wastewater from paper mill	Dumpsite leachate, Phenolic wastewater from paper mill, oily wastewater, metal finishing industry

3.2.3   NF

An NF membrane, first used in the late 1980s, has properties between RO and UF membranes. It is sufficient for removing ions that greatly add to osmotic pressure and thus requires lower operating pressures than Ros. Highly contaminated waters require successful pretreatment prior to NF, although soluble fractions cannot be removed by it. Free chlorine in feed water affects the membranes (Wang et al., 2016). NF membranes have been used for dairy, medicine and wastewater treatment and in desalination applications.

3.2.4   RO

RO, a pressure-driven procedure, is used to eliminate dissolved substances and smaller particles and is only permeable by water molecules. The pressure applied to RO must be sufficient to allow water to pass the osmotic pressure. The efficiency of the RO membrane usually benefits from higher penetrability, greater selectivity and higher resistance to fouling. The drawbacks include the use of high pressure, being costly compared with other membrane processes and being often vulnerable to fouling. In certain situations, a high pretreatment is essential (Liu et al., 2017).

3.2.5 FORWARD OSMOSIS

Forward osmosis (FO) is a mechanism in which water is driven across a semipermeable membrane from a feed solution to a drawing solution due to the osmotic pressure gradient (Ong et al., 2017). The obvious benefit over traditional pressure-driven membrane technology is that the FO mechanism does not rely on high hydraulic pressure, thereby offering an incentive to conserve electricity and membrane maintenance costs (low fouling potential).

3.3 MBRS

MBR is a method that combines the biodegradation of pollutants by activated sludge with direct solid–liquid separation by membrane filtration, that is, by means of an MF or UF membrane (Judd, 2010). Of late, MBR technology is commonly recognized as an alternative core technology to treat wastewater containing micropollutants. The effectiveness of MBR therapy is greater than other biological processes due to a very large microbial population. In addition, the sieving effect of the membrane sorts according to the size of the contaminant and holds them to the membrane, thereby bringing it in contact with the degrading microorganisms within the MBR for their complete degradation (Ahmed et al., 2017). The widespread use of MBRs has been due to their significant advantages, such as high-quality produced water and the high biodegradation ability of pollutants with a lower cumulative footprint.

3.3.1 MBR CONFIGURATIONS

Configuration wise MBRs can be classified into two: side-streams and submerged MBRs (Figure 3.1). However, side-stream MBRs have the benefit of providing more durable physical ability, more robust crossflow control and hydraulic loading and simpler chemical washing (summed up in Table 3.2). They are still mainly found in commercial and small-scale waste water treatment plants (WWTPs). As suggested by the name, in the case of submerged MBRs, the membrane is placed within the bioreactor but is pumped into the membranes placed outside the bioreactor for side-stream MBRs.

3.3.2 MBR PERFORMANCE DETERMINATION AND AFFECTING FACTORS

Considering the wastewater quality and level, separate MBR operations are deemed ideal for the treatment of water, which includes membrane properties and sludge characteristics (Mutamim et al.,

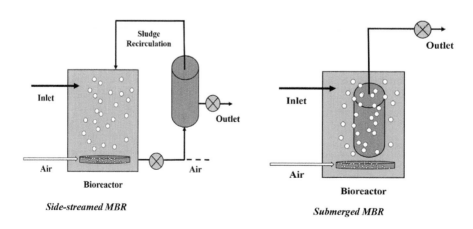

FIGURE 3.1 Configuration of side-stream and submerged MBRs.

Source: Adapted from Jiang (2007).

TABLE 3.2

Submerged MBRs and Side-Stream MBRs (adapted from Hamedi et al., 2019; Jiang, 2007)

Properties	Submerged MBR	Side-stream MBR
Complexity	Simple	Complex
Versality	Less versatile	Versatile
Operation mode	Dead-end filtration	Crossflow filtration
Robustness	Less robust	Robust
Flux	Low	High
Shear provided by	Aeration	Pump
Energy consumption	Significantly less	High
Pressure	Low	High
Fouling probability	Low	High
Fouling reducing methods	• Air bubble agitation • Backwashing • Chemical cleaning	• Crossflow • Airlift • Backwashing • Chemical cleaning
Application	• Municipal-scale systems • Solids in activated sludge	• Industrial systems • Feed with high • Temperature

2012). Factors that are considered for optimum MBR performance are (Basile et al., 2015; Pronk et al., 2019):

- Membrane intrinsic resistance
- Membrane and membrane module configuration
- The hydrodynamic regime of the solution at the membrane surface
- Temperature
- Transmembrane pressure
- Shear conditions
- Intermittent operation
- Fouling tendency
- Integration with other processes

However, factors that are of most importance to the behavior of the sludge are the following:

- Hydraulic residence time (HRT) – A significant operating determinant that denotes the time of residence of the feed stream in an MBR prior to treatment by it (Basile et al., 2015). The relation between HRT and removal performance is proportional.
- Mixed liquor suspended solids (MLSS) – Another crucial operating factor for aerobic systems is the concentration of total suspended solids of the wastewater in the aeration tank (Lousada-Ferreira et al., 2010). It is used to monitor suspended growth processes in treatment plants.
- Sludge retention time – This shows how often sludge removal occurs from the system. It influences mixed liquor properties and changes microbial physiological conditions (Chang & Lee, 1998; Massé et al., 2006).
- Organic loading rate (OLR) – This is the ratio of feed to microorganism (F/M) and is of significant importance. It is classified as kilograms of chemical oxygen demand (COD) divided by kilograms of MLSS times a day. The relation between OLR and MBR performance is inversely proportional (Basile et al., 2015).

3.3.3 Types of MBRs

3.3.3.1 Moving Bed Biofilm Reactors

Moving bed biofilm reactor (MBBR) and integrated fixed-film activated sludge (IFAS) are correlated with growth secondary biological treatment in WWTPs. Contaminated water can be biologically treated through adequate analysis and environmental control. However, knowledge of biological pathways is of prime importance to ensure adequate conditions (Ødegaard et al., 1994). Small plastic carrier materials support biofilm growth in MBBRs (Rusten et al., 1999). The performance of the reactor has been shown in many coupled operations for the elimination of biochemical oxygen demand (BOD) and nutrients. The key benefit of the process relative to the activated sludge reactors is its compactness and it does not involve the recirculation of sludge. Flexibility is the advantage over most biofilm systems.

3.3.3.2 Anaerobic MBRs

Two most effective anaerobic technologies in use for wastewater treatment are up flow anaerobic sludge blankets (UASBs) and expanded granular sludge bed reactors (EGSBs; Tauseef et al., 2013; van Lier et al., 2015). Anaerobic processes in industrial wastewater treatment are beneficial due to lower sludge generation and the conversion of organic matter into useful biogas without energy consumption (Gouveia et al., 2015).

3.3.3.3 Membrane-Biofilm Reactors

The membrane-biofilm reactor (MBfR) or membrane-aerated biofilm reactor (MABR) is an emerging treatment technology. MBfR is centered on gas-permeable membranes that offer a gaseous substrate to biofilms naturally formed on the outer surface of the membrane in a counter-diffusional manner. This technology presents distinct benefits over traditional biofilm treatment methods and allows advanced treatment for a broad range of reduced, oxidized and organic compounds (Martin & Nerenberg, 2012).

3.3.3.4 Nanomaterial MBRs

The idea of nanomaterial membranes (NMs) promises to be a sustainable route to improve membrane characteristics and enhance the efficiency of membrane bioreactors (MBRs) in wastewater treatment (Pervez et al., 2020). NM-based membranes are more efficient than traditional membranes in terms of hydrophilicity, surface roughness, thermal stability, hydraulic stability, fouling, higher water permeability and higher selectivity due to their tiny pore size (Kim & Van der Bruggen, 2010; Qu et al., 2013). Different types of NF-MBRs are being used in wastewater treatment, including nanofiber membrane bioreactors (NF-MBRs), nanoparticle membrane bioreactors (NP-MBRs), nanotube membrane bioreactors (NT-MBRs), nanocrystal membrane bioreactors (NC-MBRs), nanowire membrane bioreactors (NW-MBRs) and nanosheet membrane bioreactors (NS-MBRs; Pervez et al., 2020).

3.4 APPLICATION OF MEMBRANE TECHNOLOGY FOR WASTEWATER TREATMENT

3.4.1 Industrial Wastewater Treatment

Industrial wastewater is also generated on a discontinuous basis and the composition of the streams can differ greatly. In this case, therefore, a large processing heterogeneity is needed to accommodate inherent variation. The features of industrial wastewater can generally be represented by specific parameters, including COD, BOD, suspended solids (SS), ammonium nitrogen (NH_4^+-N), heavy metals, pH, color, turbidity and biological parameters. Since the properties of industrial wastewater are highly dependent on the form of industrial wastewater and industrial processes (Lin et al., 2012). Membrane methods are commonly used for the handling of municipal wastewater leading to higher costs for treated water and wastewater discharge.

Membrane technologies used in industrial wastewater treatment to

- directly recover recycled materials, by-products and solvents;
- partial flow recirculation;
- prevent massive, high-polluted wastewater flows; and
- reuse of concentrated streams as raw or low-cost raw materials.

3.4.1.1   Food Industries

The food industry covers a diverse number of subsidiaries, such as fish, dairy, livestock, vegetable and beverage manufacturing sectors. As a consequence, the wastewater of each branch varies in its quality with high organic loads. In addition, these wastewaters contain high-added-value compounds (e.g., phenols, carotenoids, pectin, lactose, proteins) that can be extracted. Successful implementation of membrane technology includes wastewater from potato starch production, fruit juice, seafood industries and so on (Zirehpour & Rahimpour, 2016).

3.4.1.2   Pulp and Paper Industries

The processes in the pulp and paper industries are focused on the use of water and an incredible amount of wastewater can be generated. Membrane filtration makes it possible to increase the performance of the existing wastewater treatment system in the pulp and paper industry. Usually, MBR systems will extract 82–99% of COD, approximately 100% of suspended solids at a HRT of 0.12–2.5 days (Lin et al., 2012). The NF treatment process decreased the COD and the color of the effluent by about90% (Zirehpour & Rahimpour, 2016).

3.4.1.3   Textile Industry

The textile processing industry (TPI) is a water-intensive field, as water is used as the primary medium for the application of coloring, finishing agents and the elimination of impurities. In recent trends in industrial wastewater treatment for energy recovery and reuse, the combination anaerobic MBR (AnMBR) and aerobic MBR method will be a viable technique for TPI wastewater treatment. The AnMBR method is used for energy recovery and the subsequent use of aerobic MBR will accomplish color reduction in order to generate the effluent for subsequent reuse (Lin et al., 2012).

3.4.1.4   Tannery Industries

Tanning is a water-consuming process, and as a result, wastewater disposal is one of the biggest issues of tanneries. A hybrid system of low-cost MBR minerals and found that the combined system could easily remove chromium, while the additional minerals mitigated fouling (Malamis et al., 2009). The aerobic MBR is a viable technology for tannery wastewater treatment; however, pilot and full-scale implementations are minimal. More attention needs to be given to the possible role of AnMBR in tannery wastewater treatment (Lin et al., 2012).

3.4.1.5   Landfill Leachate

Leachate is a high organic matter and ammonia nitrogen-strong wastewater produced as a result of rainwater percolation and moisture from waste in landfills. The chemical constituent of the leachate depends on the age and maturity of the dump site. For a young leachate, the organic components are much higher as compared to old or matured ones (Klimiuk & Kulikowska, 2006). Successful reduction of leachate contaminants can be done using stripping accompanied by flocculation, MBR and reverse osmosis therapy. A combination of MBR and electrooxidation methods can reduce COD and NH^{+4}-N and were followed by substantial detoxification (Lin et al., 2012).

3.4.1.6 Pharmaceutical Wastewaters

The pharma industry disposal contains a wide-ranging class of compounds with significant structural heterogeneity, function, actions and operation. Cephalosporin-containing pharmaceutical wastewater after treatment with MBR causes increased degradation by bioaugmentation (Saravanane & Sundararaman, 2009). MBRs implementing special microorganisms can serve as potential contenders to current pharmaceutical wastewater treatment processes.

3.4.1.7 Oily and Petrochemical Wastewaters

Oil and petrochemical wastewater are amongst the most troubled sources of pollutants due to their poisonous and refractory traits that originate from a number of sources, such as crude oil extraction, oil refining, petrochemical industry, metal manufacturing, lubricants and coolants, and car wash. A modified full-scale facility from chemical de-emulsification to a UF process accompanied by an MBR method was used to treat oil-contaminated wastewater and was able to remove 90% COD and full tar, grease and phenolic (Kim et al., 2006).

3.4.2 MUNICIPAL WASTEWATER TREATMENT

The quantity and type of wastewater and contaminants from the municipality vary by country due to climate change, socioeconomic conditions, household infrastructure and other factors (Henze & Comeau, 2008). Municipal wastewater is typically treated to eliminate unwanted contaminants by bacterial biodegradation of organic matter to smaller molecules (CO_2, NH_3, PO_4 etc.) in presence of oxygen (Assayie et al., 2017).

3.5 INTEGRATED MEMBRANE SYSTEMS

As no particular treatment procedure can meet all of the treatment goals, generally shuffling the use of several procedures is done to hit the bull's-eye. This advancement benefits by reducing membrane fouling, for high organic matter containing feed water (Schäfer et al., 1998). Traditional methods such as flocculation, coagulation and so on are added as pretreatment prior advancing to membrane treatment. Without adequate pretreatment, pollutants such as suspended and dissolved solids can obstruct NF/RO membranes and reduce their efficiency. Combining MBR with advanced oxidation processes (AOPs) and electrocoagulation was found to be efficient in the reduction of membrane fouling, recalcitrant compounds, colored compounds and metal removal from pharmaceutical and textile wastewater. Of late, for its diverse advantages, namely superior water quality, low energy requirement, osmotic membrane bioreactor (OMBR) has drawn increasing interest. Combining thermophilic bioprocess with membrane distillation, membrane distillation bioreactor (MDBR), continuous pumping and recirculation of mixed liquor from the bioreactor to the MD machine and bioreactor provides high-quality water (Neoh et al., 2016). In biofilm membrane bioreactor (BF-MBR), carriers are incorporated within the MBR that decreases the accumulation of suspended solids and reduces membrane fouling without limiting the efficiency of the process (Leyva-Díaz et al., 2013). Using an entrapping method, carriers are formed by immobilizing cells in bio entrapped membrane bioreactor (BE-MBR). The investigators observed that the fouling in BF-MBR was extreme relative to traditional MBR at the later stage. However, it has shown the best output in both phenol removal efficiency and membrane fouling (Rafiei et al., 2014).Aerobic granular sludge systems with spherical-shaped granules, tested for higher organic load, can simultaneously nitrify or denitrify within it. It contains a diverse microbial community instead of a specific type and thus functions effectively (Zhao et al., 2014).

The benefits and inconvenience of the different integrated technologies follow (adapted from Neoh et al., 2016):

Membrane distillation – This integrated system helps in higher recalcitrant biodegradation; thus, less sludge is produced and causes lowered footprint from this process in spite of providing better effluent quality. For its outstanding stability, it is cheaper than RO-MBR. It has limited potential in COD removal from the feed water.

FO-MBR – This process is more energy efficient than other conventional methods, can recover phosphorus from the feed and can produce decent quality effluent. It also can remove trace organic contaminants successfully from high TSS-containing wastewater (better than RO-MBR). The fouling is mostly reversible and less than RO-MBR. Its drawbacks include membrane stability uncertainty and, with rising salinity, can reduce microbial kinetics and water flow.

RO-MBR – It is a cheaper alternative to FO-MBR as it consumes less energy compared to conventional MBRs and possesses a lower fouling tendency. It shows lower effectiveness for high-saline wastewater treatment than FO-MBR.

AOPs/electrocoagulation-MBR – It is an easy-to-handle system and can remove colors and recalcitrant such as pharmaceutical contaminants. During operation, less sludge is generated and possesses a lower fouling potential. Its main setback is that it is not really effective in treating high-TSS-contaminated wastewater. High operation cost has also limited its application.

Granular MBR – This process has higher rate of nitrification and denitrification, and it is more shock-resistant. It possesses less fouling potential and leaves a smaller footprint during operation. Although fouling can become a severe concern during later stage of operation, it takes a longer time during the start-up to form granules.

BF-/bio-entrapped MBR – This system has considerably good nitrification and denitrification rates, has less fouling tendency and can reduce the concentration of suspended solids. But severe fouling can be a drawback after a long time of operation.

3.6   MEMBRANE FOULING

Considering the benefits of MBRs with respect to a reduced footprint and improved treated water quality, the process is constrained by membrane fouling, which results in flux degradation and membrane-cleaning downtime. Fouling is a decrease in the permeability of the membrane. Usually, the transmembrane pressure (TMP) has to be raised to keep the flow steady. Membrane fouling decreases efficiency by increasing TMP, which in turn raises maintenance and operating costs. Physical and biological factors can cause membrane fouling (Judd, 2004). It is caused by the accumulation of bacteria and bacterial products, suspended particles, and colloidal particles, as well as inorganic dissolved chemical compounds on the membrane surface and results in a reduction in flow and permeability.

Fouling is of two types, that is, reversible and irreversible. Reversible fouling happens in the shape of a crust of cake on the membrane, which can be extracted by physical processes, such as backwash or hydrodynamic scouring. Irreversible fouling is induced by chemisorption and pore-plugging mechanisms can only be expelled by chemical washing or high-temperature decomposition (Guo et al., 2012; Huang et al., 2018).

3.6.1   FACTORS THAT INFLUENCE MEMBRANE FOULING

Membrane fouling relies on different aspects of the setup, namely, feed properties (pH and ion strength), membrane features (roughness, hydrophobicity, etc.) and processing parameters (cross-flow rate, TMP and temperature). Several of these variables combine in one form or another to intensify membrane fouling. Factors that can be held responsible for fouling are summed up in the following sections (Iorhemen et al., 2016).

3.6.1.1   Membrane Characteristics

- *Membrane materials* –Hydrophilic such as ceramic membranes are less prone to fouling whereas hydrophobic membranes like polymeric membranes are more prone to fouling.
- *Membrane surface roughness* –As the rough surface creates a groove for colloidal particles to gather on the membrane surface during the operation, fouling keeps increasing with rising surface roughness.
- *Membrane pore size* –The higher the membrane pore sizes, the higher chance of blocking by contaminant, and thus a greater chance of fouling.
- *Water affinity* –Increased hydrophilicity implies less membrane fouling, while hydrophobicity associates with enhanced membrane fouling tendency.
- *Membrane surface charge* –Membranes get negatively charged due to the dumping of colloidal particles and thus can accumulate positively charged ions such as Ca^{2+}, Al^{3+} from MLSS, causing inorganic membrane fouling.

3.6.1.2   Operating Conditions

- *Operating mode* – Running in crossflow filtration mode causes less cake-layer formation on the membrane, thus lowering the chance of fouling of the membrane.
- *Aeration* – Higher aeration rates lead to lower rates of membrane fouling.
- *Temperature* –Low temperatures enhance the potential for membrane fouling as more bacterial extracellular polymeric substances (EPSs) are released and higher the load of filamentous bacteria.
- *COD/N ratio* – Higher COD/N ratio in feed lowers the membrane fouling rate and has a better membrane efficiency and a longer operating time. However, reports also suggest that a low COD/N ratio implies lower fouling.
- *HRT* –Fouling increases with declining HRT. However, excessive HRT results in the aggregation of fouling agents.
- *Solids retention time (SRT)* – Low EPS production by operating at high SRT limits fouling. Fouling increases at extremely high SRT as it incorporates MLSS and high sludge viscosity.
- *Organic loading rate (OLR)* –Fouling increases with an increase in OLR.
- *F/M ratio* –Increased EPS production from increasing F/M ratio through high biomass intake results in spiked fouling.

3.6.1.3   Feed/Biomass Properties

- *Floc size* – Fouling of the membrane rises with smaller floc sizes.
- *Salinity* –Bound EPSs released with rising salinity cause more membrane fouling.
- *pH* –A reduction in pH leads to an increase membrane fouling rates.
- *MLSS* –High fouling is caused by higher MLSS. However, studies also suggest that no or very little impact of it on fouling.
- *EPSs* –The higher the EPS concentration in the feed, the higher the chance of fouling.
- *Sludge apparent viscosity* –Increased viscosity leads to increased membrane fouling.

3.6.2   Control of Membrane Fouling

In spite of being one of the promising candidates, fouling is an obstacle to the development of membranes with high flux and permeability. Rapid action is needed to solve the problem. Some particles move through the membrane, and some of them may stick the pores of the membrane, thereby causing membrane blockage, or within the surface of the membrane by the feed elements, which, in turn, adversely affects membrane permeability, flux and lifespan (Chen et al., 2018). Antifouling methods can be divided into multiple types depending on the terms used to manage membrane fouling (Table 3.3).

TABLE 3.3

Control of Membrane Fouling (summarized from the review of Bagheri & Mirbagheri, 2018)

Strategies	Features
Air sparging	• Lowers concentration polarization and fouling • Fouling management by turbulence fluctuations and putting shear stress on the membrane surface • High aeration rate can enhance the fouling of the membrane
Mechanical cleaning	• Control of fouling by applying shear stress to the membrane surface • Scouring of the membrane
Ultrasonic mitigation	• Ultrasound-assisted aqueous medium to remove soluble and insoluble particles • Essentially reduce the concentration polarization and eliminate the biofilm covering on the membrane surface
Chemical cleaning	• Include the use of acids, bases, oxidants, surfactants and chelates, and the recent introduction of nitrite and rhamnolipids • Acids eliminate fouling through solubilization and neutralization • Bases are responsible for hydrolysis, solubilization and saponification of the foulant • Oxidation and disinfection of foulant with oxidants • Regulation of hydrophilic/hydrophobic interactions and chelation by surfactants and chelates • Solubilization, disinfection and hydrophilic/hydrophobic associations with rhamnolipid and free nitrous acid of the foulant • Known to exhibit bad effects in membrane characteristics and adversely affect the microbial process in MBRs
Fouling release surfaces and nanomaterials	• Membrane fouling can be controlled by preparing membranes with antifouling surfaces with specific physical and chemical surface properties • Hydrophilic surfaces have demonstrated tremendous usefulness to regulate different forms of foulants by suppressing non-specific interactions. • Postmodification of membranes by polymeric anti-fouling materials or inorganic nanomaterials can reduce fouling. • Mussel-inspired surface modification is a recent method of bioadhesion for the formation of antifouling surfaces, based on the multi-functionality of dopamine and its derivatives.
Cell entrapment (CE)	• Cell immobilization (passive immobilization and CE) restricts the free movement of cells by confining them into, or attaching them to, a solid support • Artificially entraps cells in a porous polymer matrix • Cannot be solely reliable with removal of pathogens and Large particles • Good alternative for conventional biological treatment systems
Biological mitigation	• Newer approach with high capabilities in biofouling control • Different molecules (AHL), enzyme (acylase) functions as quorum quencher to help in mitigation • Microbial attachment or biofilm formation inhibition through inhibition of adenosine triphosphate synthesis • Nitric oxide favors planktonic growth by stimulating phosphodiesterase activity and degrading cyclic diguanylate monophosphate • Enzymes (proteinase K, trypsin, subtilisin etc.) which targets EPSs, can prevent initial microbial attachment than disrupt established biofilm. • Protease is much better than traditional chemicals for the control of irreversible membrane in spite of drawbacks (instability, temperature and pH) • Disruption of fouling layers by hydrolysis of microbial macromolecules using supplementing exogenous hydrolases • bacteriophage as antifoulant has received enormous attention
Electrically based mitigation	• Electrophoresis (EP) and electrostatic repulsion and the forces exerted by electric fields on the charged particles can inhibit membrane fouling • Electrical methods to control fouling in MBRs are mainly external such as electrocoagulation and EP or internal such as microbial fuel cells

3.7 CONCLUDING REMARKS AND FUTURE DIRECTION

With the ever-rising human population, agriculture and industrial needs, water resource is getting depleted exponentially. On the other hand, daily anthropogenic activities produce a significant amount of wastewater that cannot be used directly. Wastewater mitigation and reuse are keys to the solution of this issue. Wastewater treatment plants employ different methods to make it reusable and membrane technology is the most advanced and efficient one in their arsenal. Membrane technology is dramatically improving the management of water and wastewater. This chapter was an effort to summarize the major membrane technologies, focusing on bio-membrane technology; the MBR configuration, types and their application; the integration of MBR systems, quoting their merits and demerits; and the major drawbacks of membranes, that is, fouling and their antifouling strategies. A lot of research work has been done in this area for many years. There is still space for reform in many ways, though. Regarding the previous success of conventional MBRs, NMs-MBR technology can also be used in other emerging areas.

New researchers are being carried out till date for application and development of new, more efficient membrane materials, copolymers like polyvinylid-enefluoride–hexafluoropropylene and polyvinylid-enefluoride–tetrafluoroethylene. Also the application of carbon nanotubes and buckey-paper membrane is being tested.

For the last decade, most efforts have concentrated primarily on the use of modern and innovative approaches to solve the issue of membrane fouling in MBRs. Most recent experiments have worked on the use of NMs, CE, biological principles, and electrically based approaches to manage membrane fouling. These novel membrane-fouling management techniques have demonstrated high efficiency. However, the introduction of these for large-scale MBRs needs further study and investigation. Moreover, the regulation of membrane fouling requires more than one solution. Membrane fouling is still a significant problem in the area of membrane methods, especially bio-membrane technology, which must be addressed in the coming years.

Wastewater treatment technologies generally skip the potential of resource recovery. However, resource recovery technologies coupled with wastewater treatment can generate commercial products from wastes, namely, biofuels, biopolymers, single-cell proteins and others, receiving more and more attention lately. Future studies should focus on a full economic analysis of all innovations addressed, taking into account all running costs and energy recovery from biogas processing. Hopefully, this chapter would be helpful in providing good knowledge for future study into membrane technology developments in wastewater treatment.

ACKNOWLEDGMENT

We acknowledge financial assistance from Water Technology Initiative (WTI) program of Department of Science and Technology, Govt. of India.

BIBLIOGRAPHY

Ahmed, M. B., Zhou, J. L., Ngo, H. H., Guo, W., Thomaidis, N. S., & Xu, J. (2017). Progress in the biological and chemical treatment technologies for emerging contaminant removal from wastewater: A critical review. *Journal of Hazardous Materials, 323*, 274–298. https://doi.org/10.1016/j.jhazmat.2016.04.045

Aliyu, U. M., Rathilal, S., & Isa, Y. M. (2018). Membrane desalination technologies in water treatment: A review. *Water Practice and Technology, 13*(4), 738–752. https://doi.org/10.2166/wpt.2018.084

Assayie, A. A., Gebreyohannes, A. Y., & Giorno, L. (2017). Municipal wastewater treatment by membrane bioreactors. In A. Figoli & A. Criscuoli (Eds.), *Sustainable membrane technology for water and wastewater treatment* (pp. 265–294). Springer. https://doi.org/10.1007/978-981-10-5623-9_10

Bagheri, M., & Mirbagheri, S. A. (2018). Critical review of fouling mitigation strategies in membrane bioreactors treating water and wastewater. *Bioresource Technology, 258*, 318–334. https://doi.org/10.1016/j.biortech.2018.03.026

Basile, A., Cassano, A., & Rastogi, N. K. (2015). *Advances in membrane technologies for water treatment: Materials, processes and applications.* Elsevier.

Chang, I.-S., & Lee, C.-H. (1998). Membrane filtration characteristics in membrane-coupled activated sludge system: The effect of physiological states of activated sludge on membrane fouling. *Desalination, 120*(3), 221–233. https://doi.org/10.1016/S0011-9164(98)00220-3

Chen, W., Qian, C., Zhou, K.-G., & Yu, H.-Q. (2018). Molecular spectroscopic characterization of membrane fouling: A critical review. *Chem, 4*(7), 1492–1509. https://doi.org/10.1016/j.chempr.2018.03.011

Gouveia, J., Plaza, F., Garralon, G., Fdz-Polanco, F., & Peña, M. (2015). Long-term operation of a pilot scale anaerobic membrane bioreactor (AnMBR) for the treatment of municipal wastewater under psychrophilic conditions. *Bioresource Technology, 185*, 225–233. https://doi.org/10.1016/j.biortech.2015.03.002

Guo, W., Ngo, H.-H., & Li, J. (2012). A mini-review on membrane fouling. *Bioresource Technology, 122*, 27–34. https://doi.org/10.1016/j.biortech.2012.04.089

Hamedi, H., Ehteshami, M., Mirbagheri, S. A., Rasouli, S. A., & Zendehboudi, S. (2019). Current status and future prospects of membrane bioreactors (MBRs) and fouling phenomena: A systematic review. *The Canadian Journal of Chemical Engineering, 97*(1), 32–58. https://doi.org/10.1002/cjce.23345

Henze, M., & Comeau, Y. (2008). Wastewater characterization. *Biological Wastewater Treatment: Principles Modelling and Design*, 33–52.

Huang, S., Ras, R. H. A., & Tian, X. (2018). Antifouling membranes for oily wastewater treatment: Interplay between wetting and membrane fouling. *Current Opinion in Colloid & Interface Science, 36*, 90–109. https://doi.org/10.1016/j.cocis.2018.02.002

Iorhemen, O., Hamza, R., & Tay, J. (2016). Membrane bioreactor (MBR) technology for wastewater treatment and reclamation: Membrane fouling. *Membranes, 6*(2), 33. https://doi.org/10.3390/membranes6020033

Jiang, T. (2007). *Characterization and modelling of soluble microbial products in membrane bioreactors.* [s.n.].

Judd, S. (2004). A review of fouling of membrane bioreactors in sewage treatment. *Water Science and Technology: A Journal of the International Association on Water Pollution Research, 49*(2), 229–235.

Judd, S. (2010). *The MBR book: Principles and applications of membrane bioreactors for water and wastewater treatment.* Elsevier.

Judd, S. (2016). The status of industrial and municipal effluent treatment with membrane bioreactor technology. *Chemical Engineering Journal, 305*, 37–45. https://doi.org/10.1016/j.cej.2015.08.141

Kamali, M., & Khodaparast, Z. (2015). Review on recent developments on pulp and paper mill wastewater treatment. *Ecotoxicology and Environmental Safety, 114*, 326–342. https://doi.org/10.1016/j.ecoenv.2014.05.005

Kim, B. R., Anderson, J. E., Mueller, S. A., Gaines, W. A., Szafranski, M. J., Bremmer, A. L., Yarema, G. J., Guciardo, C. D., Linden, S., & Doherty, T. E. (2006). Design and startup of a membrane-biological-reactor system at a ford-engine plant for treating oily wastewater. *Water Environment Research, 78*(4), 362–371. https://doi.org/10.2175/106143006X98778

Kim, J., & Van der Bruggen, B. (2010). The use of nanoparticles in polymeric and ceramic membrane structures: Review of manufacturing procedures and performance improvement for water treatment. *Environmental Pollution, 158*(7), 2335–2349. https://doi.org/10.1016/j.envpol.2010.03.024

Klimiuk, E., & Kulikowska, D. (2006). Organics removal from landfill leachate and activated sludge production in SBR reactors. *Waste Management, 26*(10), 1140–1147. https://doi.org/10.1016/j.wasman.2005.09.011

Krüger, R., Vial, D., Arifin, D., Weber, M., & Heijnen, M. (2016). Novel ultrafiltration membranes from low-fouling copolymers for RO pretreatment applications. *Desalination and Water Treatment, 57*(48–49), 23185–23195. https://doi.org/10.1080/19443994.2016.1153906

Kumar, C. M., Roshni, M., & Vasanth, D. (2019). Treatment of aqueous bacterial solution using ceramic membrane prepared from cheaper clays: A detailed investigation of fouling and cleaning. *Journal of Water Process Engineering, 29*, 100797. https://doi.org/10.1016/j.jwpe.2019.100797

Leyva-Díaz, J. C., Martín-Pascual, J., González-López, J., Hontoria, E., & Poyatos, J. M. (2013). Effects of scale-up on a hybrid moving bed biofilm reactor: Membrane bioreactor for treating urban wastewater. *Chemical Engineering Science, 104*, 808–816. https://doi.org/10.1016/j.ces.2013.10.004

Lin, H., Gao, W., Meng, F., Liao, B.-Q., Leung, K.-T., Zhao, L., Chen, J., & Hong, H. (2012). Membrane bioreactors for industrial wastewater treatment: A critical review. *Critical Reviews in Environmental Science and Technology, 42*(7), 677–740. https://doi.org/10.1080/10643389.2010.526494

Liu, G., Han, K., Ye, H., Zhu, C., Gao, Y., Liu, Y., & Zhou, Y. (2017). Graphene oxide/triethanolamine modified titanate nanowires as photocatalytic membrane for water treatment. *Chemical Engineering Journal, 320*, 74–80. https://doi.org/10.1016/j.cej.2017.03.024

Lousada-Ferreira, M., Geilvoet, S., Moreau, A., Atasoy, E., Krzeminski, P., van Nieuwenhuijzen, A., & van der Graaf, J. (2010). MLSS concentration: Still a poorly understood parameter in MBR filterability. *Desalination, 250*(2), 618–622. https://doi.org/10.1016/j.desal.2009.09.036

Malamis, S., Katsou, E., Chazilias, D., & Loizidou, M. (2009). Investigation of Cr(III) removal from wastewater with the use of MBR combined with low-cost additives. *Journal of Membrane Science, 333*(1–2), 12–19. https://doi.org/10.1016/j.memsci.2009.01.028

Martin, K. J., & Nerenberg, R. (2012). The membrane biofilm reactor (MBfR) for water and wastewater treatment: Principles, applications, and recent developments. *Bioresource Technology, 122*, 83–94. https://doi.org/10.1016/j.biortech.2012.02.110

Massé, A., Spérandio, M., & Cabassud, C. (2006). Comparison of sludge characteristics and performance of a submerged membrane bioreactor and an activated sludge process at high solids retention time. *Water Research, 40*(12), 2405–2415. https://doi.org/10.1016/j.watres.2006.04.015

Mutamim, N. S. A., Noor, Z. Z., Hassan, M. A. A., & Olsson, G. (2012). Application of membrane bioreactor technology in treating high strength industrial wastewater: A performance review. *Desalination, 305*, 1–11. https://doi.org/10.1016/j.desal.2012.07.033

Neoh, C. H., Noor, Z. Z., Mutamim, N. S. A., & Lim, C. K. (2016). Green technology in wastewater treatment technologies: Integration of membrane bioreactor with various wastewater treatment systems. *Chemical Engineering Journal, 283*, 582–594. https://doi.org/10.1016/j.cej.2015.07.060

OboteyEzugbe, E., & Rathilal, S. (2020). Membrane technologies in wastewater treatment: A review. *Membranes, 10*(5), 89. https://doi.org/10.3390/membranes10050089

Ødegaard, H., Rusten, B., & Westrum, T. (1994). A new moving bed biofilm reactor: Applications and results. *Water Science and Technology, 29*(10–11), 157–165. https://doi.org/10.2166/wst.1994.0757

Ong, C. S., Al-anzi, B., Lau, W. J., Goh, P. S., Lai, G. S., Ismail, A. F., & Ong, Y. S. (2017). Anti-fouling double-skinned forward osmosis membrane with zwitterionic brush for oily wastewater treatment. *Scientific Reports, 7*(1), 6904. https://doi.org/10.1038/s41598-017-07369-4

Pervez, M. N., Balakrishnan, M., Hasan, S. W., Choo, K.-H., Zhao, Y., Cai, Y., Zarra, T., Belgiorno, V., & Naddeo, V. (2020). A critical review on nanomaterials membrane bioreactor (NMs-MBR) for wastewater treatment. *Npj Clean Water, 3*(1), 43. https://doi.org/10.1038/s41545-020-00090-2

Pronk, W., Ding, A., Morgenroth, E., Derlon, N., Desmond, P., Burkhardt, M., Wu, B., & Fane, A. G. (2019). Gravity-driven membrane filtration for water and wastewater treatment: A review. *Water Research, 149*, 553–565. https://doi.org/10.1016/j.watres.2018.11.062

Qu, X., Brame, J., Li, Q., & Alvarez, P. J. J. (2013). Nanotechnology for a safe and sustainable water supply: Enabling integrated water treatment and reuse. *Accounts of Chemical Research, 46*(3), 834–843. https://doi.org/10.1021/ar300029v

Rafiei, B., Naeimpoor, F., & Mohammadi, T. (2014). Bio-film and bio-entrapped hybrid membrane bioreactors in wastewater treatment: Comparison of membrane fouling and removal efficiency. *Desalination, 337*, 16–22. https://doi.org/10.1016/j.desal.2013.12.025

Rusten, B., Johnson, C. H., Devall, S., Davoren, D., & Cashiont, B. S. (1999). Biological pretreatment of a chemical plant wastewater in high-rate moving bed biofilm reactors. *Water Science and Technology, 39*(10), 257–264. https://doi.org/10.1016/S0273-1223(99)00286-3

Saravanane, R., & Sundararaman, S. (2009). Effect of loading rate and HRT on the removal of cephalosporin and their intermediates during the operation of a membrane bioreactor treating pharmaceutical wastewater. *Environmental Technology, 30*(10), 1017–1022. https://doi.org/10.1080/09593330903032865

Schäfer, A. I., Fane, A. G., & Waite, T. D. (1998). Nanofiltration of natural organic matter: Removal, fouling and the influence of multivalent ions. *Desalination, 118*(1–3), 109–122. https://doi.org/10.1016/S0011-9164(98)00104-0

Tauseef, S. M., Abbasi, T., & Abbasi, S. A. (2013). Energy recovery from wastewaters with high-rate anaerobic digesters. *Renewable and Sustainable Energy Reviews, 19*, 704–741. https://doi.org/10.1016/j.rser.2012.11.056

van Lier, J. B., van der Zee, F. P., Frijters, C. T. M. J., & Ersahin, M. E. (2015). Celebrating 40 years anaerobic sludge bed reactors for industrial wastewater treatment. *Reviews in Environmental Science and Bio/Technology, 14*(4), 681–702. https://doi.org/10.1007/s11157-015-9375-5

Wang, N., Liu, T., Shen, H., Ji, S., Li, J.-R., & Zhang, R. (2016). Ceramic tubular MOF hybrid membrane fabricated through *in situ* layer-by-layer self-assembly for nanofiltration. *AIChE Journal, 62*(2), 538–546. https://doi.org/10.1002/aic.15115

Werber, J. R., Deshmukh, A., & Elimelech, M. (2016). The critical need for increased selectivity, not increased water permeability, for desalination membranes. *Environmental Science & Technology Letters*, *3*(4), 112–120. https://doi.org/10.1021/acs.estlett.6b00050

Zhang, L., Zhang, P., Wang, M., Yang, K., & Liu, J. (2016). Research on the experiment of reservoir water treatment applying ultrafiltration membrane technology of different processes. *Journal of Environmental Biology*, *37*(5), 1007.

Zhao, X., Chen, Z., Wang, X., Shen, J., & Xu, H. (2014). PPCPs removal by aerobic granular sludge membrane bioreactor. *Applied Microbiology and Biotechnology*, *98*(23), 9843–9848. https://doi.org/10.1007/s00253-014-5923-0

Zirehpour, A., & Rahimpour, A. (2016). Membranes for wastewater treatment. In P. M. Visakh& O. Nazarenko (Eds.), *Nanostructured polymer membranes* (pp. 159–207). John Wiley & Sons, Inc. https://doi.org/10.1002/9781118831823.ch4

4 Membrane and Membrane-Based Processes for Wastewater Treatment

Har Lal Singh, Sarita Khaturia, Mamta Chahar, and Anjali Bishnoi

CONTENTS

DOI: 10.1201/9781003165019-4

4.1 INTRODUCTION

Water plays many important roles in the body including flushing waste from the body, regulating body temperature, and transporting nutrients and is necessary for digestion. Accessible amount of water is highly contaminated by industrial and agricultural waste and cannot be consumed; therefore, water quality and quantity are the main parameters that need to be improved [1]. Contaminant removal from water is very much essential to avoid negative effects on human health environment degradation [2]. For the treatment of wastewater, different techniques have been employed like reverse osmosis (RO) [3], ion exchange [4], gravity [5] and adsorption [6]. For removing water contaminants, among all these methods, one of the superior methods is adsorption due to its less price, easy operation and diversely available adsorbents. Various types of adsorbents can be used in this method are magnetic nanoparticles [7] activated carbon [8], nanotubes [9] and polymer nanocomposites [10]. Nearly all the contaminants can be removed by these adsorbents including heavy metals, which are very harmful even at low concentrations. As discussed earlier, almost all the contaminants are removed by employing adsorption, but due to some limitations like insufficiency of suitable adsorbents that possess a high adsorption capacity and the limited commercial availability of these adsorbents [11], there is a necessity for more efficient techniques like membrane technology. The development of membranes started in the 1960s when the first water desalination plants based on RO technology were designed and built and progressively this process is accepted as a cost-effective method for treating wastewater. Due to its diversified applications in various sectors US Environmental Protection Agency (EPA) has recognized membrane processes such as RO as a 'best available technology' (BAT) because it follows all the Safe Drinking Water Act (SDWA) regulations.

A membrane is a permeable or semipermeable solid phase (polymer, inorganic or metal) that controls the relative rates of transport of certain species present in the source waters and restricts their motion [12–13]. Generally, membranes work by selectively allowing some constituents to pass through the membrane while blocking the passage of others. For this to happen, the movement of material across a membrane requires a driving force. Therefore, membrane processes can normally be classified based on the type of driving force that causes components in the water to separate. The different type of driving force that initiates solute separation includes a pressure differential [micro-, ultra-, nanofiltration (NF), RO], a concentration difference across the membrane that initiates diffusion of a species between two solutions (dialysis) and a potential field applied to an ion-exchange membrane that initiates migration of ions through the membrane [electrodialysis (ED), electro-electrodialysis, electrochemical devices]. Membranes processes applicable to water/wastewater are summarized according to the driving force

4.1.1 Membrane Composition

All the membranes, either cellulosic or non-cellulosic membranes, that are utilized for municipal water treatment are prepared from synthetic organic polymers. Cellulosic membranes are usually asymmetric (made of one material but with a dense 'barrier layer' and porous support), whereas non-cellulosic membranes are either asymmetric or composites (barrier and support layers made of different materials). ED and electrodialysis reversal use synthetic polymers consisting of either crosslinked sulfonated copolymers of vinyl compounds (cation transfer type) or crosslinked copolymers of vinyl monomers with quaternary ammonium anion exchange groups (anion transfer type). Microfiltration (MF) and ultrafiltration (UF) membranes are made from a wide variety of materials, like polypropylene, polyvinyl difluoride (PVDF), polysulfone, polyethersulfone and cellulose acetate. The various membrane materials have different properties and characteristics such as pH, surface charge and hydrophobicity. This can affect the exclusion characteristic of a membrane.

4.1.2 TYPES AND CONFIGURATIONS

Water and wastewater treatment polymer membranes are typically prepared in two forms, namely, flat sheet and hollow fiber (capillary). With the exception of cellulose triacetate and the DuPont™ polyamides, all of the RO and NF polymer types are normally 'cast' in flat sheets onto woven or nonwoven backing materials. Flat-sheet castings are made in either one or two steps. Polyamide membranes are formed by casting a base membrane first (typically polysulfone) onto which the polyamide is cast in a process called in situ polymerization. The two-step process allows for the optimization of the membrane's properties by keeping the rejecting layer very thin to maximize productivity. Most commercially available MF and UF membranes currently used for drinking water treatment are made in a hollow-fiber configuration. Hollow-fiber membranes are operated in either an inside-out or outside-in mode.

4.1.3 CHALLENGES OF MEMBRANE TECHNOLOGY

- High penetration rates
- Low activation energy of transport mechanisms
- High-resolution separation
- Design of membrane reactors
- Upscaling of technology
- Process integration
- Low membrane degradation

4.2 MEMBRANE TECHNOLOGY IN INDUSTRIES

In the last two decades, membrane technology is the major contributor for solving water-related problems. Today, industries, municipalities and water companies treat approximately 60 million m³/day of wastewater using several membrane plants [14].

Pore size and molecule size are the main criteria, and on that basis, we can separate the pollutants through the membranes, shown in Figure 4.1 and Table 4.1 [15, 16], and for wastewater treatment, the most common membrane separation methods, such as MF, UF, NF and RO, are used in industries.

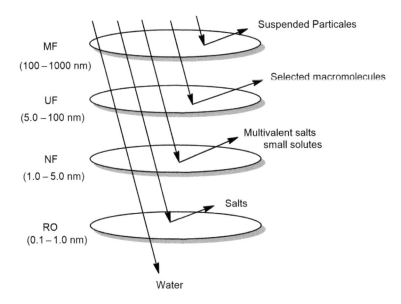

FIGURE 4.1 Ranges for pore sizes (nm) of various membranes.

TABLE 4.1

Various Types of Membranes

Membranes	Pore size (micrometer)	Separation mechanism	Mol. wt (kilo Dalton)	Pressure (Bar)	Average permeability (L/m² h Bar)	Material retained	Membrane types
MF membranes	large pore size 10^{-1} to 10	Molecular Sieve	100–500	1–3	500	Bacteria, fat, oil, grease, colloids, organics, micro-particles.	Symmetric Polymer or ceramic membranes
UF Membranes	Pore size smaller than MF membrane 10^{-3} to 1	Molecular Sieve	20–150	2–5	150	Oils, Pigments sugar, organics, Proteins, microplastics	Asymmetric Polymer composite or ceramic membrane
NF membranes	Porous, thin film composite 10^{-3} to 10^{-2}	Solution diffusion	2–20	5–15	10–20	Divalent ions, lactose, sucrose, sodium chloride, Pigments	Asymmetric polymer or thin-film composite membrane
RO membranes	nonporous, thin-film composite	Solution diffusion	0.2–2	15–75	5–10	All contaminants including monovalent ions	thin-film composite membrane

For the water treatment, several types of membranes are used as follows:

1. MF
2. UF
3. NF
4. RO
5. ED
6. Pervaporation
7. Hybrid membrane process

In these methods, hydraulic pressure is used for separation.

4.2.1   MF

This method filters remove mainly sediment, algae, protozoa and bacteria while water (H_2O); monovalent ions like Na^+, Cl^-; dissolved or organic matter; and small colloids and viruses can pass through the filter [17]. In MF, the membrane material can be organic or inorganic. Organic membranes are composed of different types of polymers such as polyvinylidine fluoride, polyamide, polysulfone, cellulose acetate and others, while inorganic membranes are made up of porous alumina and metals. It is a pretreatment for UF and a post-treatment for granular media filtration to reduce the fouling.

4.2.1.1   Principle of the MF Process

MF is a physical separation process that contains a porous membrane. It removes dissolved solids, turbidity and microorganisms by a sieving mechanism, based on the pore size of the membrane. If the particle size is larger than the pore size of the membranes, they can be fully removed while particles smaller than the pores of the membrane are partially removed.

An MF system is depicted in Figure 4.2; in this setup, the suspended liquid is passed through a semipermeable membrane, and the pump is allowed the liquid to pass through the membrane. Two

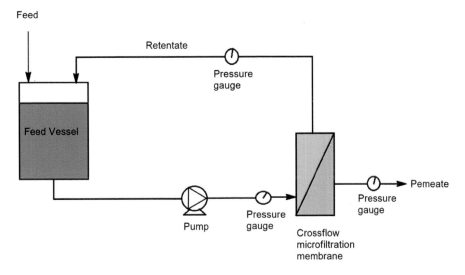

FIGURE 4.2   Schematic diagram of MF process [18].

pumps are adjusted for measure the pressure between the outlet and inlet streams or vacuum [19]. MF membranes operate in two configurations: crossflow filtration and dead-end filtration. After treatment, micro-filters are used with a recovery rate of about 90–98% [20].

Applications of the process: These membranes are used in the water, beverage and bioprocessing industries.

4.2.2   UF

For the separation of particles, a pressure or concentration gradient is required through membranes. High-molecular-weight substances and suspended particles separated, and low-molecular-weight particles dissolve in solute.

These membranes retain proteins, endotoxins, viruses and silica. This method is applied in industries like the pharmaceutical, dairy, beverage and food-processing industries, among others, and in research for purifying and converting raw water to portable water. UF also used for the protection of RO membranes as the prefiltration in RO (Figure 4.3).

4.2.2.1   Principle of the Process

In this method, a pressure-induced separation of solutes from a solvent through a semipermeable membrane. The relation between the applied pressure and the flux through the membrane is expressed by the Darcy equation:

$$J = \frac{TMP}{\mu R},$$

where
J = flux (flow rate per membrane area),
TMP = transmembrane pressure,
μ= solvent viscosity and
R = total resistance (sum of membrane and fouling resistance).

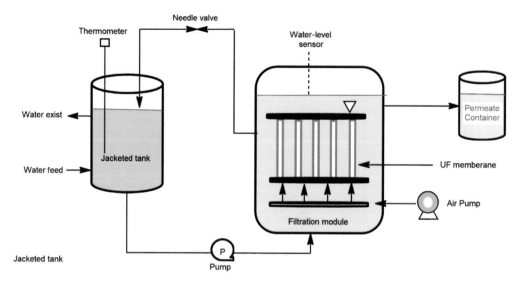

FIGURE 4.3   UF process.

4.2.2.2   Application of the Process

Conventionally, proteins and colloids have been removed by UF membranes. Mainly UF membranes are applied in the food industry to recover milk proteins and eliminate lactose and salts, as well as in the metal finishing industry to concentrate oil emulsions. [21, 22]. MF/UF plants are also applied for wastewater.

4.2.3   NF

Water softening and removal of by-products from surface water and fresh groundwater use this method. Membranes used for NF are composed of cellulose acetate blends or polyamide composites, or they could be modified forms of UF membranes like sulfonated polysulfone [23, 24].

4.2.3.1   Principle of the Process

In this process, hydrostatic pressure is applied to transport a molecular mixture to the surface of a membrane. The solvent and some low-molecular-weight solutes permeate the membrane while other components are retained.

4.2.3.2   Applications

1. It removes multivalent ions, synthetic dyes, sugars and specific salts.
2. The NF technique has been extensively used for milk and juice production, pharmaceuticals, fine chemicals, paper industry, and flavor and aroma industries as shown in Table 4.2 [25].

4.2.3.3   Advantages and Disadvantages

1. This method is used for the softening of water.
2. The main benefit associated with NF is gentle molecular separation that is often not included with other forms of separation processes (centrifugation). NF has a very favorable benefit of being able to process large volumes and continuously produce streams of products. In NF, the membrane's pore size is limited to only a few nanometers; if the pore size

TABLE 4.2
Industrial Applications of NF Technique

S. No.	Type of Industry	Use
1.	Oil and petroleum chemistry	Removal of tar Purification of gas condensates
2.	Pharmaceuticals	Maintain temperature Solvent recovery and management
3.	Medicinal	Amino acids and lipids extraction takes place from blood and other cells
4.	Natural essential oils and similar products	For fractionation of crude extracts, enrichment of natural compounds and smooth separations
5.	Bulk chemistry	Product polishing Continuous recovery of homogeneous catalysts

is smaller than NF, then RO is used, and if larger than NF, then UF is used. UF can also be used in cases in which NF can be used, due to it being more conventional.

3. A disadvantage of this method is that NF membranes are very expensive, and repair and replacement are dependent on total dissolved solids (TDS), flow rate and feed components.

4.2.4 RO

In 1867, the first synthetic membrane was reported by Moritz Traube. Cellulose acetate membranes were prepared by Loeb and Sourirajanin 1963, which showed relatively high flux and removal of salt [26]. It is the finest separation membrane process available. Dupont de Nemours used a polyamide-based membrane in the 1970s for the improvement in the mechanical resistance of the membrane. RO is a high pressure–driven process for the desalting of salt water. Both RO and NF are fundamentally different because the flow goes against the concentration gradient and because those systems use pressure to force water so that it goes from the low-pressure side to the high-pressure side. In this process, contaminants are removed by a semipermeable membrane as shown in Figure 4.4.

4.2.4.1 Principle

In RO systems, applied pressure to the salt solution side reverses the osmotic water flow, so that movement of water from the salt solution side to the pure water side of the membrane takes place. The applied pressure must be higher than the osmotic pressure difference. Here, water is put under pressure and forced through a membrane that filters out the minerals and nitrate. RO retains nearly all molecules except water, and due to the size of the pores, the required osmotic pressure is significantly greater than that for MF.

4.2.4.2 Advantages of RO

1. Nearly all contaminant ions and most dissolved non-ions are removed.
2. RO is suitable for small systems with a high degree of seasonal fluctuation in water demand.
3. It is insensitive to flow and TDS levels.
4. RO operates immediately without any minimum break-in period.
5. Low effluent concentrations are possible.
6. It removes bacteria and particles.
7. Simplicity and automation operations allow for less operator attention, which makes them suitable for small-system applications.

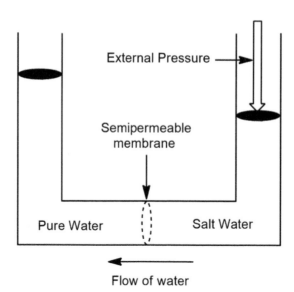

FIGURE 4.4 RO process.

4.2.4.3 Limitations of RO Method
1. High operating costs and capital
2. Potential problem with managing the wastewater brine solution
3. Pretreatment at high levels
4. Fouling of membranes

4.2.5 ELECTRODIALYSIS

For separating dissolved ions from water, the ED process combines electricity and ion-permeable membranes. It is effective in removing TDS, fluoride and nitrate from water. This process also uses membranes, but direct electrical currents are used to attract ions to one side of the treatment chamber as shown in Figure 4.5. This system includes a source of pressurized water, a direct current power supply and a pair of selective membranes.

4.2.5.1 Principle of the Process
In this process, the membranes adjacent to the influent steam are charged either positively or negatively and this charge attracts counter-ions toward the membrane. These membranes are designed to allow the positive or the negative charged ions to pass through the membrane, where the ions move from the product water stream through a membrane to the two reject water streams.

4.2.5.2 Advantages of Electrodialysis
1. All the contaminant ions and many of the dissolved non-ions are removed.
2. It is insensitive to flow and TDS levels.
3. Low effluent concentrations are possible.

4.2.5.3 Limitations of Electrodialysis
1. Operating costs and capital are high.
2. The level of pretreatment required is high.

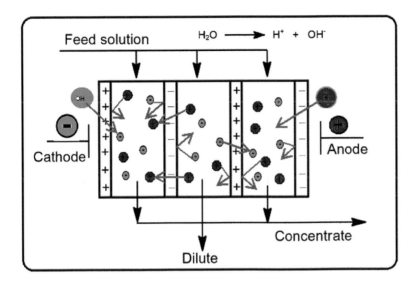

FIGURE 4.5 Electrodialysis process.

4.2.6 PERVAPORATION

This method couples membrane permeation and evaporation for the separation of liquid mixture on the basis of their preference. In the upstream, the liquid mixture fed from the one side and permeate vanishes on the other side (Figure 4.6).

The liquid mixture's more permeable material is occluded on the membrane and then spread through the membrane under the influence of a concentration gradient of diffusing, and after that, it evaporates downstream of the membrane. Finally, the vapor is condensed and collected as liquid due to which it is also known as the solution–diffusion model [27, 28]. This method is used for the separation of ethanol–water mixture.

Wijmans et al. found that removal of various organic solvents like benzene, toluene, naphtha, butane, ethyl ether and others from dilute aqueous streams takes place using organophilic membranes [29]. Kondo and Sato reported [30] on the use of polyether block amide (PEBA) membrane, which is an aromatic hydrocarbon selectively used to remove phenol from industrial wastewater discharged from a phenolic resin process. The wastewater contains up to 10% phenol and other contaminants.

4.2.7 HYBRID MEMBRANE PROCESS

A major drawback of membrane methods is membrane fouling. To overcome this, hybrid processes have been introduced to increase water quality and reduce operating costs. The hybrid process, as shown in Figure 4.8, mainly integrates (1) two or more membrane processes and (2) membrane processes with other water treatment processes, including coagulation, ozonolysis and sand filtration.

4.2.7.1 Coagulation Membrane Process

Combining coagulation with membrane filtration increases the removal of pollutants and reduces membrane fouling. Many researchers have combined coagulation with membrane filtration for the treatment of surface water and coagulants such as chitosan [31], aluminum sulfate, aluminum chloride, polyaluminum chloride, ferric chloride [32] and ferric sulfate [33]. In coagulation membrane process, they found that permeate quality increased and membrane fouling was diminished. Moreover, coagulation combined with a UF membrane for the removal of heavy metal ions like As, Sb.

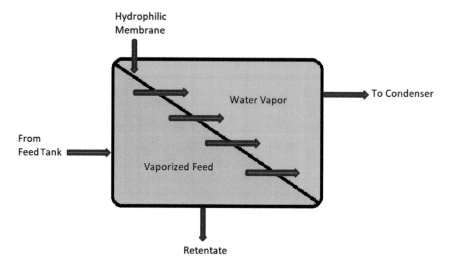

FIGURE 4.6 Schematic diagram of pervaporation.

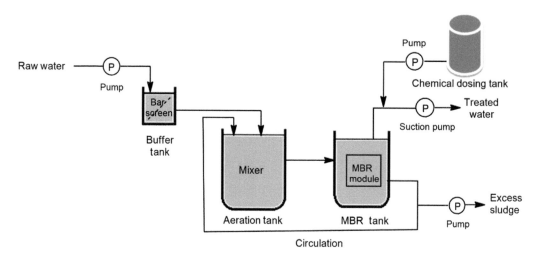

FIGURE 4.7 MBR process.

4.2.7.2 Adsorption Membrane Process

Adsorption technology is mainly used for the treatment of water. Organic compounds can be removed by powdered activated carbon (PAC). Hybrid adsorption–membrane process reduced the membrane fouling rate [34, 35]. Khan et al. [36] reported the effect of particle size on membrane fouling at a PAC-UF system.

4.2.7.3 Ion-Exchange Membrane Process

In this process, fluidized ion exchange and magnetic ion exchange are combined. Removing NOM (nominal organic matter) is done by a nanoporous anion exchanger. Cornelissen et al. [37] used fluidized ion exchange treatment before UF and NF treatment for the surface water treatment.

4.2.7.4   Prefiltration Membrane Process

In this method, for removing coarse materials and microorganisms, sand and packed bed materials are used as preliminary barriers [38]. By using granular media filters, both membrane surface fouling and pore clogging can be reduced.

4.2.7.5   Membrane Bioreactors

A membrane bioreactor (MBR) is a hybrid system of membrane filtration and biological treatment, which is used for wastewater treatment (Tables 4.3 and 4.4). UF and MF can be used in MBR systems. In this process, the membrane (Figure 4.7) acts as an absolute barrier to suspended matter. So, the system is capable of removing suspended solids concentration (MLSS up to 15 g/l).

4.3   APPLICATION OF MEMBRANE TECHNOLOGY IN VARIOUS SECTORS

In food industries, the use of membrane technology as a purifying and separation method is gaining wide application, as shown in Figure 4.8, because the introduction of membrane technology into the food processing industry shows enormous improvement in processing techniques [45, 46]. Due to this, it is used as an alternative to conventional techniques or as novel technology for processing new ingredients and foods. Membrane processes are more advantageous than traditional technologies, and separations done by this technology are considered as green, for example, using cold pasteurization and sterilization with suitable membranes. In this technology, for removing microorganisms, instead of using high temperatures, appropriate membranes are used, which are more economical in terms of energy consumption. In the processing procedure, using membrane filtration to remove microorganisms for shelf-life extension of foods instead of using additives and preservatives also creates a green image. This technique preserves the natural taste of food products and the nutritional value of heat-sensitive components by using a concentration of membrane filtration. Most

TABLE 4.3
Some Combinations of Membrane Processes in Wastewater Treatment

Combinations	Type of wastewater	Removal	References
MF-RO	Urban wastewater	Pesticides and pharmaceuticals	[39]
NF-RO	Dumpsite leachate	95% water recovery	[40]
UF-RO	Metal finishing industry	contaminants	[41]
UF-RO	Oily wastewater	Oil and grease (100%), TOC (98%), COD (98%), TDS e (95%), Turbidity (100%)	[42]
UF-NF/RO	Phenolic wastewater from paper mill	COD (95.5%), phenol (94.9%)	[43]

TABLE 4.4
Conventional Membrane Bioreactors and Hybrid Membrane Bioreactors [44]

	Conventional MBR	NF-MBR	OMBR	MDBR
Membrane type	MF/ UF	NF/RO	Forward osmosis	Porous MF
Driving force	HP	HP	HP	VP
NaCl Rejection (%)	Very less	40–90	~100	100
TOC in permeate (mg/L)	3–10	1–4	<3	<0.8
Water flux (L/m².h)	10–30	<2.5	<10	1.2–15

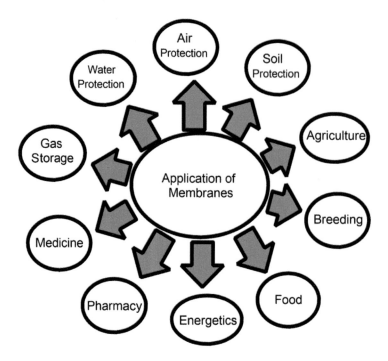

FIGURE 4.8 Applications of MT in various sectors.

useful aspects of membrane technology are the recovery of valuable components from diluted effluents in wastewater treatment. Pressure-driven membrane processes, namely, MF, UF, NF and RO, facilitate the separation of components with a large range of particle sizes and find a wide range of applications in the food-processing industry. For the treatment of wastewater, especially for permeated water, the use of membrane separations is a very good alternative, and this treated water can be reused in production activities. The membrane separation technologies of RO, UF and MF have been used to concentrate and purify both small and large molecules in pharmaceutical production processes [47] This technology also shows its importance in the medical field [48, 49] in various lifesaving treatment methods such as in drug delivery, artificial organs, tissue regeneration, diagnostic devices and for performing coatings on medical devices. Novel membrane technologies like forward osmosis (FO) and membrane distillation (MD) are successfully employed for agricultural water production and the recovery of nutrients from saline water and wastewater. Membranes are not only used for filtration, extraction and distillation; they can also be applied for gas storage [50] in biogas plants or act as catalysts [51] in syntheses.

4.4 CONCLUSION AND FUTURE PERSPECTIVES

Membrane technology shows extensive applications and has been observed to be a very beneficial method for wastewater treatment. Due to its versatile and multifaceted character, this technology is used as a separation process in several chemical and pharmaceutical, food and beverages, metallurgical and biotechnological wastewater treatment industries. It has been proved by literature that from different activities several membrane technologies can be efficiently used to treat wastewater. Along with these applications, membrane fouling and membranes' sensitivity to toxicity are the main limitations of this technology. Today, the advancement in technology overcomes some of these membrane limitations, such as short life, expensiveness, fouling and permeate quality. Still, there are many technical challenges to optimizing and making membrane technology more competitive in the market, large-scale industries and communities. There is a need for further research on advanced membrane materials, which may be resistant to both chemical and mechanical barriers. These may be helpful in prolonging membrane life span and induce long-term performance.

BIBLIOGRAPHY

1. Bethi, B., Sonawane, S. H., Bhanvase, B. A., & Gumfekar, S. P. (2016). Nanomaterials-based advanced oxidation processes for wastewater treatment: A review. *Chemical Engineering and Processing-Process Intensification*, *109*, 178–189.
2. Moussa, D. T., El-Naas, M. H., Nasser, M., & Al-Marri, M. J. (2017). A comprehensive review of electrocoagulation for water treatment: Potentials and challenges. *Journal of Environmental Management*, *186*, 24–41.
3. Yang, Y., Pignatello, J. J., Ma, J., & Mitch, W. A. (2016). Effect of matrix components on UV/H2O2 and UV/S2O82− advanced oxidation processes for trace organic degradation in reverse osmosis brines from municipal wastewater reuse facilities. *Water Research*, *89*, 192–200.
4. Beita-Sandí, W., & Karanfil, T. (2017). Removal of both N-nitrosodimethylamine and trihalomethanes precursors in a single treatment using ion exchange resins. *Water Research*, *124*, 20–28.
5. Carr, S. A., Liu, J., & Tesoro, A. G. (2016). Transport and fate of microplastic particles in wastewater treatment plants. *Water Research*, *91*, 174–182.
6. Hatton, T. A., Su, X., Achilleos, D. S., & Jamison, T. F. (2017). Redox-based electrochemical adsorption technologies for energy-efficient water purification and wastewater treatment.
7. Lai, G. S., Lau, W. J., Goh, P. S., Ismail, A. F., Yusof, N., & Tan, Y. H. (2016). Graphene oxide incorporated thin film nanocomposite nanofiltration membrane for enhanced salt removal performance. *Desalination*, *387*, 14–24.
8. Saleh, T. A., Sarı, A., & Tuzen, M. (2017). Optimization of parameters with experimental design for the adsorption of mercury using polyethylenimine modified-activated carbon. *Journal of Environmental Chemical Engineering*, *5*(1), 1079–1088.
9. Saleh, T. A. (2016). Nanocomposite of carbon nanotubes/silica nanoparticles and their use for adsorption of Pb (II): From surface properties to sorption mechanism. *Desalination and Water Treatment*, *57*(23), 10730–10744.
10. Lofrano, G., Carotenuto, M., Libralato, G., Domingos, R. F., Markus, A., Dini, L., & Meric, S. (2016). Polymer functionalized nanocomposites for metals removal from water and wastewater: An overview. *Water Research*, *92*, 22–37.
11. Gaouar, M. Y., & Benguella, B. (2016). Efficient and eco-friendly adsorption using low-cost natural sorbents in waste water treatment. *Indian Journal of Chemical Technology*, *23*(3), 204–209.
12. Mukiibi, M., Feathers, R., & exclusion Nanofiltration, S. (2009). A breakthrough in water treatment. *Water Conditioning & Purification*.
13. Staude, E. (1992). *Marcel Mulder: Basic Principles of Membrane Technology*. Kluwer Academic Publishers, Dordrecht, Boston, London, 1991. ISBN 0-7923-0978-2, 363 Seiten, Preis: DM 200.
14. Bouhabila, E. H., Aim, R. B., & Buisson, H. (1998). Microfiltration of activated sludge using submerged membrane with air bubbling. *Desalination*, *118*, 315–322.
15. Singh, R., & Hankins, N. (2016). *Emerging Membrane Technology for Sustainable Water Treatment*. Elsevier, Amsterdam, The Netherlands.
16. Muro, C., Riera, F., & del Carmen Díaz, M. (2012). Membrane separation process in wastewater treatment of food industry. *In Food Industrial Processes – Methods and Equipment; In Tech, Rijeka: Rijeka, Croatia*, 253–280.
17. Crittenden, J., Trussell, R., Hand, D., Howe, K., & Tchobanoglous, G. (2012). *Principles of Water Treatment*, 2nd edn. John Wiley and Sons, Hoboken, NJ.
18. Baker, R. (2000). *Microfiltration, in Membrane Technology and Applications*. John Wiley & Sons Ltd, Newark, CA, 279.
19. Perry, R. H., & Green, D. W. (2007). *Perry's Chemical Engineers' Handbook*, 8th edn. McGraw-Hill Professional, New York, 2072.
20. Kenna, E., & Zander, A. (2000). *Current Management of Membrane Plant Concentrate*. American Waterworks Association, Denver, 14.
21. Zeman, & Zydney, A. L. (1996). *Microfiltration and Ultrafiltration: Principles and Applications*. CRC Press, New York.

22. Cheryan, M. (1998). *Ultrafiltration and Microfiltration Handbook*. CRC Press, Boca Raton, FL.

23. Rahimpour, A., et al. (2010). Preparation and characterisation of asymmetric polyethersulfone and thin-film composite polyamide nanofiltration membranes for water softening. *Applied Surface Science*, *256*(6), 1657–1663.

24. Mason, E. A. (1991). From pig bladders and cracked jars to polysulfones: An historical perspective on membrane transport. *Journal of Membrane Science*, *60*, 125–145.

25. Nunes, S. P., & Peinemann, K. V. (2001). *Membrane Technology in the Chemical Industry*. Wiley-VCH, Weinheim.

26. Loeb, S., & Sourirajan, S. (1963). Seawater dimineralization by means of an osmotic membrane. *Advances in Chemistry Series*, *38*, 117–132.

27. Nagai, K. (2010). Fundamentals and perspectives for pervaporation. *In Comprehensive Membrane Science and Engineering; Elsevier Inc.: Amsterdam, The Netherlands*.

28. Zhang, S., & Drioli, E. (1995). Pervaporation membranes. *Sep. Sci. Technol.*, *30*, 1–31.

29. Wijmans, J. G., Kaschemekat, J., Davidson, J. E., & Baker, R. W. (1990). Treatment of organic-contaminated wastewater streams by pervaporation. *Environ. Prog.*

30. Kondo, M., & Sato, H. (1994). Treatment of wastewater from phenolic resin process by pervaporation. *Desalination*, *98*, 147–154.

31. Bergamasco, R., Konradt-Moraes, L. C., Vieira, M. F., Fagundes-Klen, M. R., & Vieira, A. M. S. (2011). Performance of a coagulation-ultrafiltration hybrid process for water supply treatment. *Chemical Engineering Journal*, *166*(2), 483–489.

32. Fiksdal, L., & Leiknes, T. (2006). The effect of coagulation with MF/UFmembrane filtration for the removal of virus in drinking water. *Journal of Membrane Science*, *279*(1–2), 364–371.

33. Konieczny, K., Sakol, D., Płonka, J., Rajca, M., & Bodzek, M. (2009). Coagulation-ultrafiltration system for river water treatment. *Desalination*, *240*(1–3), 151–159.

34. Tomaszewska, M., & Mozia, S. (2002). Removal of organic matter from water by PAC/UF system. *Water Research*, *36*(16), 4137–4143.

35. Matsui, Y., Hasegawa, H., Ohno, K., et al. (2009). Effects of superpowdered activated carbon pretreatment on coagulation and trans-membrane pressure buildup during microfiltration. *Water Research*, *43*(20), 5160–5170.

36. Khan, M. M. T., Takizawa, S., & Lewandowski, Z., et al. (2011). Membrane fouling due to dynamic particle size changes in the aerated hybrid PAC-MF system. *Journal ofMembrane Science*, *371*(1–2), 99–107.

37. Cornelissen, E. R., Chasseriaud, D., Siegers, W.G., Beerendonk, E. F., & Van der Kooij, D. (2010). Effect of anionic fluidized ion exchange (FIX) pre-treatment on nanofiltration (NF) membrane fouling. *Water Research*, *44*, 3283–3293.

38. Singh, R. (2005). Hybrid membrane systems: Applications and case studies. In *Hybrid Membrane Systems for Water Purification*, R. Singh, Ed., chapter 3, 131–196. Elsevier Science, Amsterdam, The Netherlands.

39. Rodriguez-Mozaz, S., Ricart, M., Köck-Schulmeyer, M., Guasch, H., Bonnineau, C., Proia, L., de Alda, M. L., Sabater, S., & Barceló, D. (2015). Pharmaceuticalsandpesticidesinreclaimedwater: Efficiencyassessmentofamicrofiltration – reverse osmosis (MF – RO) pilot plant. *Journal of Hazardous Materials*, *282*, 165–173.

40. Rautenbach, R., Linn, T., & Eilers, L. (2000). Treatment of severely contaminated waste water by a contamination of RO, high-pressure RO and NF-Potential and limits of the process. *Journal of Membrane Science*, *174*, 231–241.

41. Petrinic, I., Korenak, J., Povodnik, D., & Hélix-Nielsen, C. (2015). A feasibility study of ultrafiltration/ reverse osmosis (UF/RO)-based wastewater treatment and reuse in the metal finishing industry. *Journal of Clean. Production*, *101*, 292–300.

42. Salahi, A., Badrnezhad, R., Abbasi, M., Mohammadi, T., & Rekabdar, F. (2011). Oily wastewater treatment using a hybrid UF/RO system. *Desalination of Water Treatment*, *28*, 75–82.

43. Sun, X., Wang, C., Li, Y., Wang, W., & Wei, J. (2015). Treatment of phenolic wastewater by combined UF and NF/RO processes. *Desalination*, *355*, 68–74.

44. Luo, W., Hai, F. I., Price, W. E., Guo, W., Ngo, H. H., Yamamoto, K., & Nghiem, L. D. (2014). High retention membrane bioreactors: Challenges and opportunities. *Bioresource Technology*, *167*, 539–546.

45. Cuperus, F. P., & Nijhuis, H. H. (1993). Applications of membrane technology to food processing. *Trends in Food Science & Technology*, *4*(9), 277–282.
46. Li, N. N., Fane, A. G., Ho, W. W., & Matsuura, T. (Eds.). (2011). *Advanced Membrane Technology and Applications*. John Wiley & Sons, Hoboken, NJ.
47. He, H., Pham-Huy, L. A., Dramou, P., Xiao, D., Zuo, P., & Pham-Huy, C. (2013). Carbon nanotubes: applications in pharmacy and medicine. *BioMed Research International*.
48. Xue, G., Hu, X. L., Chen, X. P., & Zheng, X. (2009). Applications of membrane separation technology in the production of medicine and medical treatment [J]. *Chemical Industry and Engineering*, 2.
49. Wang, L., & Wang, L. (2008). Applications of membrane technology in biotransformation [J]. *Chemical Industry and Engineering Progress*, 6.
50. Tabe-Mohammadi, A. (1999). A review of the applications of membrane separation technology in natural gas treatment. *Separation Science and Technology*, *34*(10), 2095–2111.
51. Galiano, F., Castro-Muñoz, R., Mancuso, R., Gabriele, B., & Figoli, A. (2019). Membrane technology in catalytic carbonylation reactions. *Catalysts*, *9*(7), 614.

5 Developments of Nanomaterial-Incorporated Polymeric Membranes for Water and Wastewater Treatment

Sinu Poolachira and Sivasubramanian Velmurugan

CONTENTS

DOI: 10.1201/9781003165019-5

5.1 INTRODUCTION

Water is the core essential and basic necessity for all known forms of life and has become the significant workhorse of industries worldwide as a functional fluid, transport medium, heat-transfer fluid and surface cleaning agent. The industrial effluent from process plants comprises highly toxic chemical compounds and their direct discharge to the environment causes water quality deterioration and results in water pollution. Moreover, uncontrolled wastewater streams have initiated the pollution of some prevailing freshwater bodies also. This water pollution and water scarcity are predicted as crucial problems in the coming years.

Industries followed various treatment techniques to separate the water contaminants as it is mandatory to a great extent before effluent discharges to the environment. Therefore, several research works have focused on suitable arrangements to acquire freshwater by eliminating water contaminants (chemicals, organics, biological compounds), purifying polluted water using available technologies and reusing it for the same process or recycled for another [1]. Reusing or recycling can reduce freshwater costs, wastewater flows and water footprint size. Besides, clearwater availability can also improve the production capacity, sustainability and operation efficiency of the plant. Every industry has its wastewater that must be handled carefully and managed with the proper treatment method. Various methods for water treatment depend on the water quality requirements, budgetary considerations and space constraints. By considering all these things, membrane separation has become the most successful process to complete purification in a low-cost, environmentally friendly, and energy-saving manner. Advanced membranes developed from novel materials have become excellent options for industrial water management applications.

5.1.1 INDUSTRIAL WASTEWATER COMPONENTS

Inorganic components present in wastewater have a central role in determining the characteristics of receiving water and the aquatic ecosystem. Every industry has different types of wastewater that must be analyzed carefully to find the proper treatment and reuse solutions. The composition of the wastewater depends on the usage process of the particular industry. However, wastewater from industries generally comprises significant organic matter levels, heavy metals and the like. These are responsible for many acute diseases such as cancer risk, skin diseases, anemia, low blood platelets and headache. In addition, humans are threatened with exposure to chemical toxins directly through water or indirectly by consuming plants watered by polluted water or consuming aquatic organisms. The chemical toxins have been bioaccumulated and harm human health. Hence, removing the major microchemical pollutants is of utmost importance even at low concentrations. Even though the daily discharge level of any pollutant may not be high, its continuous discharge for a long time brings them to accumulate in the groundwater and finally reached into water bodies through surface runoff or leaching into the groundwater [2]. Environmental researchers should consider these pollutants, which exhibit organic or inorganic nature, according to their environmental impacts. Among them, heavy metals are significant because of their toxicity and antagonistic known effects on the human body.

The oxidation number determines the toxicity and removal efficiency of the particular heavy metal ions. Also, heavy metal ions cannot degrade or be destroyed, and they tend to accumulate in nature in their pure form. Heavy metal contaminants are usually found in every step of industrial production, from mining metal ore to final finishing and even up to end use of metal [3]. World Health Organization (WHO) indicates some specific heavy metal pollutants are potentially hazardous to the surroundings when they accumulate above than the concentration confines. According to WHO, toxic heavy metals include cadmium (Cd), lead (Pb), mercury (Hg), chromium (Cr), nickel (Ni), arsenic (As), zinc (Zn) and copper (Cu) [4]. The disposal of these waste streams within water bodies lacking sufficient treatment become exceedingly dangerous for humans and other organisms.

Some of the wastewater characteristics also depend on weak acids, bases and salts, along with heavy metal ions. These materials can be subjected to disinfection methods for separation purposes.

The waste generated from metal finishing operations is generally in slurry form, containing metals dissolved in liquid, such as hydroxides of ferric, magnesium, nickel, zinc, copper and aluminum, and must be treated to fulfill all pertinent regulations. Conventional pollutants like suspended solids, chromium, ammonia, phenols, sulfide, oil and grease are the major pollutants from petroleum and petrochemical industries. Many power plants release wastewater with substantial amounts of heavy metals, notably lead, cadmium, mercury, chromium, arsenic, selenium, and nitrogen compounds. The wastewater produced from commercial textile services is filled with organic dyes, sand, grit, heavy metals, oil, grease and volatile organic carbons (VOCs). Chemical industries seem to challenge environmental supervisory policies in handling their wastewater discharges. Today, only 60% of the industrial effluent is treated, and the remaining part is directly discharged in itself to the environment, even if it is a large or small-scale industry. Furthermore, establishing a treatment plant is difficult and expensive for small-scale industries, and hence, it again reduces the removal percentage, resulting in a large amount of wastewater.

Effective treatment technology, stringent pollution control acts and measures and appropriate execution play essential roles in preventing such issues [5]. Appropriate separation mechanisms and target pollutants contained in the effluent narrow the ranges of the membrane. Yet the combination of a membrane with other separation processes can be utilized to achieve complete separation. Moreover, the separation performance of the membranes can be improved by modifying the active membrane surface with a target compound [6]. Hence, modified membranes are an essential part of the industrial effluents treatment plants for on-site solutions across various industries.

5.1.2 Membrane Separation Process

Recently, membrane-based processes have advanced from simple operations in the research workroom to industrial applications with remarkable methodological and commercial impact. During the last decades, an appreciable effort of environmentalists has been committed to developing better membranes and extending their application range. Generally, a membrane is a specific barrier that permits individual components and retains other components in the liquid or gas mixture. Figure 5.1 represents the membrane process in which raw water fed into the membrane is called the feed; the stream that goes through the membrane is known as permeate, while the stream that comprises the remained components is called retentate or concentrate.

According to their ability to sequestrate industrial effluent components, pressure-driven membranes are the most accessible and straightforward among the various membrane technologies. Moreover, the isolation of the components also depends on the pore size of the components. Accordingly, the pressure-driven membranes are classified as microfiltration (MF), ultrafiltration (UF), nanofiltration (NF) and reverse osmosis (RO) membranes. The largest pore size characterizes MF membranes, and they usually reject various microorganisms and large particles. Compared to MF membranes, UF membranes have smaller pore sizes.

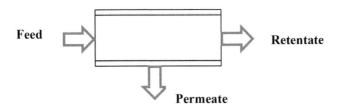

FIGURE 5.1 Membrane process.

TABLE 5.1

Summary of the Pressure-Driven Membrane Process

Membrane system	Size range (μm)	Operating pressure (Pa)	Particles removed
Microfiltration (MF)	1.0–0.01	$< 2 \times 10^5$	Suspended solids, large colloids and bacteria
Ultrafiltration (UF)	0.01–0.001	$1 \times 10^5 - 7 \times 10^5$	Viruses, colloids and macromolecules
Nanofiltration (NF)	0.01–0.0001	$3 \times 10^5 - 20 \times 10^5$	Organic molecules, salt ions, metal ions, humic acids and lactose
Reverse osmosis (RO)	< 0.0001	$15 \times 10^5 - 70 \times 10^5$	Dissolved salts and metal ions

Consequently, they can reject bacteria and soluble macromolecules along with microbes and large particles. In all pressure-driven membranes, RO membranes are effectually nonporous structures, and thus, it can eliminate particles and several low–molar mass compounds, such as ions, salts and organics. In comparison with all membranes, NF membranes are moderately advanced ones. They are occasionally named loose RO membranes and have a porous nature with the pore size are around 10 Å; they display good performance between RO and UF membranes. Classification of the pressure-driven process is shown in Table 5.1.

5.1.3 MEMBRANE SELECTIVITY AND PERMEABILITY

The central part of the membrane separation process is the membrane itself. Hence, membranes should be made with good mechanical stability that can retain high productivity and good selectivity for the desired permeate. Furthermore, the ideal physical structure of the membrane material mainly depends on the topmost thin layer of material with a good pore size distribution and surface porosity. The separation is carried out at the dense top layer. In contrast, the bottom support layer provides a very near resistance-free path for water and the unseparated solutes present in the permeate water stream. The driving force (hydrostatic pressure, concentration, or electric field) across the membrane surface facilitates the mass transfer across the membranes. In addition, the connection between the force applied and the flow generated in the membrane is overseen by factors that depend on membranes and the chemical species. Besides, the flow of components through a membrane can be related to applied force, solubility and mobility.

Membrane selectivity limits the quality of the separation process and membrane permeability limits the flow rates through the membrane. The selectivity and permeability trade-off between membranes plays a significant role in maximizing the right stuff. And this is the reason for the limitation of membranes in the wastewater industry. Selectivity represents how the desired molecules are separated from the rest and permeability represents how fast the molecules permeate through the membrane material. More efforts have been conducted to overcome selectivity and permeability limitations by adjusting the pore structure, including pore size, porosity and skin layer thickness. Even so, novel strategies have to be developed to breach this trade-off limitation because of the complexity of adjusting pore structure [7].

Moreover, the selectivity terms in membrane applications are generally described as the difference between the rate ratios of two species mixed in the solutions. They can flow through the membrane because of their permeability behavior, but they only allow noncomplex species and water to pass through its surface while the rest of the complex species are retained. A membrane with enhanced selectivity is a fundamental necessity instead of a membrane with improved permeability [8], especially water purification membranes. Membrane separation with nanotechnology has been chosen as an efficient technology for processing separation faster than any other conventional separation technique [9]. Still, the decline in permeate flux due to membrane fouling is a critical problem associated with the membrane process. Membrane fouling mainly depends on its surface properties like hydrophobicity, roughness, and filtration mode (dead-end/normal-flow filtration or crossflow filtration). The foulant adhesion and pore-blocking during the filtration process lead to a decline in permeability. An increase in membrane hydrophilicity can strengthen its fouling resistance by considering that most contaminants are naturally hydrophobic. Several modification methods have been observed to increase the hydrophilicity of polymeric membranes, such as blending with hydrophilic polymers, introducing nanoparticles into the casting solution and chemical modification and so on. An extra repulsive force can also help retain the foulants, which results in an improvement in selectivity and facilitates the trade-off limitation [7].

5.1.4 Mode of Filtration

One of the best selections to determine while using a membrane is feeding and concentrating components through the membrane system. Mainly, two flow geometries of membrane processes are used: dead-end and crossflow (or) tangential flow filtrations.

5.1.4.1 Dead-End Flow

Dead-end flow or normal-flow pattern is a batch process in which the feed is slowly subjected to the membrane sample. It allows some components based on the driving force over the membrane surface while retaining the other components and results in high permeate yield in a simple way. As the filtration time prolongs, the membrane is subjected to clogging and decreases the filtration capacity. So that pressure should be increased to maintain the flux value. The significant advantages of dead-end filtration include the easiness of fabrication and implementation and inexpensive operation cost. The major disadvantage of dead-end filtration is that the concentration polarization is followed by extensive membrane fouling. Higher driving forces are induced for a faster fouling rate, and the process requires a subsequent step to remove the accumulated matter [10].

5.1.4.2 Crossflow

In crossflow mode, the raw water is introduced tangentially over the active membrane surface under pressure rather than directly onto the membrane. The tangential flow generates turbulence over the surface and that reduces the buildup of solute particles. The smaller particles (smaller than the membrane pore) permeate through the membrane, but others are retained. In contrast, larger suspended particles persist in the retentate stream with minimum solid buildup and constant low flow resistance. Figure 5.2 represents the mode of the filtration process.

Either crossflow or dead-end flow geometries propose some advantages and disadvantages. The membrane is considered a consumable component in dead-end mode. The crossflow can hinder

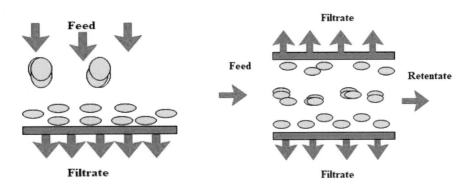

FIGURE 5.2   Normal-flow and crossflow filtration process.

the accumulation of materials at the surface and the membrane is not like a consumable component. Considering the crossflow mechanism of mixed matrix membrane is vital for the design and operational perspectives because of the obligation of a constant mode of operation in combined membrane processing applications [11]. Hence, the dead-end mode is more likely to recommend flux-limited applications and crossflow mode for flux stability and fouling reduction applications.

5.1.5 TRANSPORT MECHANISM

A membrane's superior property is considered as its competence to allow the selected species to interact with kinetics and thermodynamic state. The components penetrate through the membrane based on their mass diffusivity and concentration gradient. Somehow, it is also affected by thermodynamic contribution. Besides, the transport mechanism in a membrane should be varied based on the structure of the membrane. Commonly, three transport processes occur through the membranes and can be named passive, facilitated, and active. For passive transport, the feed-stream flow is due to the electrochemical gradient concerning the chemical potential in the membrane phase, and it is the same as the driving force. A constant interaction was found between the membrane molecules and the permeant molecules in facilitated transport, followed by a three-step mechanism; that is, carrier molecules were associated at the membrane's internal surface and passed through the membrane. Finally, it dissociated from the internal membrane surface.

Moreover, the transport is from a higher chemical potential to a lower chemical potential with a third component in the membrane phase. However, the feed flow is against its driving force for active transport, representing that the feed side chemical potential is lower than the permeant side potential. For nonporous membranes, solution–diffusion flow theory can explain the transport mechanism, whereas, for porous membranes, it depends on the sieving mechanism. The Donnan exclusion can explain the sorption and desorption process in nonporous membranes, in which the membrane surface charge determines the attraction and repulsion of the solute particles. In a porous membrane, the size and shape of the solute particles depend on the separation process [12]. Furthermore, the selection of additives plays a significant part in the transport mechanism of composite membranes.

5.1.6 MEMBRANE MATERIALS

On the whole, polymeric materials provide a broad range of structures and properties. Hence, almost all organic membranes are built from polymeric materials. Furthermore, water flux, solute rejection, thermal and chemical stability, mechanical strength, antimicrobial activity and cost-effectiveness are notable characteristics of membrane material. Because of the immense diversity of polymer membrane materials, sound knowledge in membrane classifications such as membrane materials, cross section, casting procedure, membrane shape and structure help with selecting a suitable material for technical applications.

Today, researchers concentrate on polymeric materials to produce membranes with various features because of their excellent control in pore generation and comparatively lesser cost than any other inorganic nanomaterials. All polymeric materials have definite characteristics that produce a membrane with different characteristics and behave appropriately for different water treatment applications [13]. In polymeric membranes, the components are separated by the sorption–diffusion mechanism. Generally, sorption depends on the chemical nature of the molecules and diffusion depends on the size and shape of the molecules.

Cellulose acetate (CA), polysulfone (PSF), polyethersulfone (PES), polyvinyl alcohol, polyacrylonitrile, polyvinylidene fluoride (PVDF), polypropylene, polytetrafluoroethylene and polyimide are the increasingly being used organic membrane materials. Today, many investigative works are taking place on modified-membrane developments to increase the process efficacy [14]. Some of the leading polymers utilized for membrane development and their benefits and drawbacks are listed in Table 5.2.

TABLE 5.2

Membrane Polymers with Their Advantages and Disadvantages

Polymer	Advantages	Disadvantages
Cellulose acetate	Excellent film-forming properties, high hydrophilicity, eco-friendly and suitable cost	Poor abrasion resistance, rapidly loses strength when wet and limited chemical resistance and thermal performance, not appropriate for aggressive cleaning
Polysulfone (PSF)	High strength, rigidity, creep resistance and dimensional stability due to its replicating phenylene rings, Excellent chlorine resistance, Resilience in membrane production over a wide range of pore size	Poor limits of operating pressure, hydrophobic nature and lack of solvent resistance
Polyethersulfone	Strict pore size distribution, high flux, low protein adsorption and high mechanical and thermal performance	Hydrophobic nature, Low resistance to ultraviolet (UV) light, Attacked by polar solvents such as ketones and aromatic hydrocarbons
	More polar moiety, making it slightly more hydrophilic than PSF	
Polyvinylidene difluoride	High mechanical strength and chemical resistance, high thermal stability (up to 75 °C), resistance to Cl_2 and easy fabrication	Hydrophobic nature and coating sensitive to high pH
Polyamide	Broad pH tolerance, high thermal stability and high mechanical properties	Low resistance to Cl_2 and microbial attack
Polytetrafluoroethylene	Excellent thermostability, strong chemical inertness, high mechanical strength and great insulating performance	Hydrophobic nature and require frequent cleanings

Every membrane-forming polymer has more than one characteristic, but at the same time, the polymer shows these characteristics instantaneously. Consequently, substantial exertions are conducted through membrane modification methods to improve the functions such as permeate flux and membrane life. Mainly, membrane modifications are successfully performed by combining nanoparticles (NP) onto the polymer matrix. NPs demonstrate improved selectivity, good adsorption capability, greater specific surface area, excessive dynamic groups, higher binding capacity, reduced price and easiness in reusability. Most marketable membranes are produced from synthetic polymers and inorganic materials, resulting in new nanocomposite membranes with combined properties of polymers and NPs. On the other hand, inorganic materials on massive productions are restricted due to higher operational costs and characteristic mechanical stability [15, 16].

5.2 NANOCOMPOSITE MEMBRANES

5.2.1 MIXED MATRIX MEMBRANES

Comparing polymeric membranes with inorganic molecular sieving materials, they have low separation performance characteristics. Hence, the generation of mixed matrix membranes is considered an alternative strategy in membrane research. In this field, the excellent separation characteristics of molecular sieve materials and appropriate mechanical properties of polymers were combined for better performance and economical processability. In mixed matrix membranes (MMMs), the bulk phase typically is a polymer, and the distributed phase represents the inorganic materials, which may be metal oxides, carbon-based materials or mineral-based materials. When NPs are infused into the polymer matrix, it is anticipated that the following membrane features turn out to be better than normal polymer membranes. Simultaneously, the inherent fragility of the inorganic membranes can be averted by the continuous polymer phase. The addition of NPs with superior separation characteristics results in MMMs having the ability to accomplish more permeability and selectivity with trivial loss in membrane flexibility expected for the resultant MMMs.

The MMM fabrication difficulty was noticed due to the weak particle contact and the low particle distribution in the polymer matrix. Additionally, average particle size, pore size, and polymer features can influence the mixed matrix properties. The polymer matrix's continuous distributed phase comprises porous (activated carbon, zeolite, carbon nanotubes) or nonporous material (TiO_2, silica, fullerene). Applying NPs on the topmost layer of nanofibers scaffolds (fibers having a diameter less than 100 nm) encourages higher porosity, resulting in excellent permeability and offers more economical energy demands. The basic principle behind the membrane production process is to control the pore size and pore size distribution at the top layer and decrease the surface layer thickness.

Conversely, few studies in the past three decades reported that conventional polymers could also increase gas separation membranes' performance. To fully exploit the budding openings in gas separation, solid awareness is desirable in identifying new membrane materials with suitable inorganic NPs that can comply with current requirements. The precise particle size distribution and shape ensure outstanding selectivity for the membranes. In the same way, the instantaneous application of inorganic membranes is delayed by the lack of expertise to develop a continuous and defect-free membrane, along with the higher production cost and handling issues. However, as may be understood, the progress on the production and application of MMMs comprising inorganic particle–infused polymer matrices for gas separation is still relatively small and provides an opportunity for future developments to modify gas separation membranes.

5.2.2 Thin-Film Composite Membranes

A thin-film composite (TFC) membrane is a layered membrane that contains two or more layered materials constructed in a thin film. The two classifications are the NF membrane and the RO membrane. Both membranes can be prepared on a thin-film polyamide layer (thickness less than 200 nm) placed on the top of a PSF or PES porous sublayer of approximately 50 microns. A TFC membrane's advantages are that it provides a high filtration rate, mechanical strength, and rejection rate of unwanted materials like salt ions. A high percentage of rejection is achieved on account of the top polyamide layer. The primary reason for selecting this membrane is its ability to permeate water and impermeability to the solution's impurities. A TFC membrane may also offer good water permeability, selectivity and higher pure water flux. Figure 5.3 represents a TFC membrane.

Thin-film polyamide is coated over the base membrane along with the nanomaterials is achieved by the interfacial polymerization reaction. Interfacial polymerization is a type of polycondensation reaction. Similarly, RO and TFC membranes are prepared via interfacial polymerization reaction amongst a polyfunctional amine (1, 3-phenylenediamine [MPD]) and an acid chloride (1,3,5 benzene

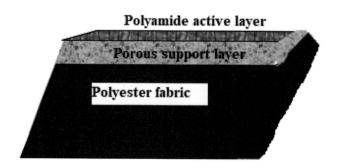

FIGURE 5.3   TFC membranes.

tricarbonyl trichloride [TMC]). These amine monomers and acid chlorides are completely dissolved in water and a hydrocarbon solvent. In the meantime, water and the organic solvent are immiscible; polymerization reaction occurs at the water–hydrocarbon interface – those consequences in forming a thin skin layer on the surface of the porous substrate membrane. During interfacial polymerization, nanoparticles are introduced into an organic phase of the solution.

5.3 MEMBRANE MODIFICATION

The surface properties of the membrane critically influence the performance as it associates with the feed. Fouling or the adsorption of unwanted species on top of the membrane surface disturbs the membrane and compromises the performance. Accordingly, excessive effort is needed to minimize the unwanted buildup of molecules on the membrane surface. Generally, hydrophobicity is one of the significant disadvantages of any membrane material and is directly related to fouling behavior. So the scaling up of an additional modification and its implementation are needed to reduce the fouling rate and biocompatibility. Some researchers adopted several strategies such as surface modification, bulk modification and multi-bore configuration to alleviate the fouling resistance to tackle the fouling problem. Handling the membrane surface with water-soluble solvent, termed hydrophilization, is a viable option for its surface modification – moreover, the chemical treatment methods through covalent bonding result in a better-modified membrane. Bulk modification using responsive nanomaterials is a forward-looking approach to overcome the limitations of existing membranes. Blending polymers with organic or inorganic molecules results in a more hydrophilic membrane, reducing fouling and enhancing water flux [17–20]. By adding porogen or NPs, blending methods and surface-coating methods are significantly affect the membrane field. A wide variety of surface-modifying agents has been used to modify the membrane surface by covalent bonding. That connects the essential components by carboxylation, sulfonation, amination or epoxidation. Finally, the measurement of hydrophilicity, homogeneity of modification and surface roughness are needed as these are the crucial factors indicating the total success of the modification process.

A study was performed by Roy and Raghunath et al. (2018) based on developing a membrane material in water and energy production reliability. They highlighted that in making a membrane for a sustainable solution to water pollution, proper design and novel membrane fabrication with customized separation properties is mandatory [21]. According to Ahamed Yusuf et al. (2020), the recent developments in membrane science and technology for endurable water treatment involve reusing membranes, reusing waste brine or sludge and energy recovery and waste minimization by membrane antifouling approaches [22]. B. Chakrabarty et al. (2008) considered the influence of polyvinyl pyrrolidone (PVP) on PSF membrane structure and permeation properties, and it is interrelated with morphological parameters and flux performance of the membranes [23]. A L Ahmad et al. (2013) studied the performance of modified PES membranes by hydrophilic, amphiphilic and inorganic materials to intensify the membrane flux and hydrophilicity [24]. Heidi Lynn Richards et al. (2012) studied a metal NP-modified polymer matrix to reduce fouling capability and increase membrane performance. The results showed that the addition of TiO_2, Al_2O_3 and ZrO_2 NPs enriched the tensile strength and hydrophilicity and alleviated fouling of organic matter [25]. The literature concludes that membrane modification is considered one of the most effective approaches to minimize membrane fouling by improving the membranes' surface hydrophilicity.

5.4 MEMBRANE FABRICATION: IMMERSION PRECIPITATION

The choice of polymer and its desired structure are the main criteria for selecting a polymer membrane fabrication technique. The most common method to produce polymeric membranes is nonsolvent-induced phase separation (NIPS) via immersion precipitation. The membrane-forming polymer is dissolved in an organic solvent, which is miscible with the nonsolvent. By immersing a casted film in the nonsolvent coagulation bath, solvent and nonsolvent exchanges are started. Hence,

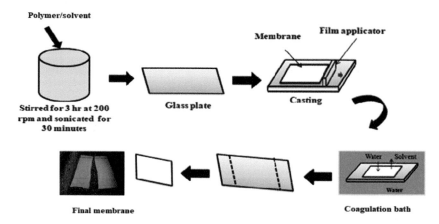

FIGURE 5.4 Immersion precipitation.

the demixing of solvent and nonsolvent results in two phases, a polymer-rich and polymer-lean phase. Hence, the phase separation technique is a demixing of solvents in a controlled manner in which dissolved polymer is converted into a solid state. During the demixing process, the solvents in which the polymer is dissolved (dope solution) are segregated by either evaporating into the atmosphere or employing another nonsolvent, namely, water.

Phase inversion can be encouraged via several methods such as (1) dipping of polymer solution in nonsolvent, (2) coming into contact with a polymer solution and vapors of a nonsolvent, (3) solvent evaporation and (4) solution quenching to a lesser temperature [26]. When the glass plate is submerged in nonsolvent (water), the nonsolvent may diffuse into the polymer solution. If the diffusion coefficient of polymer material is far lesser than solvent, polymer molecules can move a small distance only. A ternary system between the polymer–solvent–nonsolvent can predict the membrane formation by this immersion precipitation method. Similarly, the diffusion of the particles can be understood with the help of mass transfer models. The prediction of the dope solution rheology, including viscoelasticity and thermodynamics, is necessary to fabricate an asymmetric membrane with anticipated features. Comprehensive knowledge from rheology can envisage the solidification process and instantaneous demixing and better understand the phase separation method's phenomena. The one constraint of the phase inversion process is that it necessitates polymers' good solubility in a suitable solvent. Additionally, the process efficiency is limited due to insufficient quantitative knowledge concerning the system's thermodynamics and kinetics.

5.5   TYPES OF INORGANIC NANOADDITIVES

A nanomaterial-infused polymeric membrane is considered a promising method for enhancing membrane performance. The NPs can enhance membrane characteristics such as permeability, hydrophilicity, mechanical stability, conductivity, selectivity and antimicrobial activity. However, there is a slight chance of membrane deterioration due to the presence of NPs. Hence, extreme care should be taken while selecting an NP and its composition in a polymer dope solution. Metal oxide, carbon-based materials and minerals-based materials are the most widely used and successfully demonstrated nanoadditives to get the desirable membrane properties.

5.5.1   METAL AND METAL OXIDES

The incorporation of metal or metal hydroxide nanomaterials has attracted in membrane research based on the synergistic effects on produced membranes. The particle dimensions show a prominent role in the practical implementation of the membrane, which contains metal or metal oxide as nanoadditives. The particle size is an identifiable factor along with the distribution dimensions and mean particle size. Moreover, metal/metal oxide particles can upgrade polymeric membranes' permanence because of the perm selectivity with the varying temperature conditions [2, 27]. The NPs can simply be infused into the polymer dope solution by probe sonication techniques. First of all, dispersed NPs in the organic solvent are used, and then a polymer is added to the mixture, which is followed by an ultrasonication treatment results in a uniformly dispersed dope solution. Finally, the glass plate in which the dope solution is cast is dipped in the coagulation bath of nonsolvent (water) at room temperature meant for manufacturing flat-sheet membranes over a wet-phase inversion process. Otherwise, these NPs can be coated on the surface of the supporting membranes. In that case, the rate of filtration depends on the nature of the surface film. The growth of membranes in catalytic, optical, electronic and magnetic applications mainly depends on these nanosized metal particles. The countless metal NPs that may be familiarized in the polymeric membranes are TiO_2, silica, ZnO and silver NPs, and detailed explanations are included in the following sections.

5.5.1.1 Titanium Dioxide

TiO_2 was developed with the Ti element as the forerunner because of its outstanding stability, super hydrophilicity, photocatalytic activity and nontoxicity. It exhibits an excellent self-cleaning and pliable nature in chemical and thermal environments. Simultaneously, the homogeneous suspension of TiO_2 in water is engaged as a catalyst and can effectively destroy several bacteria and instantaneously support organic chemical disintegration. Moreover, TiO_2 has publicized competencies in numerous additional environmental difficulties apart from water and air pollution cases. The self-assembly behavior of TiO_2 has been utilized to fabricate ultra-thin films without using any solvent at high-temperature conditions.

TiO$_2$ is mixed with the PVDF membrane; it improves the hydrophilicity and averts the membrane from biofouling. TiO_2 can move the membrane action more by relocating the polymer outline or by repolymerization. The alleviation of PSF fouling during humic acid separation can also improve by introducing TiO_2 in the membrane matrix. The photocatalytic activity of TiO_2 negotiates a noble agreement in self-cleaning property and succeeding in membrane fouling drawbacks. Since TiO_2 is not toxic to humans, it favors castoff as the best potential photocatalyst in ultraviolet (UV) light. The powder form may be used in coatings and coverings of the membrane to reduce the element cost. Some studies reported that the application of TiO_2 NPs shows a reduction in membrane porosity. For such cases, NF helps avoid the failure of TiO_2-based porous structure and results in a reduction in flux drop, and it toughens the membrane mechanically along with an improvement in hydrophilicity.

L. Penboon et al. (2019) studied the consequence of TiO_2 on PVDF membrane for dye wastewater treatment by a photocatalytic membrane. TiO_2 particles were witnessed at the top surface while the porosity of the coated membrane is substantially decreased. However, the TiO_2 photocatalytic membrane is developed as a reasonable technology for the decolorization of wastewater [28]. Jing Guo et al. (2017) fabricated a PES composite membrane mixed with sulfated-TiO_2 nanoparticles. The sulfated-TiO_2 upgraded the hydrophilicity, whereas dropping the fouling rate due to the acidic nature caused by the assimilation of sulfate groups with PES [29]. K A Gebru et al. (2016) prepared a composite membrane from CA and TiO_2 NPs utilizing an electrospinning technique. The TiO_2 addition brings about vastly organized fiber networks, which substantially improve the membrane's pore organizations [30].

5.5.1.2 Zinc Oxide

Wide-reaching researchers have utilized zinc oxide (ZnO) NPs identified as hydrophilic, inexpensive and environmentally safe inorganic nanoparticles. ZnO NPs can offer a polymeric membrane with excellent antifouling performance and photocatalytic self-cleaning capacity as necessitated by the methodical works. Therefore, ZnO-infused composite membranes were deemed to be a key priority in the membrane field. Over the last decades, they have been meaningfully developed by an enormous quantity of reported work in the literature. The existing assessment points out the up-to-date conclusions in polymeric membranes infused with ZnO NPs for fouling moderation [31].

The presence of hydrophilic functional groups such as –OH, –SO$_3$H, and –COOH enables ZnO NPs to own their strong hydrophilic nature [32]. ZnO-modified PVDF membranes have been comprehensively employed in separation processes for several industrial applications [33]. However, the fouling behavior shortens the PVDF membrane lifetime and proliferates the operational cost. Hong and He (2014) fabricated a PVDF–ZnO composite membrane using a simple blending method that exhibited typical asymmetric cross-sectional structures. It appeared that the accumulation of ZnO NPs results in a decline in water contact angles, which represents enhanced hydrophilicity, and, therefore, promoted antifouling nature [34].

Similarly, Zhao S. et al. (2015) fabricated a PES–ZnO composite membrane and found that the composite membrane exhibited a more porous structure. The decline in contact angle value represents the better hydrophilic nature of the new membrane. They concluded that the composite membrane's improved surface hydrophilicity and thermal stability were accredited with accumulating these ZnO NPs [35]. Kim et al. (2018) established the polyurethane (PU)–ZnO composite material using the collective superficial modification method. The composite membrane exposed unusual antimicrobial activity with photocatalytic ability supported a potential implementation in the degradation of organic pollutants and wastewater decontamination [36].

5.5.1.3 Iron Oxide

Iron is the most abundant transition metal and element in the Earth's crust. The reactivity of iron particles is essential in macroscopic applications, especially rusting, but it has a leading concern about nanoscale applications as the backbone of our modern infrastructure. Iron NPs have not been entirely studied due to their fine form and high reactivity due to their pyrophoric character. Nevertheless, iron can offer countless benefits at the nanoscale, together with effective magnetic and catalytic properties. As discussed, the iron metal origin's high reactivity is inappropriate for use as pure metal NPs. Therefore, iron NPs were combined into the polymeric membranes instead of iron compounds and used effectively.

Negin Ghaemi et al. (2015) prepared a modified PES membrane with functionalized Fe$_3$O$_4$ NPs for removing copper from water. Blending a PES polymer with surface-modified Fe$_3$O$_4$ NPs increases the hydrophilicity and pure water flux of membranes. Moreover, the mean pore radius and the overall porosity of membranes are also enhanced by these NPs [37]. Nasim Barati et al. (2021) invented a modified ceramic membrane with in situ–grown iron oxide NPs and applied them for oily wastewater treatment. A homogeneous distribution of the NPs onto the membrane surface, as well as within the active layer, was observed. Besides, almost no major agglomeration was noticed; hereafter, low-NP content can almost preserve the membrane morphology [38].

5.5.1.4 Silver

Silver and silver-based compounds are expansively used to make coatings, wound and burn dressings and antimicrobial plastics, among others, based on better-quality biocidal properties. The accumulative usage of silver (Ag) NPs in recent materials ensure that they discover their way to environmental systems. Ag NPs show powerful inhibitory and biocidal possessions in contradiction of various types of microorganisms with long periods. Substantial hard work has been dedicated to introducing Ag NPs into polymeric membrane substrates to form Ag NPs/membrane nanocomposites. Further studies need to be conducted to introduce silver into the membrane matrix to progress the antibiofouling nature in wastewater treatment. Li X (2013) conducted an in situ development of Ag NPs in PVDF ultrafiltration membranes to alleviate organic and bacterial fouling. The modified PVDF membrane showed organic antifouling nature with antibacterial properties measured using a halo zone test [39]. Jabran Ahmad et al. (2019) infused triangular-shaped Ag NPs for membrane modification to connect the potential of NPs based on their different shapes. The modified membrane exhibited notable shape-dependent antifouling, antibacterial and antiadhesion performance contrary to *E. coli*. The enhanced antiadhesion performance and reduced biofouling result from the negative membrane surface charge, which conveyed electrostatic repulsive force between the membrane surface and bacteria [40].

5.5.2 CARBON-BASED MATERIALS

Recently, motivated research works are carried out on the discoveries of different carbon nanostructures and their applications in various fields. The intensive examinations of carbon nanolayers and carbon/polymer nanocomposites encouraged curiously by the innovation of graphene layers and carbon nanotubes (CNTs) stemmed from the inference that the eccentric character of the carbon atom in the carbon layer has an essential significance for the structure and the properties of carbon NPs and their thin films. Nano-carbon-based materials, that is, graphene and CNTs, can obstruct bacterial progression upon straight interaction through the cells [41].

5.5.2.1 Graphene Oxide

Graphene oxide (GO), an able two-dimensional (2D) nanomaterial with excellent properties, has been used as an additive for water purification. It has shown increased membrane hydrophilicity and pure water fluxes and superior pollutant adsorption due to its large surface area. With a significant quantity of oxygen-carrying functional groups (hydroxyl and epoxy), carboxyl surface groups and large surface area, GO nanosheets can be easily used for chemical functionalization and composite materials production [42]. Because of the excellent adsorption capacity of GO, it can be used as an efficient adsorbent for toxic metal particles like Pb(II), Cu(II) and Cd(II) [41]. However, there is a chance of GO NPs leaching into the water due to their high affinity toward water molecules [43]. Hence, GO-impregnated MMMs are an option that results in an attractive alternative for membranes and prevents the leaching behavior of GO. Along with the GO's ability to improve the membrane hydrophilicity, it enhances the polymer membrane's surface roughness and strength, influencing antifouling properties and performance [44].

Generally, GO is a nanomaterial with an amphiphilic nature and builds up with a water channel that progresses permeation flux. GO displays numerous appealing properties like high strength, low thickness, high flexibility and a negatively charged surface, proposing water dispersibility and excessive miscibility with polymers [45]. Usually, the by-product of graphite oxide is provided by scattering graphite oxide in critical solutions or in polar solvents to produce a monomolecular layer identified as GO. Generally, graphite oxide, created by reacting graphite with strong oxidizers. Lee et al. (2013) arranged polysulfone/GO mixed matrix membranes (MMMs) for managing wastewater with membrane bioreactors and revealed a higher conflict from both fouling and biofouling distinguished with the pure polymer as validated by the upsurge in transmembrane pressure [46]. Chang et al. (2014) analyzed the influence of GO and polyvinylpyrrolidone (PVP) on the PVDF layers. Indeed, it has been shown that the membrane antifouling properties with good hydrophilicity are improved by combining GO into PVP. Furthermore, the authors declared that the enhancement is ascribed to hydrogen bonds' enlargement amongst PVP and GO [47]. Sinu and Sivasubramanian (2022) fabricated GO-modified PES membranes that resulted in higher water flux with good removal for lead ions [41].

5.5.2.2 CNTs

CNTs are allotropes of carbon, made of cylinder-shaped graphite sheets in a tube-like structure. Lately, CNTs have pulled researchers' attention because of their exceptional electrical, mechanical, and thermal properties and their fractional antibacterial activity. Being significant, the most recent property of CNTs is appropriated for essential water purification applications. CNTs are made out of a single graphene sheet, named single-walled carbon nanotubes (SWCNTs). Alternately, multilayers of graphene sheets are identified as multiwalled carbon nanotubes (MWCNTs).

CNT has recently fascinated significant consideration for producing innovative membranes with striking water purification features and electrical applications. CNT-based composite membranes are highly demanded because they offer enhanced membrane properties owing to the collective advantages of CNTs and membrane separation. For instance, CNTs' composite membranes possess excellent virus removal and antimicrobial action toward gram-positive and gram-negative bacteria [48]. Consequently, such membranes were possibly developed as membrane filters for drinking water management. Moreover, CNTs can revise the membranes' physicochemical possessions, which invigorate their capabilities for numerous solicitations. Customarily, the inner pores of CNTs give priority to turn as discriminatory nanopores. Thus CNT-filled membranes have a habit of displaying an improved permeability deprived of the reduction in their selectivity. In contrast, developments in mechanical and thermal properties can be gained as thriving [49]. Regardless of the limitations of its original qualities, research still has an interest in CNTs and their applications in membrane preparation; for instance, an increasingly recognized focus on CNT functionalization focuses on the involvement and hydrophobicity of membranes in solution [50].

5.5.3   Minerals-Based Materials

The usage of minerals-based materials such as clays, mineral feedstocks, cement, sands and ash can be used as a base for producing ceramic membranes and propose a favorable passageway headed for procurement of effective filtration systems sparingly executed in large volumes. The challenge in producing low-cost membranes from naturally available raw materials and waste products relates to the achievement of a membrane that displays suitable pore structures for efficient pollutant separation. Layered double hydroxides and silica nanoparticles are commonly used minerals-based materials for modifying membranes for water treatment.

5.5.3.1   Layered Double Hydroxides

A promising layered material, namely, layered double hydroxides (LDHs), otherwise called anionic clays or very often known as hydrotalcite (HT)-like materials, has found different applications in the medical field, as polymer additives, in composite nanomaterials formation, as mixed metal oxide catalyst precursors and in the expulsion of ecological threats [51, 52]. HT is having brucite-type octahedral sheets, in which hydroxyl groups octahedrally organize and the water molecules and interchangeable anions like nitrate, carbonate and sulfate invade the interlayer space. Shedding of this HT brings about 2D nanosheets with exceptional properties; thus, the exfoliated hydrotalcite (EHT) nanosheets could have tremendous opportunities in various functional nanomaterials [53].

Jindun Liu et al. (2014) prepared Mg-Al hydrotalcite-based hybrid membranes in dimethylacetamide solvent. The in-situ exfoliation produces a nanocomposite membrane with a positive charge. The formed membranes are suitable for nanofiltration and ultrafiltration applications with enhanced water flux, good rejection and hydrophilicity [54]. Poolachira et al. (2019) fabricated a PES membrane modified with Mg-Al hydrotalcite, which has been efficiently used to remove lead ions from aqueous solutions with a rejection percentage as high as 50.2%. The presence of nanoadditives in the membrane has improved its characteristics features, like tensile strength, porosity and hydrophilicity. A better percentage of rejection and permeability has been achieved, that is, almost 1.7 times higher than membranes without the additives, for metal ions [55]. Yu Zhao (2016) modified thin-film polyamide nanocomposite membranes with Al-Zn LDHs to remove organic matter naturally. Here, the interfacial polymerization reaction is conducted between the monomers of MPD and TMC [56].

5.5.3.2 Silica

Silica NPs, an ideal protein host with a large surface area, hydrophilic nature, environmental inertness and high chemical and thermal stability, are investigated extensively in the area of nanotechnology and membrane separation. The fine miscibility in an aqueous solution is due to the excellent dispersion of NPs because of its electrostatic stabilization. The lower toxicity and environmental inertness of silica NPs nominate it for incorporating into the membranes used in drinking water applications. Additionally, silica nanoparticles embedded in membranes may help to produce gases which free from impurities. Silica nanoparticles could be further used in processes such as environmental remediation, seawater desalination and petroleum chemicals and fuel production due to their ability to trap molecular-sized impurities

SiO_2 nanoparticles are now promising candidates as additives to fabricate inorganic–organic hybrid composite membranes due to the low-cost synthesis technique and low toxicity levels when used in aqueous solutions. The NPs need to form a stable and robust hydrophilic surface for which the particles need to have good monodispersity [57]. These NPs are also said to be excellent additives to create super-hydrophilic films using SiO_2 polymer nanocomposites, which helps relieve the disadvantage of membrane fouling [58]. Aftab Ahmad Khan et al. (2021) modified PES membranes using fluorinated silica and perfluorodecyl triethoxysilane and polydimethylsiloxane solutions (which are omniphobic agents) for the application of oily wastewater treatment. The observations that have been noticed on modifying are that the membrane's performance has improved noticeably and its antifouling properties when used in oily systems [59]. Antonio Martin et al. (2015) fabricated mesostructured silica SBA-15 particles and loaded them onto a PES membrane. The functionalization of the silica particle is achieved by the co-condensation technique between amine and carboxylic groups [60]. An asymmetrical structure with distributed open macro voids was observed from the membranes fabricated using the immersion precipitation technique.

The feasible and desirable incorporation of modified silica in PSF membrane improves the antifouling nature and good tolerance to tensile force and enhanced gas permeability. Modified PVDF membrane with silica NPs showed excellent selectivity, thermal stability and diffusivity. Lately, hybrid membranes made from mesoporous silica comprising sulfuric acid groups with Nafion by sol-gel processes showed improved proton conductivity.

5.6 CONCLUSION AND FUTURE SCOPE

The demand for clean water is rapidly increasing as its sources are dwindling and ever-increasing water demand requires water conservation, recycling or treatment of polluted water to create a clear one. Recycling and reuse dominate a circular economy tactic and suggest improving water supply by managing wastewater healthier. Efficient treatment techniques are needed to remove the toxic contaminants from wastewater as they can be reused or recycled. Moreover, the opted water purification technique should become more reliable, cost-effective and without compromise; it should be environmentally friendly. Upgradation and modifying polymeric nanocomposite membranes could be suitable since they exhibit larger surface areas and enhance the water treatment process. Due to the operation flexibility, high removal percentage, and economic analysis, polymer membranes have been verified as powerful technologies in the water management process.

Significant problems associated with water purification membranes are their fouling nature during long-term operation, resulting in the deterioration of the membrane by the accession of particles on the superficial surface and the inner pores. The deposition and subsequent growth of microbes, precipitation of inorganic compounds (salts) and organic matter are typical in any filtration process and affect the permeability, selectivity and reduction the lifetime of membranes, and finally, it causes membrane damage. Some traditional methods like physical cleaning are mainly applied to prevent fouling, which adds to operational expenditures. However, nanocomposite membranes can expose better results. Besides, a considerable capital investment with low repossession rates is still a problem faced by desalination membranes. An additional constraint of using nanocomposite membrane on large scale applications are

a. its difficulty of fabrication techniques, which still require further modifications to over-awed issues like leaching of nanomaterial;
b. improper accumulation of NPs in an organic solvent; and
c. the risk of the inconsistency of the polymer together with the NPs.

The improper dispersion control is complicated due to the surface interactions, especially for the NPs smaller than 100 nm in size. Nevertheless, scientists comprehend the surface interaction theories; on the other hand, the aspects that would promote augmenting or further encouraging the agglomerations remain indistinct. This is a struggle in dispersing NPs in the course of membrane fabrication. Still, some processing plants use inefficient technologies and have not improved, leading to adverse environmental impacts and process inefficiency. The future of membrane research appears to be encouraging the development of innovative NPs and their implementation. Membrane modification should be carried out to afford them good antifouling features in a certain way since there have been various signs of progress in the membranes' mechanical strength and robustness.

BIBLIOGRAPHY

[1] A. Zirehpour, A. Rahimpour, Membranes for Wastewater Treatment, 2016. https://doi.org/10.1002/9781118831823.ch4.
[2] L. Bashambu, R. Singh, J. Verma, Metal/metal oxide nanocomposite membranes for water purification, Mater. Today Proc. (2020). https://doi.org/10.1016/j.matpr.2020.10.213.
[3] S. Chowdhury, M.A.J. Mazumder, O. Al-Attas, T. Husain, Heavy metals in drinking water: Occurrences, implications, and future needs in developing countries, Sci. Total Environ. 569–570 (2016) 476–488. https://doi.org/10.1016/j.scitotenv.2016.06.166.
[4] E. Hanhauser, M.S. Bono, C. Vaishnav, A.J. Hart, R. Karnik, Solid-phase extraction, preservation, storage, transport, and analysis of trace contaminants for water quality monitoring of heavy metals, Environ. Sci. Technol. 54 (2020) 2646–2657. https://doi.org/10.1021/acs.est.9b04695.
[5] P. Chowdhary, R.N. Bharagava, S. Mishra, N. Khan, Role of industries in water scarcity and its adverse effects on environment and human health, Environ. Concerns Sustain. Dev. (2020) 235–256. https://doi.org/10.1007/978-981-13-5889-0_12.
[6] M. Kárászová, M. Bourassi, J. Gaálová, Membrane removal of emerging contaminants from water: Which kind of membranes should we use?, Membranes (Basel). 10 (2020) 1–23. https://doi.org/10.3390/membranes10110305.
[7] Y. Zhang, W. Yu, R. Li, Y. Xu, L. Shen, H. Lin, B.Q. Liao, G. Wu, Novel conductive membranes breaking through the selectivity-permeability trade-off for Congo red removal, Sep. Purif. Technol. 211 (2019) 368–376. https://doi.org/10.1016/j.seppur.2018.10.008.
[8] P.S. Goh, A.F. Ismail, S.M. Sanip, B.C. Ng, M. Aziz, Recent advances of inorganic fillers in mixed matrix membrane for gas separation, Sep. Purif. Technol. 81 (2011) 243–264. https://doi.org/10.1016/j.seppur.2011.07.042.
[9] A.C. Enten, M.P.I. Leipner, M.C. Bellavia, L.E. King, T.A. Sulchek, Optimizing flux capacity of dead-end filtration membranes by controlling flow with pulse width modulated periodic backflush, Sci. Rep. 10 (2020) 1–11. https://doi.org/10.1038/s41598-020-57649-9.
[10] S. Mondal, S. Chatterjee, S. De, Theoretical investigation of cross flow ultrafiltration by mixed matrix membrane: A case study on fluoride removal, Desalination. 365 (2015) 347–354. https://doi.org/10.1016/j.desal.2015.03.017.
[11] A. Abdelrasoul, H. Doan, A. Lohi, C.-H. Cheng, Mass transfer mechanisms and transport resistances in membrane separation process, Mass Transf. – Adv. Process Model. (2015). https://doi.org/10.5772/60866.
[12] L.Y. Ng, A.W. Mohammad, C.P. Leo, N. Hilal, Polymeric membranes incorporated with metal/metal oxide nanoparticles: A comprehensive review, Desalination. 308 (2013) 15–33. https://doi.org/10.1016/j.desal.2010.11.033.

[13] A. Lee, J.W. Elam, S.B. Darling, Membrane materials for water purification: Design, development, and application, Environ. Sci. Water Res. Technol. 2 (2016) 17–42. https://doi.org/10.1039/c5ew00159e.

[14] T.A. Saleh, V.K. Gupta, Application of nanomaterial-polymer membranes for water and wastewater purification, Nanomater. Polym. Membr. (2016) 233–250. https://doi.org/10.1016/b978-0-12-804703-3.00009-7.

[15] Y. Feng, G. Han, L. Zhang, S.B. Chen, T.S. Chung, M. Weber, C. Staudt, C. Maletzko, Rheology and phase inversion behavior of polyphenylenesulfone (PPSU) and sulfonated PPSU for membrane formation, Polymer (Guildf). 99 (2016) 72–82. https://doi.org/10.1016/j.polymer.2016.06.064.

[16] S. Bano, A. Mahmood, S.J. Kim, K.H. Lee, Graphene oxide modified polyamide nanofiltration membrane with improved flux and antifouling properties, J. Mater. Chem. A. 3 (2015) 2065–2071. https://doi.org/10.1039/c4ta03607g.

[17] A.A.R. Abdel-Aty, Y.S.A. Aziz, R.M.G. Ahmed, I.M.A. ElSherbiny, S. Panglisch, M. Ulbricht, A.S.G. Khalil, High performance isotropic polyethersulfone membranes for heavy oil-in-water emulsion separation, Sep. Purif. Technol. 253 (2020) 117467. https://doi.org/10.1016/j.seppur.2020.117467.

[18] D. Rana, T. Matsuura, Rana-D._Surface-modifications-for-antifouling-membranes_2010.pdf, (2010) 2448–2471.

[19] M. Heijnen, R. Winkler, P. Berg, Optimisation of the geometry of a polymeric Multibore® ultrafiltration membrane and its operational advantages over standard single bore fibres, Desalin. Water Treat. 42 (2012) 24–29. https://doi.org/10.1080/19443994.2012.682968.

[20] D. Gille, W. Czolkoss, Ultrafiltration with multi-bore membranes as seawater pre-treatment, Desalination. 182 (2005) 301–307. https://doi.org/10.1016/j.desal.2005.03.020.

[21] S. Roy, S. Ragunath, Emerging membrane technologies for water and energy sustainability: Future prospects, constraints and challenges, Energies. 11 (2018). https://doi.org/10.3390/en11112997.

[22] A. Yusuf, A. Sodiq, A. Giwa, J. Eke, O. Pikuda, G. De Luca, J.L. Di Salvo, S. Chakraborty, A review of emerging trends in membrane science and technology for sustainable water treatment, J. Clean. Prod. 266 (2020) 121867. https://doi.org/10.1016/j.jclepro.2020.121867.

[23] B. Van Der Bruggen, Chemical modification of polyethersulfone nanofiltration membranes: A review, J. Appl. Polym. Sci. 114 (2009) 630–642. https://doi.org/10.1002/app.30578.

[24] A.L. Ahmad, A.A. Abdulkarim, B.S. Ooi, S. Ismail, Recent development in additives modifications of polyethersulfone membrane for flux enhancement, Chem. Eng. J. 223 (2013) 246–267. https://doi.org/10.1016/j.cej.2013.02.130.

[25] H.L. Richards, P.G.L. Baker, E. Iwuoha, Metal nanoparticle modified polysulfone membranes for use in wastewater treatment: A critical review, J. Surf. Eng. Mater. Adv. Technol. 02 (2012) 183–193. https://doi.org/10.4236/jsemat.2012.223029.

[26] A. Kausar, Phase inversion technique-based polyamide films and their applications: A comprehensive review, Polym. – Plast. Technol. Eng. 56 (2017) 1421–1437. https://doi.org/10.1080/03602559.2016.1276593.

[27] C. Cummins, R. Lundy, J.J. Walsh, V. Ponsinet, G. Fleury, M.A. Morris, nano today enabling future nanomanufacturing through block copolymer self-assembly: A review, Nano Today. 35 (2020) 100936. https://doi.org/10.1016/j.nantod.2020.100936.

[28] L. Penboon, A. Khrueakham, S. Sairiam, TiO2 coated on PVDF membrane for dye wastewater treatment by a photocatalytic membrane, Water Sci. Technol. 79 (2019) 958–966. https://doi.org/10.2166/wst.2019.023.

[29] J. Guo, J. Kim, Modifications of polyethersulfone membrane by doping sulfated-TiO2 nanoparticles for improving anti-fouling property in wastewater treatment, RSC Adv. 7 (2017) 33822–33828. https://doi.org/10.1039/c7ra06406c.

[30] C. Das, K.A. Gebru, Cellulose acetate modified titanium dioxide (TiO2) nanoparticles electrospun composite membranes: Fabrication and characterization, J. Inst. Eng. Ser. E. 98 (2017) 91–101. https://doi.org/10.1007/s40034-017-0104-1.

[31] L. Shen, Z. Huang, Y. Liu, R. Li, Y. Xu, G. Jakaj, H. Lin, Polymeric membranes incorporated with ZnO nanoparticles for membrane fouling mitigation: A brief review, Front. Chem. 8 (2020) 1–9. https://doi.org/10.3389/fchem.2020.00224.

[32] L. Shen, X. Bian, X. Lu, L. Shi, Z. Liu, L. Chen, Z. Hou, K. Fan, Preparation and characterization of ZnO/polyethersulfone (PES) hybrid membranes, Desalination. 293 (2012) 21–29. https://doi.org/10.1016/j.desal.2012.02.019.

[33] F. Liu, N.A. Hashim, Y. Liu, M.R.M. Abed, K. Li, Progress in the production and modification of PVDF membranes, J Membr. Sci. 375 (2011) 1–27. https://doi.org/10.1016/j.memsci.2011.03.014.

[34] G. Kang, Y. Cao, Application and modi fi cation of poly (vinylidene fl uoride) (PVDF) membranes: A review, J Membr. Sci. 463 (2014) 145–165. https://doi.org/10.1016/j.memsci.2014.03.055.

[35] J. Hong, Y. He, Polyvinylidene fluoride ultrafiltration membrane blended with nano-ZnO particle for photo-catalysis self-cleaning, Desalination. 332 (2014) 67–75. https://doi.org/10.1016/j.desal.2013.10.026.

[36] M. Giagnorio, B. Ruffino, D. Grinic, S. Steffenino, L. Meucci, M.C. Zanetti, A. Tiraferri, Achieving low concentrations of chromium in drinking water by nanofiltration: Membrane performance and selection, Environ. Sci. Pollut. Res. 25 (2018) 25294–25305. https://doi.org/10.1007/s11356-018-2627-5.

[37] J.H. Kim, M.K. Joshi, J. Lee, C.H. Park, C.S. Kim, Polydopamine-assisted immobilization of hierarchical zinc oxide nanostructures on electrospun nanofibrous membrane for photocatalysis and antimicrobial activity, J. Colloid Interface Sci. 513 (2018) 566–574. https://doi.org/10.1016/j.jcis.2017.11.061.

[38] X. Li, R. Pang, J. Li, X. Sun, J. Shen, W. Han, L. Wang, In situ formation of Ag nanoparticles in PVDF ultrafiltration membrane to mitigate organic and bacterial fouling, Desalination. 324 (2013) 48–56. https://doi.org/10.1016/j.desal.2013.05.021.

[39] G.M. Fabri-, H.M. Hegab, L. Zou, Author's accepted manuscript graphene oxide-assisted Membranes: Fabrication and potential applications, J. Memb. Sci. (2015). https://doi.org/10.1016/j.memsci.2015.03.011.

[40] M. Fathizadeh, W.L. Xu, F. Zhou, Y. Yoon, M. Yu, Graphene oxide: A novel 2-dimensional material in membrane separation for water purification, Adv. Mater. Interfaces. 4 (2017) 1–16. https://doi.org/10.1002/admi.201600918.

[41] S. Poolachira, S. Velmurugan, Efficient removal of lead ions from aqueous solution by graphene oxide modified polyethersulfone adsorptive mixed matrix membrane, Environ. Res. 210 (2022) 112924. https://doi.org/10.1016/j.envres.2022.112924.

[42] A.M. Dimiev, L.B. Alemany, J.M. Tour, Graphene oxide: Origin of acidity, its instability in water, and a new dynamic structural model, ACS Nano. 7 (2013) 576–588. https://doi.org/10.1021/nn3047378.

[43] A. Abdel-Karim, S. Leaper, M. Alberto, A. Vijayaraghavan, X. Fan, S.M. Holmes, E.R. Souaya, M.I. Badawy, P. Gorgojo, High flux and fouling resistant flat sheet polyethersulfone membranes incorporated with graphene oxide for ultrafiltration applications, Chem. Eng. J. 334 (2018) 789–799. https://doi.org/10.1016/j.cej.2017.10.069.

[44] A. Abdel-Karim, S. Leaper, M. Alberto, A. Vijayaraghavan, X. Fan, S.M. Holmes, E.R. Souaya, M.I. Badawy, P. Gorgojo, High flux and fouling resistant flat sheet polyethersulfone membranes incorporated with graphene oxide for ultrafiltration applications, Chem. Eng. J. 334 (2018) 789–799. https://doi.org/10.1016/j.cej.2017.10.069.

[45] J. Lee, H.R. Chae, Y.J. Won, K. Lee, C.H. Lee, H.H. Lee, I.C. Kim, J. Min Lee, Graphene oxide nano-platelets composite membrane with hydrophilic and antifouling properties for wastewater treatment, J. Memb. Sci. 448 (2013) 223–230. https://doi.org/10.1016/j.memsci.2013.08.017.

[46] X. Chang, Z. Wang, S. Quan, Y. Xu, Z. Jiang, L. Shao, Exploring the synergetic effects of graphene oxide (GO) and polyvinylpyrrodione (PVP) on poly(vinylylidenefluoride) (PVDF) ultrafiltration Membrane performance, Appl. Surf. Sci. 316 (2014) 537–548. https://doi.org/10.1016/j.apsusc.2014.07.202.

[47] F. Ahmed, C.M. Santos, R.A.M. V. Vergara, M.C.R. Tria, R. Advincula, D.F. Rodrigues, Antimicrobial applications of electroactive PVK-SWNT nanocomposites, Environ. Sci. Technol. 46 (2012) 1804–1810. https://doi.org/10.1021/es202374e.

[48] R. Castro-Muñoz, The strategy of nanomaterials in polymeric membranes for water treatment: Nanocomposite membranes, Tecnol. y Ciencias Del Agua. 11 (2020) 410–436. https://doi.org/10.24850/j-tyca-2020-01-11.

[49] M. Sianipar, S.H. Kim, Khoiruddin, F. Iskandar, I.G. Wenten, Functionalized carbon nanotube (CNT) membrane: Progress and challenges, RSC Adv. 7 (2017) 51175–51198. https://doi.org/10.1039/c7ra08570b.

[50] G. Cual, N.S. Ahmed, R. Menzel, Y. Wang, A. Garcia-gallastegui, S.M. Bawaked, A.Y. Obaid, S.N. Basahel, N.S. Ahmed, R. Menzel, Y. Wang, S.M. Bawaked, A.Y. Obaid, N. Sulaiman, Author's accepted manuscript reference: Graphene-oxide-supported CuAl and CoAl layered double hydroxides as enhanced catalysts for carbon-carbon coupling via Ullmann reaction, J. Solid State Chem. (2016). https://doi.org/10.1016/j.jssc.2016.11.024.

[51] L. Li, R. Ma, Y. Ebina, N. Iyi, T. Sasaki, Positively charged nanosheets derived via total delamination of layered double hydroxides, Chem. Mater. 17 (2005) 4386–4391. https://doi.org/10.1021/cm0510460.

[52] P. Lu, Y. Liu, T. Zhou, Q. Wang, Y. Li, Author's accepted manuscript recent advances in layered double hydroxides (LDHs) as two-dimensional membrane materials for gas and liquid separations, J. Memb. Sci. (2018). https://doi.org/10.1016/j.memsci.2018.09.041.

[53] J. Liu, L. Yu, Y. Zhang, Fabrication and characterization of positively charged hybrid ultra fi ltration and nano fi ltration membranes via the in-situ exfoliation of Mg/Al hydrotalcite, DES. 335 (2014) 78–86. https://doi.org/10.1016/j.desal.2013.12.015.

[54] S. Poolachira, S. Velmurugan, Exfoliated hydrotalcite: Modified polyethersulfone-based nanofiltration membranes for removal of lead from aqueous solutions, Environ. Sci. Pollut. Res. 27 (2020) 29725–29736. https://doi.org/10.1007/s11356-019-06715-5.

[55] Y. Zhao, N. Li, S. Xia, Polyamide nano fi ltration membranes modi fi ed with Zn e Al layered double hydroxides for natural organic matter removal, Compos. Sci. Technol. 132 (2016) 84–92. https://doi.org/10.1016/j.compscitech.2016.06.016.

[56] J. Lin, W. Ye, K. Zhong, J. Shen, N. Jullok, A. Sotto, B. Van der Bruggen, Enhancement of polyethersulfone (PES) membrane doped by monodisperse Stöber silica for water treatment, Chem. Eng. Process. Process Intensif. 107 (2016) 194–205. https://doi.org/10.1016/j.cep.2015.03.011.

[57] H. Dong, P. Ye, M. Zhong, J. Pietrasik, R. Drumright, K. Matyjaszewski, Superhydrophilic surfaces via polymer-SiO2 nanocomposites, Langmuir. 26 (2010) 15567–15573. https://doi.org/10.1021/la102145s.

[58] A.A. Khan, M.I. Siyal, J.O. Kim, Fluorinated silica – modified anti – oil-fouling omniphobic F – SiO2@ PES robust membrane for multiple foulants feed in membrane distillation, Chemosphere. 263 (2021) 128140. https://doi.org/10.1016/j.chemosphere.2020.128140.

[59] A. Martín, J.M. Arsuaga, N. Roldán, J. de Abajo, A. Martínez, A. Sotto, Enhanced ultrafiltration PES membranes doped with mesostructured functionalized silica particles, Desalination. 357 (2015) 16–25. https://doi.org/10.1016/j.desal.2014.10.046.

[60] S. Karimi, Y. Mortazavi, A. A. Khodadadi, A. Holmgren, D. Korelskiy, J. Hedlund, Functionalization of silica membranes for CO2 separation, Sep. Purif. Techno. 235 (2020) 116207. https://doi.org/10.1016/j.seppur.2019.116207

6 Nanocomposite Membrane–Based Water Treatment Processes

Sougata Ghosh and Bishwarup Sarkar

CONTENTS

6.1 INTRODUCTION

Increasing urbanization and industrialization have resulted in the predominance of contaminants like heavy metals, dyes, pesticides, oil and other hazardous chemicals in the soil and water (Rathoure and Dhatwalia, 2016). The indiscriminate release of untreated industrial effluents loaded with toxic pollutants affects the quality of water and the health of the aquatic flora and fauna. These harmful chemicals can enter the body of aquatic animals and plants and accumulate in tissues. Thus, when these organisms are eaten by higher animals, the contaminants enter the food chain and results in biomagnifications (Ghosh, 2020). The refractory pollutants can be carcinogenic, teratogenic and mutagenic and can affect various metabolic processes resulting in acute or chronic toxicity (Ghosh et al., 2021a).

Industrial effluents are initially subjected to various physical, chemical and biological treatments as depicted in Figure 6.1. Several conventional methods, shown in Figure 6.2, like chemical precipitation, coagulation/flocculation, froth floatation, chemical oxidation, adsorption, ion exchange, incineration and electrochemical techniques are employed to remove various types of contaminants from the industrial effluents (Crini and Lichtfouse, 2019). However, all the conventional water treatment processes are not always effective for diverse types of contaminants as they are often slow and cost-intensive. Furthermore, each and every treatment method comes up with its specific set of limitations, the most critical being the consumption of chemicals like lime, oxidants, H_2S, and others that incurs high cost (Ghosh et al., 2021b). Continuous physicochemical monitoring of the effluent pH and additional oxidation steps for metal complexes are time-consuming and labor-intensive. Other major drawbacks are high sludge production, handling and disposal (Bratby, 2006; Ghosh and Webster, 2021a).

Nanotechnology-driven solutions have led to the development of efficient and rapid strategies for removing refractory pollutants from the environment. Table 6.1 shows several nanoparticles (NPs) that are impregnated and incorporated in membranes for water purification by ultrafiltration (UF), microfiltration, nanofiltration (NF) and reverse osmosis (RO; Ghosh and Webster, 2021b). This chapter elaborates recent advances in the fabrication of membranes by introducing nanoscale carbon

DOI: 10.1201/9781003165019-6

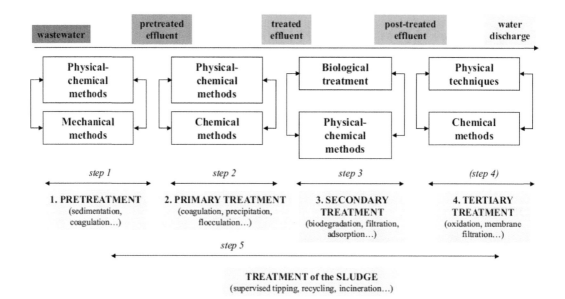

FIGURE 6.1   Main processes for the decontamination of industrial wastewater.

Source: Reprinted with permission from Crini, G., Lichtfouse, E., 2019. Advantages and disadvantages of techniques used for wastewater treatment. Environ. Chem. Lett. 17, 145–155. Copyright © 2018 Springer Nature Switzerland AG.

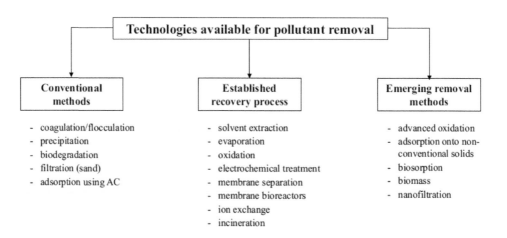

FIGURE 6.2   Classification of technologies available for pollutant removal and examples of techniques.

Source: Reprinted with permission from Crini, G., Lichtfouse, E., 2019. Advantages and disadvantages of techniques used for wastewater treatment. Environ. Chem. Lett. 17, 145–155. Copyright © 2018 Springer Nature Switzerland AG.

nanotubes (CNTs), graphene oxide (GO), titania (TiO_2), zinc oxide (ZnO), silver NPs (AgNPs) and copper NPs (CuNPs). Furthermore, the use of NP-impregnated membranes for microfiltration, UF, NF and reverse osmosis is discussed. Eventually, the scope of using nanobiotechnology for improving the biocompatibility of the membrane in order to design a green water treatment process is highlighted.

6.2 MICROFILTRATION

Refractory pollutants are often difficult to remove and hence alternative technologies like microfiltration are employed for removing hazardous pollutants, including dyestuffs. Various nanostructure-based microfiltration techniques are discussed in this section. Mulopo (2017) reported treatment of bleach effluent using CNT/polysulfone (PSf) nanocomposites that were integrated into an anaerobic membrane bioreactor (AMBR). CNTs were produced using the chemical vapor deposition (CVD) method wherein; acetylene was used as a source of carbon while PSf membranes were prepared using phase inversion and immersion technique wherein PSf was used as a polymer and 1-methyl-2-pyrrolidinone (NMP) was used as a solvent to prepare 20% w/w casting solution. CNTs were functionalized (fCNTs) via immersion in HNO_3 followed by reflexing at 110 °C for 4 h. fCNT/PSf nanocomposites were created using 0.04 wt.% of fCNTs. Scanning electron microscopy (SEM) images showed the porous and permeable nature of the membrane, with an average diameter of 0.164 mm, and fractures on the subsurface, respectively. The pore size of membranes in presence of CNTs was reduced to 0.659 μm. The water contact angle was also decreased from 79.832° to 72.158° after the addition of CNTs in PSf membranes indicated a more hydrophilic nature for fCNT/PSf membranes. Fourier transform infrared spectroscopy (FTIR) spectra analysis confirmed the presence of fCNTs, with an extra band from 3400 cm^{-1} to 3840 cm^{-1} that is a characteristic peak indicating the presence of hydrogen bonds. Furthermore, an AMBR was fed with bleach effluent and was coupled with fCNT/PSf for the removal of chemical oxygen demand (COD), volatile fatty acids (VFA) and suspended solids (SS) from the effluent. The COD, VFA and SS removal was similar to bare PSf membranes, as well as PSf membranes with added CNTs; however, a higher flux was achieved with fCNT/PSf membranes after mechanical cleaning that could be due to decreased fouling layer of cake. The permeability profiles showed PSf membranes containing CNTs reached a value of 0.6 to 0.7 $l/m^2/h$ indicating better flux. It was proposed that such an improvement in flux could be due to O–H bonds present in membranes modified with CNT that lead to a contact angle change and membrane roughness. Methane content in the bioreactor was found to be 57.4% with an average yield of 0.11 to 0.18 m^3 CH4 per kg of COD removed. Moreover, mixed liquor suspended solids (MLSS) and mixed liquor volatile suspended solids (MLVSS) increased from 7.562 and 4.892 mg/l to 13.600 and 8.250 mg/l, respectively. This highlights microbial growth in a bioreactor after 85 days of operation.

In another study, a PSf/graphene oxide (GO) nanocomposite membrane was prepared by Badrinezhad et al. (2018) that efficiently separated methylene blue (MB) from water. Phase inversion method was carried out to synthesize PSf/GO nanocomposite membranes wherein 0.75% nanocomposites were prepared using 16 wt.% of PSf solution and 6 mg of GO dissolved in N,N-dimethylformamide (DMF). Raman spectroscopy, FTIR spectroscopy and X-ray diffraction (XRD) were carried out to visualize and characterize the structure of GO. The XRD pattern highlighted the amorphous nature of the membrane. The singular sheets that were a part of the polymer matrix were also confirmed to be completely exfoliated of GO. The incorporation of GO was suggested to be responsible for the reduction of irregularities present in the polymer as well as contributing to significant changes in the PSf structure. FTIR spectra results revealed no chemical interaction between GO and PSf as the results of PSf/GO nanocomposites membranes were similar to neat PSf membranes. A cracked structure was observed for PSf upon SEM analysis that was observed to be subsequently reduced after the addition of GO. An increase of GO content in the PSf matrix was speculated to change the enthalpy, which may affect the process of phase separation. Hence, in presence of 2.5 wt.% of GO, a sponge-like structure was observed. An aqueous solution containing MB dye was taken for measurement of the contact angles. It was found that there is a decrease in the contact angle value in the presence of GO NPs, suggesting an increase in the hydrophilicity of the surface. The reason for this reduction in contact angle values was suggested to be due to the introduction of negatively charged carboxylic and hydroxylic functional groups that are present in GO. This subsequent addition of hydrophilic groups on the surface of the membrane may result in a decrease of energy at the interface of water and membrane, thus leading to a decrease in the contact angle measurements. Moreover, an approximate increase of 20% MB dye adsorption was observed by PSf/GO membranes compared to only-PSf membranes. Desorption efficiency was also found to be highest in PSf/GO membranes having 1.25 wt.% of GO.

Antifouling properties of polyvinylidene fluoride (PVDF) membrane were increased by Madaeni et al. (2011) using polyacrylic acid (PAA) functionalized TiO_2 NPs. Two different procedures were carried out for TiO_2 immobilization on a PVDF membrane. In the first approach, self-assembly of 0.05 wt.% TiO_2NPs on the surface of a PVDF membrane was carried out wherein the membrane was polymerized with PAA prior to addition of TiO_2 NPs using ethylene glycol as a crosslinking agent. While in the second approach, a technique called "grafting from" was performed in which 0.05 wt.% TiO_2 NPs was mixed with an acrylic acid monomer and then this mixture was added onto support PVDF membranes. This method provides minimal aggregation along with a strong interaction between the polymer and the nanofillers. It also allows an easy diffusion of monomer molecules along the membrane surface. The membranes prepared from both of these methods were then irradiated by ultraviolet (UV) light with 160-W lamp for 15 min. Proper immobilization of PAA was confirmed using FTIR spectra analysis. It was proposed that TiO_2 NPs' interaction with the surface of membrane require appropriate binding sites such as –COOH and –OH groups that are created only after polymerization of PVDF membranes with PAA. This interaction is due to coordination between Ti^{4+} and –OH groups present on the polymerized surface. A higher grafting yield of 28 wt.% was attained using the "grafting from" technique as compared to 24 wt.% using self-assembly of TiO_2 NPs. A whey solution was used as a superior foulant with a pH value of 7.0 ± 0.1 for investigating the antifouling properties of the modified membranes. A lower flux decline was obtained in a modified membrane, compared to unmodified membrane, indicating a decrease in fouling of the membrane. Bovine serum albumin (BSA) is the major protein present in whey solution that was suggested to have an overall negative charge at a pH value of 7 due to ionization. Hence, this repulsive force is present due to similarly charge molecules being present on both BSA molecules and the membrane functional groups; it was also speculated to be one of the reasons for the improvement in antifouling performance. The flux recovery ratio of modified membrane prepared using self-assembly was drastically reduced after the second filtration of a whey solution as compared to membranes prepared using grafting-from technique, suggesting better durability and stability of TiO_2 NPs of membranes prepared using the second approach. Furthermore, the exposure of modified membranes to UV light prior to the filtration process showed improvement in antifouling properties and flux recovery. It was proposed that exposure of TiO_2 NPs to UV light may create a pair of holes due to electron transfer from the capacity band to the conduction band that may react with water. Superhydrophilicity was also assumed to be one of the reasons for enhancing membrane properties.

Zinc oxide nanoparticles (ZnONPs) are considered the most biocompatible and are used in pharmaceutical industries largely. Their attractive physicochemical and optoelectronic properties have rationalized their use in various environmental and medical applications (Adersh et al., 2015; Kitture et al., 2015; Robkhob et al., 2020; Karmakar et al., 2020). Liang et al. (2012) reported a novel anti-irreversible fouling membrane made up of PVDF that was blended with ZnONPs. The fabrication of a PVDF membrane was performed using wet phase separation method wherein 1 g of polyvinylpyrrolidone (PVP) along with 3 g of glycerol and ZnONPs with a definite concentration was added into 84 g of NMP. The mixture was then kept in an ultrasonic bath for a minimum of 5 h with a temperature at 30 °C. After the proper facilitation of the dispersion of ZnONPs, 15 g of PVDF was added into the solution that was continuously stirred for 8 h at 30 °C followed by sealing and storage at room temperature for 9 h in order to remove bubbles. Water permeability was observed to be significantly improved upon modification of PVDF membranes with ZnONPs as compared to unmodified PVDF membranes. It was suggested that the hydrophilic nature of ZnONPs may be responsible for such increased water permeability. The property of anti-irreversible fouling was observed when only 78% recovery was observed after physical cleaning in unmodified membranes as compared to 100%

recovery in ZnONP-modified membranes. With an increasing dosage of ZnONPs, it was observed that the contact angle measurements were subsequently decreased, suggesting an increase in surface hydrophilicity. A concentration of 6.7% of ZnONPs was found to be optimum for best permeability of membranes. The microstructure of membrane was observed using SEM wherein; cross-sections was found to be observed having finger-like cavities and large voids on a thin upper layer of a highly inhomogeneous structure having an asymmetry as shown in Figure 6.3. A pore size range of 0.01 to 0.05 μm was observed for top surfaces. Furthermore, spherical and cuboidal particles were observed on the ZnONP-modified membranes. An elemental analysis revealed that spherical particles mainly consisted of fluorine and carbon while cuboidal particles were made up of ZnO. It was assumed that spherical particles may have formed from PVDF materials that remained undissolved. Synthetic wastewater samples were used for the analysis of membrane performance after multicycle

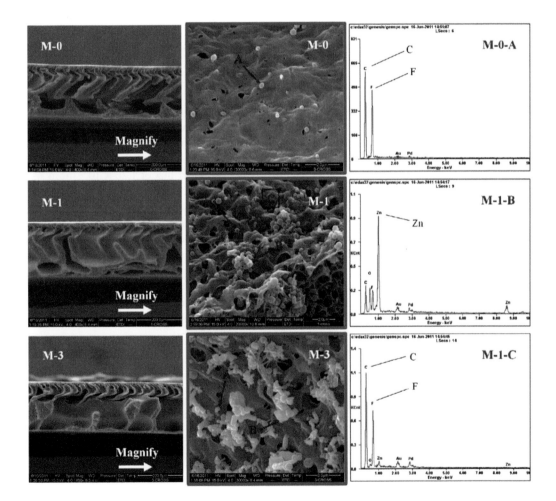

FIGURE 6.3 Magnified views (the middle column) of the internal surfaces of top membrane cavities (Mem-0, −1, −3) and EDS spectrograms (nominated by suffixing the letter A [spherical], B [cuboid] or C [control], standing for different locations, to the symbol of SEM images, e.g. M-1-B) for elemental analyses of the certain locations (marked in SEM pictures) in membrane internal surfaces.

operations of microfiltration process. After four cycles, it was observed that ZnONPs modified membranes showed the largest and steady filtration flux. A reinforcement effect was observed upon an increased ZnONP dosage on the mechanical strength of the membrane.

A PVDF-ZnO nanocomposite membrane was fabricated by Hong and He (2012) that helped improve the antifouling properties of the membrane during reclaimed water treatment. The phase inversion method was used to prepare the composite membranes. PVDF and ZnONPs were dissolved in dimethylacetamide (DMAc) in order to prepare casting dopes. An increase in viscosity of PVDF membrane was observed upon increasing the content of ZnONPs. It was speculated that unique properties of ZnONPs, such as a high specific area and surface energy, may increase the interactive linkages between the PVDF membranes, which could result in a subsequent increase in viscosity. The water contact angle was reduced from 82.33° to 70.06° when the ZnONP concentration was increased from 0% to 1%, which indicated a significant increase in the hydrophilic nature of the membrane surface. SEM images of cross sections of PVDF membranes loaded with ZnONPs showed a porous structure in the sublayer while a fingerlike structure was observed in the upper layer that increased with subsequent increase in ZnONPs. A larger number of pores were observed to form on the addition of less than 0.005 wt.% of ZnONPs as compared to an unmodified PVDF membrane that was speculated to be due to an increase in precipitation rate with subsequent increase in ZnONP concentration from 0 wt.% to 0.005 wt.%, which later on decreased when the concentration of ZnONPs was increased from 0.01 wt.% to 0.1 wt.% that led to small porous structure formation. Atomic force microscopy (AFM) images highlighted that the surface roughness of the membrane having less than 0.005 wt.% of ZnONPs was lower than unmodified membrane. However, an increase in surface roughness was seen when ZnONPs were increased from 0.005 wt.% to 0.1 wt.% because of the surface accumulation of NPs. The porosity of membrane was 75.16% with a mean pore size of 7.98×10^{-8} m along with a maximum pure water flux of 1.26×10^{-4} m^3 m^{-2} s^{-1} in presence of 0.005 wt.% of ZnONPs. However, the highest flux recovery efficiency, 67.12%, was found when 0.01 wt.% of ZnONPs was added to a PVDF membrane. In the presence of 1 wt.% of ZnONPs, the COD of effluent was reduced to 10.57 mg L^{-1}, with a removal efficiency of 70.21%, which was postulated to be due to pore size reduction and an improvement in hydrophilicity. The lowest resistance against a cake layer of 9.87×10^{12} m^{-1} was observed in PVDF membranes supplemented with 0.005 wt.% of ZnONPs. Furthermore, the tensile strength of the PVDF membrane was maximum in the presence of 0.01 wt.% of ZnONPs. Also, differential scanning calorimetry (DSC) measurements revealed the melting temperature and enthalpy of fusion of membranes to be 439.55 K and 2.75×10^3 J kg^{-1}, respectively in presence of 0.1 wt.% of ZnONPs.

6.3   UF

Among various advanced water purification techniques, UF is the most popular and wisely used. Among various metals, silver is considered to be the most bioactive due to its high bactericidal and photocatalytic dye-degrading effect, which can be further exploited in water purification (Ghosh et al., 2016a, 2016b, 2016c, 2016d). The fabrication of polyethersulfone (PES) UF membranes was reported by Zhang et al. (2014) using a biogenic Ag nanocomposite. *Lactobacillus fermentum* LMG 8900 was used for biogenic Ag^0 (Bio-Ag^0-6) NP synthesis that was subsequently dispersed in DMAc and mixed with a PES membrane. SEM analysis revealed a smooth surface with an absence of any accumulation on the membrane surface. Energy-dispersive X-ray spectroscopy (EDS) spectrum analysis confirmed the presence of Ag NPs while elemental mapping showed successful blending of evenly distributed Bio-Ag^0-6 NPs onto the surface of PES membranes. The elongation of fingerlike microvoids was highlighted with the addition of Bio-Ag^0-6 NPs, which could be due to reduced interaction between solvent molecules and polymer. AFM analysis indicated that membranes having Bio-Ag^0-6 NPs had a smoother surface compared to pure PES membranes. Also, an improvement in hydrophilicity was suggested because of a decrease in the water contact angle value compared to pure PES membranes. No change in pore size of the membrane was observed after NP addition with molecular weight cutoff (MWCO) value of 45 kDa for all membranes. Membrane performance was analyzed using UF of BSA solutions, wherein a maximum BSA flux of 85 L m^{-2}h^{-1} was obtained in presence of 1 wt.% Bio-Ag^0-6 along with 97.9% of protein rejection. A better resistance against BSA adsorption was observed for membranes containing Bio-Ag^0-6 NPs, with only 46.6 µg cm^{-2} of BSA adsorbed as compared to 57.3 µg cm^{-2} of BSA adsorption by PES membranes. Inductively coupled plasma mass spectrometry (ICP-MS) results highlighted the release of silver from membranes containing 0.5 wt.% of Bio-Ag^0-6, along with a 85.6% ratio of the remaining silver to the initial amount of silver after 80 days indicating stability and antifouling properties of membrane. The treatment of membranes with 2% HNO$_3$ showed only a slight decrease of BSA rejection from 98% to 92%, which signifies strong binding between Bio-Ag^0-6 NPs with PES matrix. Furthermore, clear inhibition zones in growth were observed when composite membranes containing varying concentrations of Bio-Ag^0-6 NPs were incubated with *P. aeruginosa* and *E. coli* highlighting efficient anti-bacterial effects of the membrane. The membranes were also investigated for its anti-biofouling properties wherein; membranes were immersed in an activated sludge tank for 9 weeks. The surface of membranes containing Bio-Ag^0-6 NPs were comparatively clean after 9 weeks as compared to PES membranes that were observed to be covered with thick biofilms. Hence, the introduction of Bio-Ag^0-6 NPs revealed significant improvement in anti-adhesion and anti-biofilm formation properties. Moreover, biofouling experiments indicated a 34% and an 8% decline in flux for membranes containing 0.3 wt. % and 1 wt. % of Bio-Ag^0-6 NPs, respectively. SEM and confocal laser scanning microscopy (CLSM) images revealed the presence of a few bacteria in the M2 membrane (0.3 wt.% of Bio-Ag^0-6 NPs) that were dead compared to an unmodified M0 membrane forming thick biofilms of live *E. coli* cells as shown in Figure 6.4. The flux recovery ratio (FRR) was found to be 56% for the M0 membrane while it reached up to 91% in the case of the M2 membrane.

Pastrana-Martinez et al. (2015) reported the preparation of UF membranes using graphene oxide TiO$_2$ (GOT) as a photocatalyst. Mixed cellulose ester (MCE) membranes were incorporated with 2g/L of GOT, TiO$_2$ and commercial TiO$_2$ material (P25) dispersions in order to photocatalytic membranes. SEM images showed homogeneous depositions of TiO$_2$ and GOT with an absence of any defects. TiO$_2$ particles were evenly distributed on the membrane with aggregates present on the top surface having a size of about 4–5 nm. The overall thickness of the membrane having TiO$_2$ was about 175 µm, with a 35-µm-thick layer of particles of TiO$_2$. The GOT particles' thickness was almost double (~65 nm) while P25 was shown to have a layer of about 39 µm. The porosity of MCE membranes containing GOT was 71%, along with a low contact angle value of 11°, indicating its potential use in enhancing antifouling properties. The removal of 12% of diphenhydramine (DP) dye and 8% of methyl orange (MO) from distilled water (DW) was observed using membranes modified with GOT under dark conditions while TiO$_2$ membranes showed 9% and 6% removal of DP and MO dyes, respectively. Simulated brackish water (SBW) showed no difference in the removal efficiency of both membranes. Near UV-Vis irradiation, photocatalytic degradation of DP was found to be equally efficient using P25 and GOT membranes having removal efficiency of approximately 73% in DW and about 60% in SBW, along with 35% removal of TOC after 240 min. While under visible light, GOT membrane was found to have maximum DP removal efficiency of 28% as compared to 5% using both TiO$_2$ and P25 membranes. It was speculated that Cl– ions present in SBW may behave as holes and ·OH scavengers that could hinder in the removal efficiency. Similar higher photocatalytic activity was observed for MO degradation with 19% and 65% removal using GOT membranes under visible light irradiation and near-UV/Vis, respectively. The second cycle of photocatalytic degradation was similar for all membranes that indicated stability and antifouling property of all membranes. The permeate flux of all three modified membranes was lower in comparison to the unmodified MCE membrane.

Hoek et al. (2011) reported the fabrication of mixed-membrane UF membranes. The membranes were composed of PSf beads along with multiple inorganic fillers such as mesoporous silica NPs, Cu and AgNPs, submicron zeolites, supra-micron zeolites as well as amorphous silica NPs. A better permeability was seen for UF membranes having sub-micron zeolites as inorganic material along

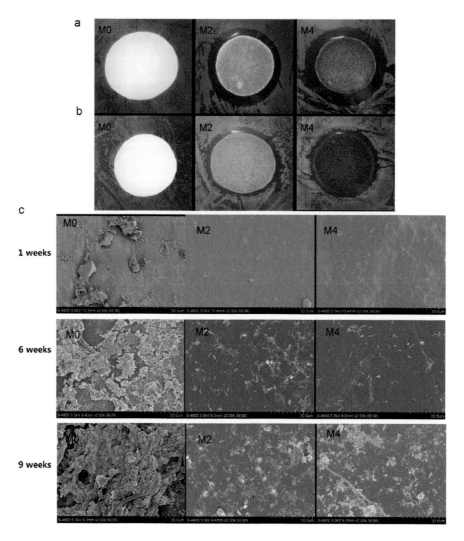

FIGURE 6.4 The antibacterial effect of nanocomposite membranes on (a) *Pseudomonas aeruginosa* and (b) *Escherichia coli* observed in the disk diffusion test and the biofilm formation on the M0, M2, and M4 membranes after immersion in an activated sludge tank for up to 9 weeks (c).

Source: Reprinted with permission from Zhang, M., Field, R.W., Zhang, K., 2014. Biogenic silver nanocomposite poly-ethersulfone UF membranes with antifouling properties. J. Memb. Sci. 471, 274–284. Copyright © 2014 ElsevierB.V.

with higher strength and a lower MWCO as compared to pure PSf membranes. The permeability of membranes containing metal NPs were also higher; however, the MWCO value was increased from 180 kDa in case of pure PSf membranes to 600 and about 400 kDa for membranes having Cu and AgNPs, respectively. Pores of all membranes were larger in size along with reduced porosity and increase in length of pore. However, a reduction in polymer stability was seen that may be due to presence of inorganic fillers. Mixed-matrix membranes that were made using metals, amorphous silica along with sub-micron zeolites showed a high ultimate strength value. Moreover, surface pores were observed in all membranes that had large macrovoids. Two well-defined layers were observed in membranes made up of silica particles wherein; one layer had a series of pores that extended from the skin layer and were lined parallel to each other while the second layer had less continuous pores and was observed to be more cellular. Zeolite fillers were properly dispersed onto the surface of membranes. White crystals of Linde type A (LTA) zeolites were visible that were not present in pure PSf membranes. Furthermore, organic-modified LTA (OMLTA) particles left voids of about 300 nm on the surface of the membrane, which led to the deterioration of the mechanical properties. BSA solutions were used for filtration and analysis of anti-fouling properties of the membranes. LTA-PSf and Ag-PSf membranes exhibited the most efficient flux recovery using hydraulic flushing, with up to 98–99% of protein rejection. Similarly, bacterial suspensions of *Pseudomonas putida* was used for dead-end filtration, and all membranes were observed to a have similar rejection of bacterial cells and flux decline.

Khalid et al. (2018) synthesized PSf membranes integrated with polyethylene glycol–functionalized CNTs (PEG-CNTs) that were used for treating wastewater. CNTs were carboxylated to avoid poor dispersion and aid in the improvement of chemical reactivity. PEG-CNTs were prepared using H_2SO_4 as a catalyst. The non-solvent-induced phase separation (NIPS) technique was used for forming PSU/PEG-CNT nanocomposite membranes. Field emission scanning electron microscopy (FESEM) images showed that the structural integrity of CNTs remained intact after PEG functionalization while the porosity was increased as compared to pristine CNTs. Thermal degradation at around 400 °C showed 27% loss of weight in PEG-CNTs that was proposed to be due to PEG chain disintegration. XRD patterns of PEG-CNTs showed a peak at $2\theta = 26.1°$, indicating the intact crystalline structure of CNT. Furthermore, addition of PEG-CNTs on PSf membranes increased hydrophilicity via a reduction in interfacial energy. PEG-CNTs entrapped membranes were shown to have enhanced water uptake capacity and better hydraulic permeability as well. Optimum concentration of PEG-CNTs was found to be 0.25 wt. % that provided a fourfold increase in membrane permeability and water permeability of 16.84 L m^{-2} h^{-1} bar^{-1}. The membrane pore size was found to increase from 12.53 nm to 21.27 nm in the presence of increasing concentrations of PEG-CNTs from 0 to 0.25 wt.%. Bulk porosity was also observed to increase in the range of 54.44% with a concomitant decrease in pore tortuosity. SEM analysis of membranes suggested presence of a dense top layer supplemented with a support layer that is porous in nature. Studies on the mechanical properties revealed that introduction of PEG-CNTs in the range of 0–0.25 wt.% led to a decrease in tensile strength. However, the higher loading of PEG-CNTs showed better membrane mechanical properties. The antifouling properties were found to be enhanced for BSA solutions with a better FFR and flux recovery of 80.33% in presence of 0.25 wt.% of PEG-CNTs as compared to only 57.14% of recovery in the case of unmodified PSf membranes. Also, the total protein resistance value was increased with increasing concentrations of PEG-CNTs.

Chung et al. (2016) reported effect of functionalized ZnONPs on PSf membranes. Zinc acetate dehydrate and oxalic acid dehydrate were used as precursors for synthesizing ZnO NPs via the sol-gel method. Ethylene glycol (EG) was added into a solution of zinc acetate that acted as a surfactant. DSC measurements of Zn-oxalate dehydrate showed an initial weight loss of 4% below 100 °C that was due to the removal of ethanol followed by a 17% weight loss below 190 °C that occurred because of the evaporation of water, and finally, a weight loss of 31% was found at about 360–400 °C, highlighting the formation of pure ZnO from Zn-oxalate. While, in the case of Zn-oxalate-containing EG, the heating profile showed an additional decomposition of EG at around 100–200 °C. The calcination temperatures for both samples were 400 °C, suggesting no effect of adding EG onto the thermal characteristics of ZnO. XRD analysis revealed a hexagonal wurtzite structure of ZnONP whose size was reduced on the addition of EG. The presence of ZnONPs was confirmed using FTIR spectra analysis wherein a characteristic peak of ZnO at 490–500 cm^{-1} was observed. Larger-sized particles with an average particle size of 50 ± 5 nm were observed in the absence of EG and were found to be accumulated with each other while smaller-sized particles having an average size of 25 ± 5 nm with reduced accumulation were obtained upon the addition of EG. The dispersive properties of NPs were much better in presence of EG as an additive. Membrane studies highlighted surface wettability when ZnONPs were introduced. The contact angle value of an unmodified PSf membrane was 69.7°, which was reduced to 53.3° in the presence of ZnO and 43.1° in presence of ZnO-EG NPs, indicating an increase in the hydrophilicity upon the incorporation of ZnONPs with EG as an additive. The water permeability of membranes was also increased subsequent to addition of ZnO-EG. Rejection studies indicated an increase of the rejection percentage from 70% to 86% in the case of ZnO-EG NPs, indicating good antifouling properties that may aid in the retention of organic pollutants within the membrane. No alterations in the structure were observed upon ZnO NP addition.

6.4   NF

Another powerful technique for effluent treatment that can effectively remove contaminants is NF. Commercial NF membrane-NF90 was modified using AgNPs by Zhang et al. (2016) in order to improve antibiofouling properties of a membrane. A thin-film composite (TFC) membrane modification was carried via in situ formation of AgNPs wherein the membrane was placed in poly (vinyl chloride) (PVC) plates and frame, and a solution of poly (vinyl alcohol) (PVA) and glutaraldehyde (GA) was added onto the membrane layer followed by the addition of an $AgNO_3$ solution to form AgNPs. Support was moisturized using a glycerol aqueous solution during the heating process. PVA was used as a crosslinking agent that reduced Ag^+ into AgNPs. A visible membrane color change from white to yellow was seen after AgNP loading, along with a new absorption spectra band observed between 400–500 nm, which is the characteristic signal of AgNPs. Water flux of modified NF90 membrane (NF90- PVA-AgNPs) decreased from 30.5 L m^{-2} h^{-1} to 20 L m^{-2} h^{-1}. Salt rejection was observed to increase from 98.87% to 99.58% with a corresponding increase in the concentration of PVA. –OH groups that are present in PVA were suggested to be responsible for the hydrogen bond formation with water that could resist the passing of water through the membrane, thus resulting in a decrease in the water flux. Also, the presence of AgNPs, along with PVA, was speculated to hinder salt ion interactions, hence giving such high salt reduction values. With the subsequent increase in temperature from 80 °C to 120 °C for in situ reaction, a decrease in water flux from 26.6 L m^{-2} h^{-1} to 16.2 L m^{-2} h^{-1} was observed. An increase in reaction time was also observed to negatively affect the water flux. Optimum conditions for NF90- PVA-AgNP fabrication was hence reported to be 1% w/v PVA, 2% w/v $AgNO_3$, 100 °C reaction temperature for 30 min. The contact angle value of a NF90-PVA-AgNP membrane was 99.6 ± 3.1°, which was higher than pristine NF90 membranes, which had a contact angle value of 44.7 ± 1.2°, suggesting poor hydrophilicity due to the presence of PVA and AgNPs. SEM analysis revealed a relatively smooth surface after PVA coating along with the presence of uniformly distributed AgNPs having a size range of 10–20 nm. EDS spectra also confirmed the presence of AgNPs with a characteristic peak of Ag^0. X-ray photoelectron spectroscopy (XPS) analysis revealed an interaction between the AgNPs and PVA, along with the presence of oxygen, carbon and nitrogen, as the main elements present on the membrane. The release rate of silver ions was about 0.73 μg/cm² day, which was reduced to 0.1 μg/cm² day after 7 days. This low silver ion release rate favors the antibiofouling properties of the membrane. Efficient antibacterial activity was observed by NF90-PVA-AgNP membrane against *E. coli* even after 14 days, indicating AgNPs' stability.

Shah and Murthy (2013) reported the synthesis of functionalized multiwalled carbon nanotube (MWCNT)/PSf membrane that was used for removing metals. Commercially available CNTs were washed and oxidized using HNO_3 and H_2SO_4 followed by a treatment with $SOCl_2$ for acyl group addition. These acylated nanotubes were further treated with NaN_3 in the presence of a DMF solvent to introduce functional azide groups. MWCNT/PSf nanocomposite membranes were prepared using a phase-inversion technique with DMF and an aqueous solution of isopropanol as a solvent and a coagulant, respectively. FTIR spectra analysis indicated the presence of –OH group in oxidized MWNTs only. Also, azide-functionalized nanotubes showed characteristic bands at 2142 cm^{-1} and 1566 cm^{-1}, indicating the presence of azide functional groups. SEM images of functionalized MWNTs showed a cleaned surface with opened tips. MWCNT/PSf membranes were 125 μm thick with a porous nature. FESEM images showed a reduction in pore size to 110–117 nm with a gradual increase in nanotube concentrations. FTIR spectra of the modified membrane confirmed the presence of functionalized MWCNTs. Thermal gravimetric analysis (TGA) highlighted an increase in the degradation temperature of 1% MWCNT/PSf membranes to 374 °C compared to 312 °C in the case of pristine PSf membranes, thus suggesting that MWCNTs may confer high heat and mechanical resistance to membranes. With an increase in MWCNT concentration from 0.1 wt.% to 1 wt.%, the hydrophobicity of PSf membranes was observed to decrease. A maximum rejection of Cr(VI) and Pb(II) in the presence of 1% functionalized CNTs was 94.2 and 90.1, respectively, under optimum conditions of 0.49 MPa pressure and a pH value of 2.6, which was reduced with an increase in pressure. It was proposed that modification of PSf membrane provides complexation sites with metal ions.

Copper (Cu) is an essential element that is generally nontoxic and biocompatible. Nanoscale copper is reported for free radical scavenging activity and several biomedical applications (Jamdade et al., 2019; Bhagwat et al., 2018). Xu et al. (2015) reported the preparation of an antimicrobial NF membrane using polycation–copper (II) complex and surface of polyacrylonitrile (PAN) as the substrate. A fixed concentration of 23.3 mM of polyethyleneimine (PEI) was used as a ligand with varying concentrations of $CuSO_4$ ranging from 0–16 mM to form PEI-Cu(II) complex. PEI PAN membranes were initially hydrolyzed using NaOH followed by the addition of a sonicated PEI-Cu(II) complex solution to form a layer via GA-mediated crosslinking. A critical ratio of (Cu^{2+})/amine group of 0.34 M was found to provide a stable complex above which precipitation of Cu^{2+} may lead to the destabilization of the PEI-Cu(II) complex. The pH of the complex solution decreased from 10 to 5 with subsequent increase in Cu^{2+} concentrations. PEI and $CuSO_4$ concentration of 2 g/L each was selected for further use in the membrane formation. Atomic absorption spectrophotometry (AAS) analysis revealed maximum Cu^{2+} sorption by hydrolyzed PAN substrate modified by the PEI-Cu(II) complex assembly (PAN_h/PEI-Cu(II) membrane) in less than 1 h of exposure to the $CuSO_4$ solution. Moreover, 1.3 times higher Cu^{2+} loading was observed in the case of a PAN_h/PEI-Cu(II) membrane compared to membranes prepared using a hydrolyzed PAN substrate and PEI and $CuSO_4$ solutions separately. The crosslinking of membranes with GA significantly helped in a strong PEI-Cu(II) complex and substrate assembly with no Cu^{2+} release after 15 min of sonication in deionized water. An efficient antibacterial efficiency of 94.6% was observed against *E. coli* using PAN_h/PEI-Cu(II) membranes compared to only 7.7% using PAN/PEI membranes as a control. Antibiofilm formation properties of PAN_h/PEI-Cu(II) membranes were also evaluated by immersing the membrane into real seawater under stirring conditions (120 rpm) for a period of 6 months. The ability of seawater softening by PAN_h/PEI-Cu(II) membrane was also identified at 1 MPa using a seawater sample as the feed solution. Flux and rejection values were found to be 32.3 L/m^2 h and 43.5%, respectively, with 98.1% rejection of Mg^{2+} ions.

Ganesh et al. (2013) prepared PSf mixed matrix membranes that were diffused with GO as seen in Figure 6.5. Graphite oxidation was carried out in presence of $KMnO_4$ as oxidizing agent to synthesize GO followed by the phase-inversion method to prepare a PSf/GO mixed matrix membrane. N-methyl 2-pyrrolidone (NMP) was used as solvent to dissolve 25 wt.% of PSf solution. Solid state ^{13}C cross-polarization/magic angle spinning (CP-MAS) nuclear magnetic resonance (NMR) spectra analysis of GO exhibited the characteristic peak at 59.5 nm, indicating the attachment of carbon to the peroxide group. Four major peaks were observed upon TGA of GO wherein the first peak at 61.6 °C indicated water evaporation while three other peaks at 231.27 °C, 269.5 °C and 651.4 °C highlight functional group dissociation present on graphite along with sublimation of carbon backbone, respectively. TEM analysis of GO showed a folding morphology with a nanometer layer of thickness. Furthermore, a prominent characteristic peak of GO at around 10.9° was highlighted in XRD analysis. The dispersion of GO in a PSf matrix membrane was confirmed using infrared (IR) spectra analysis wherein characteristic peaks at 3452 cm^{-1}, 1728 cm^{-1} and 1680 cm^{-1} were seen. Comparative XRD analysis between pristine PSf membranes showed the presence of an extra peak at around 2θ value of 11° in mixed matrix membranes, which is the characteristic peak of GO. SEM analysis of mixed matrix membrane showed alteration in the macrovoid structure due to GO addition, which could be due to the hydrophilic properties of GO. An AFM analysis showed that surface roughness was directly proportional to the concentration of GO. The water contact angle of membrane showed a decrease in the

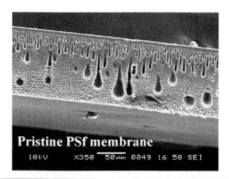

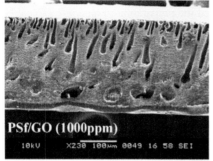 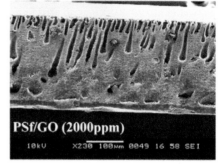

FIGURE 6.5   Cross-sectional SEM images of pristine PSf and PSf/GO mixed matrix membranes.

Source: Reprinted with permission from Ganesh, B.M., Isloor, A.M., Ismail, A.F., 2013. Enhanced hydrophilicity and salt rejection study of graphene oxide-polysulfone mixed matrix membrane. Desalination. 313, 199–207. Copyright © 2012 Elsevier B.V.

water contact value from 71° in the case of pure PSf membranes to 53° in the presence of 2000-ppm GO-doped mixed matrix membranes, hence indicating a decrease in hydrophobicity upon GO addition. Water uptake, along with water flux, was also increased with a subsequent increase in doping of GO, confirming the hydrophilic nature of mixed matrix membranes. Membranes having 2000 ppm of GO were found to reject 72% of 1000 ppm of Na_2SO_4 salt under 4 bar of applied pressure. Hence, the membranes were mechanically strong under applied pressure and moreover, salt rejection efficiency was dependent on the pH of the feed solution, with higher salt rejection in higher pH values indicating the presence of negatively charged species on the surface of the membrane.

ZnO nanofillers were dispersed in CA matrix by Khan et al. (2015) to synthesize antibacterial nanocomposites. An aqueous solution of 0.1 M of $Zn(NO_3)_2$ and 5 wt.% of carbon black was used for the synthesis of ZnO nanomaterials. CA/ZnO nanocomposites were prepared with varying weight ratios of CA/ZnO wherein cellulose and ZnO nanomaterials were dissolved in acetone and ethanol, respectively. XRD patterns indicated the presence of an amorphous CA phase and a crystalline ZnO nanosheet phase. FESEM images revealed an average thickness of 35 nm in the case of ZnO nanomaterials, with a rough, dense and compact morphology after CA dispersion. Nanocomposites were observed to have mesoporous structures. FTIR spectra analysis showed the presence of characteristic absorption for Zn-O stretching vibration at 50 cm^{-1} along with other bands, indicating CA absorption bands. TGA curve analysis highlighted that the introduction of ZnO nanofillers led to a decrease in the thermal stability of the membrane with a reduction in the decomposition temperature, which could be due to a weak association between ZnO nanosheets and CA or because of the catalytic nature of ZnO nanosheets. High antibacterial activity was observed against *E. coli* in presence of nanocomposites, which was increased with a subsequent increase in ZnO concentration. It was proposed that these nanocomposites behave as bactericidal agents and produce highly reactive oxygen species such as OH⁻, H_2O_2 and O_2^{2-} that attack the bacterial cells. The highest distribution coefficient (K_d) value was obtained in the case of Fe^{2+} metal ions, indicating the selectivity of nanocomposites toward Fe^{2+} ions. Selectivity against Fe^{2+} remained constant with varying concentrations of ZnO nanomaterials; however, a maximum uptake capacity was found in presence of 2 wt.% of ZnO. Water permeability studies were performed using membranes that were prepared using different elaboration conditions and added with 8% PVP. A range of membrane permeability was found to be related to microfiltration membranes in the presence of PVP while in the absence of PVP, the permeability was in the range of NF membranes. Furthermore, contact angle measurements showed lower hydrophobicity of membrane with increasing permeability. Isoelectric points of membranes were near 3 wherein the membranes do have any surface charge. The ester functional group present on CA was responsible for amphoteric behavior of membranes along with the repulsion of humic acid due to similar zeta potential at neutral pH that help in antifouling of membrane.

6.5 RO

Several polymer composite membranes are used for RO-mediated water purification owing to their superior resistance to chlorine, solvent and fouling. Ben-Sasson et al. (2014) reported the integration of AgNPs on TFC reverse osmosis (RO) membranes in order to enhance antibiofouling properties as illustrated in Figure 6.6. AgNPs were synthesized using in situ formation on TFC RO membrane

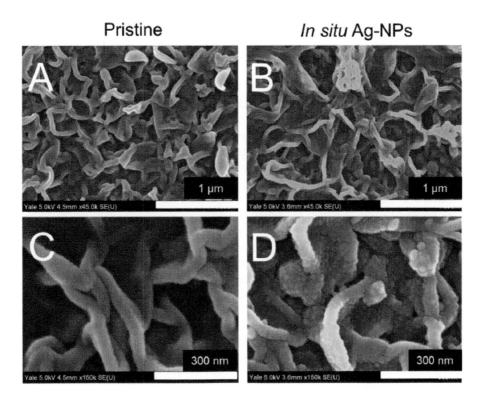

FIGURE 6.6 SEM micrographs of (A, C) pristine and (B, D) in situ AgNPs modified active layer of TFC membrane at different magnifications as indicated. Solutions of 5 mM $AgNO_3$ and 5 mM $NaBH_4$ (5:5) were used during the in situ formation reaction.

wherein AgNO$_3$ solution was mixed with the active layer of RO membrane along with NaBH$_4$ as a reducing agent. TEM images revealed discrete AgNPs that were spherical in shape, having a diameter of <15 nm. XPS analysis also confirmed the presence of Ag by providing a signal at the binding energy of 364 eV that increased with a subsequent increase in the AgNO$_3$ and NaBH$_4$ concentrations. The ratio of nitrogen/carbon was slightly reduced in the case of modified membranes, which may be due to the masking effect of AgNPs against polyamide amine functional groups present on the membrane. The highest silver loading, 3.7 ± 0.4 µg cm^{-2}, was observed on the membrane with an AgNO$_3$:NaBH$_4$ ratio of 5:5 mM. However, XPS analysis showed that the concentration of NaBH$_4$ did not influence the loading of silver ions onto the membrane. Water permeability was also decreased to 2.01 ± 0.12 L m^{-2} h^{-1} in the case of 5:5 AgNO$_3$:NaBH$_4$ in situ modified membranes as compared to 2.41 ± 0.14 L m^{-2} h^{-1} for unmodified membranes. Likewise, salt rejection was 98.33 ± 0.2% and 98.85 ± 0.26% for modified and pure membranes, respectively. The deposition of AgNPs on the membrane surface was suggested to contribute to the reduction in water permeability that aided in lowering effective membrane surface available for efficient water flow. An analysis of the silver release rate showed a release of 5.5 ± 0.6% silver ions compared to residual silver on the membrane after 7 days of dissolution. A strong antibacterial activity using 2:2 in situ modified membrane was observed after 5 h of incubation. Viable counts of *E. coli*, *P. aeruginosa* and *S. aureus* reduced to 78% ± 12%, 91% ± 8% and 96% ± 2.2%, respectively. Also, antibiofilm formation properties were also investigated for modified membranes wherein a reduction of 73% of live cells was observed compared to pristine membranes.

Ahmad et al. (2016) prepared a modified cellulose acetate/polyethylene glycol-600 (CA/PEG) membrane using AgNO$_3$ for the desalination of a water sample. A CA/PEG membrane was modified using a phase-inversion technique and further modified using in situ reduction using a 0.1 M AgNO$_3$ solution dissolved in DMF. IR spectra showed the presence of silver metal with a shift in peak to a lower frequency of 1728 cm^{-1}, which was caused due to weakening of C–O bond. Also, new peaks were obtained at 535 cm^{-1} and 370 cm^{-1}, showing the presence of Ag–O bond. SEM analysis highlighted the presence of spongy voids that could have formed because of an interaction between the polymer and Ag particles, which also facilitates an increase in flux as well as the hydrophilicity of membrane. The surface roughness of the modified membrane was higher with an average roughness of 35.47 nm compared to 10.79 nm of pristine membrane. Likewise, the contact angle value decreased from 50° to 39°, suggesting that silver may decrease the surface tension of the membrane and increase hydrophilicity. Also, modification of the membrane showed an about 18.75% increase in flux, along with 0.51% decrease in salt rejection. The hydraulic resistance of a modified membrane was less than the pristine membrane, suggesting that metal can cause the segmental movement of polymeric chains that may lead to void formation, which subsequently decreases the resistance of the membrane. Antibacterial activity against *E. coli* was investigated, with an optical density obtained that was near to zero in the case of modified membranes, indicating a strong inhibition to bacterial growth by the silver present on the membrane. Diffusion inhibition zone (DIZ) method was also carried out to observe the biocidal action of silver-loaded films on *B. subtilis*. An average diameter zone of about 0.7 mm was observed, showing clear zones of inhibition.

Biocidal CuNPs were linked with a TFC RO membrane by Ben-Sasson et al. (2016) to improve antibiofouling properties. In situ preparation of CuNPs was carried out in which RO membranes were mixed with a 50 mM-CuSO$_4$ solution along with a 50-mM NaBH$_4$ solution, which act as a reductant and aid in avoiding membrane damage during modification. The clear presence of CuNPs was observed under SEM that was not seen in pristine TFC membranes. Also, XPS analysis confirmed the presence of CuNPs, with a peak observed at 932 eV of binding energy. Surface elemental analysis revealed an increase in O/C ratio by about 10%, which was speculated to be because of the oxidation of the CuNPs. In situ modification using CuNPs was found to slightly increase water permeability from 2.53 ± 0.22 L m^{-2} h^{-1} in a pure membrane to 2.97 ± 0.32 L m^{-2} h^{-1} in the case of a CuNP-modified membrane. Salt rejection was also slightly decreased to 98.31% ± 0.32%. Moreover, zeta-potential analysis suggested no significant change in the surface charge of the membranes after modification, which could be due to the small size of CuNPs that are uniformly distributed in a thin monolayer on the surface of the membrane. The roughness parameters were also observed to remain almost constant, with a maximum roughness value of 579 ± 131 nm and 638 ± 205 nm for modified and pure membranes, respectively. The contact angle was found to increase from $45.46 \pm 1.68°$ to $59.84 \pm 3.13°$, which was attributed to the hydrophobic nature of copper oxide. CuNPs provided a strong and efficient antibacterial activity against *E. coli* as a model bacterium wherein the number of viable cells was reduced by $89.6\% \pm 8.2\%$ after 2 h of incubation. This cytotoxic effect was proposed to be due to the steady release of biocidal Cu ions that could create a toxic inhibition zone or bacterial cells engulfing CuNPs, which could lead to the formation of reactive oxygen species (ROS).

Ali et al. (2016) reported construction of GO-embedded TFC membranes using PSf as substrate and interfacial polymerization by MPD and TMC. A modified Hummers's method was used to synthesize GO in which graphite powder was mixed with cold H_2SO_4 and $NaNO_3$ followed by the addition of $KMnO_4$ under stirred conditions. FTIR spectra analyses of GO nanosheets confirmed the presence of pure GO. The PA layer present in TFC membranes was also confirmed by FTIR spectra wherein a characteristic band at 1694 cm^{-1} was observed due to C=O bond of PA active layer. XRD results showed a sharp peak at 2θ value of $10.09°$ in the case of GO, while an active layer of PA in the TFC membrane showed three peaks at 2θ values of $18.14°$, $23.25°$ and $26.55°$, respectively. SEM images showed valley morphology for TFC/GO membranes with a ridge. The contact angle was found to decrease from $64°$ to $48°$ when GO concentration was increased from 0 to 300 ppm. The increase in hydrophilicity was contributed because of the presence of carboxyl, hydroxyl and epoxy functional groups in GO. Subsequently, pure water permeability (PWP) was increased with the incorporation of GO into the membrane. For GO concentrations of more than 150 ppm, a decrease in water flux, along with a concomitant increase in salt rejection, was observed. An increase of 39% in water flux along with a 1% decrease in salt rejection was obtained in presence of 100 ppm as compared to 21.4 L m^{-2} h^{-1} water flux and 98.5% salt rejection in the case of pure TFC membranes. Furthermore, an increase in pressure to 15 bar provided an enhanced water flux of 29.6 L m^{-2} h^{-1} and a salt rejection of more than 97% for 2000 ppm of NaCl solution. TFC/GO membrane was also found to be resistant to BSA fouling with 85% of water flux recovery.

Modification of a polyamide TFC (PA(TFC)) membrane was reported by Isawi et al. (2016) using ZnONPs. A hydrothermal technique was used for ZnONPs synthesis wherein; zinc acetate was mixed with 5 M NaOH followed by an addition of sodium lauryl sulfate (SLS) surfactant that behaved as an insulator in order to prepare a homogeneous dispersion. PSf was used to synthesize the membrane substrate using a DMAc solvent while an active layer of PA was synthesized using TMC dissolved in hexane and MPD. PSf was used as a support layer on which a PA(TFC) membrane was fabricated via interfacial polymerization between MPD and TMC followed by graft polymerization using 2 wt.% of methacrylic acid (MAA) monomer and sodium metabisulfite ($Na_2S_2O_3$). It was suggested that the hydrogen atoms present on carboxylic acids and primary amine groups, as well as hydrogen present in the amide, bond on the surface of PA(TFC) membranes, providing grafting sites. PMAA-g-PA(TFC) membranes modified with ZnONPs were prepared using varying concentrations of ZnONPs ranging from 0.005 wt.% to 0.4 wt.% dissolved in MAA and $Na_2S_2O_3$ and added to active PA(TFC) membrane. ZnONPs were characterized using XRD wherein characteristic peak at 2θ value of $32.03°$, $34.64°$ and $36.51°$ was observed. FTIR spectra also revealed functional groups of ZnONPs in the range of 400–4000 cm^{-1} and TGA analysis showed successive loss of weight till 500 °C. Rod-shaped morphology with a smooth surface of NPs was seen via SEM, with a particle size range of 100–160 nm. FTIR analysis of PMAA-g-PA(TFC) membrane modified with ZnONPs showed an absorption band at 1596 cm^{-1} which may be due to COO-Zn interaction along with the presence of another peak at 3480 cm^{-1} due to intramolecular hydrogen bond formation between – COO group of MAA and –OH group of ZnONPs. XRD analysis indicated presence of ZnONPs within the grafting layer of the membrane with a slight shift in the characteristic peak, indicating an interaction between NPs and MAA. The surface morphology of

PMAA-g-PA(TFC) membrane via SEM revealed a rough surface due to the addition of ZnONPs, which were dispersed uniformly throughout the membrane. The mechanical properties of ZnO-modified PMAA-g-PA(TFC) membrane, such as tensile strength, elongation break and Young's modulus, were significantly increased compared to the PA (TFC) membrane. Furthermore, a lower water contact angle value of $50° ± 3°$ was obtained for the ZnO-modified PMAA-g-PA(TFC) membrane as compared to $63° ± 2.5°$ for the pure PMAA-g-PA(TFC) membrane, indicating improved hydrophilic properties due water attraction by NPs. Membrane performance analysis highlighted an optimum concentration of 0.1 wt.% of ZnO NPs provided a maximum salt rejection of 98% and water flux of 35 L m^{-2} h^{-1}. A pilot-scale RO unit was set up to analyze membrane desalination performance along with salt rejection. A groundwater sample from an aquifer was used as the feed solution. Almost 97% of salt rejection was achieved using a ZnO-modified PMAA-g-PA(TFC) membrane, with higher rejection rates for bivalent ions, such as Mg^{2+} and SO_4^{2-}, compared to monovalent ions, such as Na^+ and Cl^-. Release rate of Zn^{2+} ions was found to be 0.085 µg/cm^2/day during initial period of 4 days, which was reduced to less than 0.01 µg/cm^2/day after 6 days and remained constant. After 10 days, an overall 3.32% of initial Zn^{2+} ions were leached out indicating high stability of ZnONPs. Photocatalytic bactericidal activity of a ZnONP-modified PMAA-g-PA (TFC) membrane was confirmed wherein no *E. coli* cells could adhere to the membrane after 90 min of UV exposure.

TABLE 6.1
Nanocomposite Membranes for Water Treatment

Nanoparticle	Application	Polymer	Optimum concentration of filler	Reference
Microfiltration				
CNTs	Bleach effluent treatment with AMBR	PSf	0.04 wt.%	Mulopo, 2017
GO	Effluent treatment	PSf	1.25 wt.%	Badrinezhad et al., 2018
TiO$_2$NPs	Enhancing anti-fouling properties against whey solution	PVDF	0.05 wt.%	Madaeni et al., 2011
ZnONPs	Synthetic wastewater treatment	PVDF	6.7 wt.%	Liang et al., 2012
ZnONPs	Wastewater COD removal	PVDF	1 wt.%	Hong and He, 2012
Ultrafiltration				
Bio-Ag^0-6 NPs	Enhancing anti-fouling and antibacterial activity against *E. coli* and *P. putida* as model bacteria	PES	1 wt.%	Zhang et al., 2014
GOT	Photocatalytic degradation of organic pollutants	MCE	2 g/L	Pastrana-Martinez et al., 2015
Mesoporous silica, zeolites, Cu and AgNPs	Enhancing anti-fouling properties and antibacterial activity against *P. putida* as model bacterium	PSf	-	Hoek et al., 2011
PEG-CNTs	Wastewater treatment	PSf	0.25 wt.%	Khalid et al., 2018
ZnONPs	Enhancing membrane properties	PSf	0.1 wt.%	Chung et al., 2016
Nanofiltration				
Ag NPs	Enhancing salt rejection and antibacterial activity against *E. coli* as model bacterium	PA-PVA	10 mL	Zhang et al., 2016
CNTs	Removal of Cr(VI) and Pb(II)	PSf	1 wt.%	Shah and Murthy, 2013
CuSO$_4$	Seawater softening	PAN/PEI	2 g/L	Xu et al., 2015

GO	Enhancing salt rejection	PSf	–	Ganesh et al., 2013
ZnO nanomaterials	Enhancing water permeability and salt rejection	CA	2 wt.%	Khan et al., 2015
Reverse Osmosis				
AgNPs	Enhancing antibacterial activity against *E. coli*, *P. aeruginosa* and *S. aureus* as model bacteria	PA	–	Ben-Sasson et al., 2014
AgNO$_3$	Enhancing antibacterial activity against *E. coli* and *B. subtilis* as model bacteria	CA/PEG	–	Ahmad et al., 2016
CuNPs	Enhancing antibacterial activity against *E. coli* as model bacterium	PA	50 mM	Ben-Sasson et al., 2016
GO	Desalination	PSf	300 ppm	Ali et al., 2016
ZnONPs	Removal salt and metal ions	PA	0.1 wt.%	Isawi et al., 2016

6.6 CONCLUSION AND FUTURE PERSPECTIVES

The scarcity of clean water has raised global concern as the contaminated water can cause severe damage to the environment and the health of flora and fauna. Toxic metals, dyes and other hazardous chemicals cannot be removed effectively by conventional methods. Hence, innovative filtration technologies based on metallic and nonmetallic NP-impregnated membranes have emerged as potential alternative water treatment processes. The AgNPs, CuNPs, iron oxide nanoparticles (IONPs), platinum nanoparticles, palladium nanoparticles, GO and CNTs are widely being explored for efficient removal, degradation and/or detoxification of the refractory pollutants. However, several points need to be considered before nanotechnology-driven wastewater treatment techniques are implemented.

The activity of NPs is dependent on their size and shape. Hence, before impregnating NPs into a membrane, an optimized process should be developed to fabricate monodispersed nanostructures with desired size and shape. Furthermore, the stability of the NPs should be carefully monitored after incorporation into the polymeric membranes. Although AgNPs, CuNPs and their alloys can catalytically degrade dyes and other pollutants, their antimicrobial nature can also eliminate useful microbes of the water that help in the biological water treatment (Rokade et al., 2018; Rokade et al., 2017; Shende et al., 2017; Shende et al., 2018). NPs with larger surface areas can be multifunctionalized with dye-degrading enzymes, such as azoreductase, laccase, peroxidase and metal-detoxifying enzymes, such as reductases, before integrating them in the membranes. Such strategies can synergistically enhance the efficiency of the nanocomposite membranes for removing pollutants.

However, numerous NPs involve hazardous chemical agents for synthesis, which render them toxic and hence unsuitable for environment. Biogenic NPs synthesized employing bacteria, fungi, algae and green plants are more biocompatible and nontoxic (Ghosh, 2018; Shinde et al., 2018; Joshi et al., 2019; Ghosh et al., 2015; Salunke et al., 2014). Hence, biologically synthesized NPs like AgNPs, AuNPs, CuNPs, IONPs and others can be use explored for preparing membranes for microfiltration, NF, UF and RO. In view of the background, it can be concluded that nanocomposite membrane–driven wastewater treatment can emerge as a potential alternative to ensure clean water in future.

BIBLIOGRAPHY

Adersh, A., Kulkarni, A.R., Ghosh, S., More, P., Chopade, B.A., Gandhi, M.N., 2015. Surface defect rich ZnO quantum dots as antioxidant inhibiting α-amylase and α-glucosidase: A potential anti-diabetic nanomedicine. J. Mater. Chem. B. 3, 4597–4606.

Ahmad, A., Jamshed, F., Riaz, T., Waheed, S., Sabir, A., AlAnezi, A.A., Adrees, M., Jamil, T., 2016. Self-sterilized composite membranes of cellulose acetate/polyethylene glycol for water desalination. Carbohydr. Polym. 149, 207–216.

Ali, M.E., Wang, L., Wang, X., Feng, X., 2016. Thin film composite membranes embedded with graphene oxide for water desalination. Desalination, 386, 67–76.

Badrinezhad, L., Ghasemi, S., Azizian-Kalandaragh, Y., Nematollahzadeh, A., 2018. Preparation and characterization of polysulfone/graphene oxide nanocomposite membranes for the separation of methylene blue from water. Polym. Bull. 75(2), 469–484.

Ben-Sasson, M., Lu, X., Bar-Zeev, E., Zodrow, K.R., Nejati, S., Qi, G., Giannelis, E.P., Elimelech, M., 2014. In situ formation of silver nanoparticles on thin-film composite reverse osmosis membranes for biofouling mitigation. Water Res. 62, 260–270.

Ben-Sasson, M., Lu, X., Nejati, S., Jaramillo, H., Elimelech, M., 2016. In situ surface functionalization of reverse osmosis membranes with biocidal copper nanoparticles. Desalination, 388, 1–8.

Bhagwat, T.R., Joshi, K.A., Parihar, V.S., Asok, A., Bellare, J., Ghosh, S., 2018. Biogenic copper nanoparticles from medicinal plants as novel antidiabetic nanomedicine. World J. Pharm. Res. 7(4), 183–196.

Bratby, J. (ed.), 2006. Coagulation and flocculation in water and wastewater treatment. IWA Publishing, London, p. 407.

Chung, Y.T., Ba-Abbad, M.M., Mohammad, A.W., Benamor, A., 2016. Functionalization of zinc oxide (ZnO) nanoparticles and its effects on polysulfone-ZnO membranes. Desalin. Water Treat. 57(17), 7801–7811.

Crini, G., Lichtfouse, E., 2019. Advantages and disadvantages of techniques used for wastewater treatment. Environ. Chem. Lett. 17, 145–155.

Ganesh, B.M., Isloor, A.M., Ismail, A.F., 2013. Enhanced hydrophilicity and salt rejection study of graphene oxide-polysulfone mixed matrix membrane. Desalination. 313, 199–207.

Ghosh, S., 2018. Copper and palladium nanostructures: a bacteriogenic approach. Appl. Microbiol. Biotechnol. 101(18), 7693–7701.

Ghosh, S., 2020. Toxic metal removal using microbial nanotechnology. In: Rai, M., Golinska, P. (Eds.), Microbial nanotechnology. CRC Press, Boca Raton, FL.

Ghosh, S., Chacko, M.J., Harke, A.N., Gurav, S.P., Joshi, K.A., Dhepe, A., Kulkarni, A.S., Shinde, V.S., Parihar, V.S., Asok, A., Banerjee, K., Kamble, N., Bellare, J., Chopade, B.A., 2016a. *Barleria prionitis* leaf mediated synthesis of silver and gold nanocatalysts. J. Nanomed. Nanotechnol. 7, 4.

Ghosh, S., Gurav, S.P., Harke, A.N., Chacko, M.J., Joshi, K.A., Dhepe, A., Charolkar, C., Shinde, V.S., Kitture, R., Parihar, V.S., Banerjee, K., Kamble, N., Bellare, J., Chopade, B.A., 2016b. *Dioscorea oppositifolia* mediated synthesis of gold and silver nanoparticles with catalytic activity. J. Nanomed. Nanotechnol. 7, 5.

Ghosh, S., Harke, A.N., Chacko, M.J., Gurav, S.P., Joshi, K.A., Dhepe, A., Dewle, A., Tomar, G.B., Kitture, R., Parihar, V.S., Banerjee, K., Kamble, N., Bellare, J., Chopade, B.A., 2016c. *Gloriosa superba* mediated synthesis of silver and gold nanoparticles for anticancer applications. J. Nanomed. Nanotechnol. 7, 4.

Ghosh, S., Jagtap, S., More, P., Shete, U.J., Maheshwari, N. O., Rao, S. J., Kitture, R., Kale, S., Bellare, J., Patil, S., Pal, J. K., Chopade, B.A., 2015. *Dioscorea bulbifera* mediated synthesis of novel $Au_{core}Ag_{shell}$ nanoparticles with potent antibiofilm and antileishmanial activity. J. Nanomater. 2015, Article ID 562938.

Ghosh, S., Patil, S., Chopade, N.B., Luikham, S., Kitture, R., Gurav, D.D., Patil, A.B., Phadatare, S.D., Sontakke, V., Kale, S., Shinde, V., Bellare, J., Chopade, B.A., 2016d. *Gnidia glauca* leaf and stem extract mediated synthesis of gold nanocatalysts with free radical scavenging potential. J. Nanomed. Nanotechnol. 7, 358.

Ghosh, S., Selvakumar, G., Ajilda, A.A.K., Webster, T.J., 2021a. Microbial biosorbents for heavy metal removal. In: Shah, M.P., Couto, S.R., Rudra, V.K. (Eds.), New trends in removal of heavy metals from industrial wastewater. Elsevier B.V., Amsterdam, The Netherlands. pp. 213–262.

Ghosh, S., Sharma, I., Nath, S., Webster, T.J., 2021b. Bioremediation: The natural solution. In: Shah, M.P., Couto, S.R. (Eds.), Microbial ecology of waste water treatment plants (WWTPs). Elsevier, Amsterdam, The Netherlands. pp. 11–40.

Ghosh, S., Webster, T.J., 2021a. Nanotechnological advances for oil spill management: Removal, recovery and remediation. In: Das, P., Manna, S., Pandey, J.K. (Eds.), Advances in oil-water separation: A complete guide for physical, chemical, and biochemical processes. Elsevier, Amsterdam, The Netherlands. pp. 175–194.

Ghosh, S., Webster, T.J., 2021b. Nanotechnology for water processing. In: Shah M.P., Rodriguez-Couto, S., Mehta, K. (Eds.), The future of wffluent treatment plants-biological treatment systems. Elsevier, Amsterdam, The Netherlands. pp. 335–360.

Hoek, E.M., Ghosh, A.K., Huang, X., Liong, M., Zink, J.I., 2011. Physical: Chemical properties, separation performance, and fouling resistance of mixed-matrix ultrafiltration membranes. Desalination, 283, 89–99.

Hong, J., He, Y., 2012. Effects of nano sized zinc oxide on the performance of PVDF microfiltration membranes. Desalination. 302, 71–79.

Isawi, H., El-Sayed, M.H., Feng, X., Shawky, H., Mottaleb, M.S.A., 2016. Surface nanostructuring of thin film composite membranes via grafting polymerization and incorporation of ZnO nanoparticles. Appl. Surf. Sci. 385, 268–281.

Jamdade, D.A., Rajpali, D., Joshi, K.A., Kitture, R., Kulkarni, A.S., Shinde, V.S., Bellare, J., Babiya, K.R., Ghosh, S., 2019. *Gnidia glauca* and *Plumbago zeylanica* mediated synthesis ofnovel copper nanoparticles as promising antidiabetic agents. Adv. Pharmacol. Sci. 2019, 9080279.

Joshi, K.A., Ghosh, S., Dhepe, A., 2019. Green synthesis of antimicrobial nanosilver using *in-vitro* cultured *Dioscorea bulbifera*. Asian J. Org. Med. Chem. 4(4), 222–227.

Karmakar, S., Ghosh, S., Kumbhakar, P., 2020. Enhanced sunlight driven photocatalytic and antibacterial activity of flower-like ZnO@MoS$_2$ nanocomposite. J. Nanopart. Res. 22, 11.

Khalid, A., Abdel-Karim, A., Atieh, M.A., Javed, S., McKay, G., 2018. PEG-CNTs nanocomposite PSU membranes for wastewater treatment by membrane bioreactor. Sep. Purif. Technol. 190, 165–176.

Khan, S.B., Alamry, K.A., Bifari, E.N., Asiri, A.M., Yasir, M., Gzara, L., Ahmad, R.Z., 2015. Assessment of antibacterial cellulose nanocomposites for water permeability and salt rejection. J. Ind. Eng. Chem. 24, 266–275.

Kitture, R., Chordiya, K., Gaware, S., Ghosh, S., More, P.A., Kulkarni, P., Chopade, B.A., Kale, S.N. 2015. ZnO nanoparticles-red sandalwood conjugate: A promising anti-diabetic agent. J. Nanosci. Nanotechnol. 15, 4046–4051.

Liang, S., Xiao, K., Mo, Y., Huang, X., 2012. A novel ZnO nanoparticle blended polyvinylidene fluoride membrane for anti-irreversible fouling. J. Memb. Sci. 394, 184–192.

Madaeni, S.S., Zinadini, S., Vatanpour, V., 2011. A new approach to improve antifouling property of PVDF membrane using in situ polymerization of PAA functionalized TiO2 nanoparticles. J. Memb. Sci. 380(1–2), 155–162.

Mulopo, J., 2017. Bleach plant effluent treatment in anaerobic membrane bioreactor (AMBR) using carbon nanotube/polysulfone nanocomposite membranes. J. Environ. Chem. Eng. 5(5), 4381–4387.

Pastrana-Martinez, L.M., Morales-Torres, S., Figueiredo, J.L., Faria, J.L., Silva, A.M., 2015. Graphene oxide based ultrafiltration membranes for photocatalytic degradation of organic pollutants in salty water. Water Res. 77, 179–190.

Rathoure, A.K., Dhatwalia, V.K., 2016. Toxicity and waste management using bioremediation IGI Global. Hershey, PA. p. 421.

Robkhob, P., Ghosh, S., Bellare, J., Jamdade, D., Tang, I.M., Thongmee, S., 2020. Effect of silver doping on antidiabetic and antioxidant potential of ZnO nanorods. J. Trace Elem. Med. Biol. 58, 126448.

Rokade, S., Joshi, K., Mahajan, K., Patil, S., Tomar, G., Dubal, D., Parihar, V.S., Kitture, R., Bellare, J.R., Ghosh, S., 2018. *Gloriosa superba* mediated synthesis of platinum and palladium nanoparticles for induction of apoptosis in breast cancer. Bioinorg. Chem. Appl. 2018, 4924186.

Rokade, S.S., Joshi, K.A., Mahajan, K., Tomar, G., Dubal, D.S., Parihar, V.S., Kitture, R., Bellare, J., Ghosh, S., 2017. Novel anticancer platinum and palladium nanoparticles from *Barleria prionitis*. Glob. J. Nanomedicine. 2(5), 555600.

Salunke, G.R., Ghosh, S., Santosh, R.J., Khade, S., Vashisth, P., Kale, T., Chopade, S., Pruthi, V., Kundu, G., Bellare, J.R., Chopade, B.A., 2014. Rapid efficient synthesis and characterization of AgNPs, AuNPs and AgAuNPs from a medicinal plant, *Plumbago zeylanica* and their application in biofilm control. Int. J. Nanomedicine. 9, 2635–2653.

Shah, P., Murthy, C.N., 2013. Studies on the porosity control of MWCNT/polysulfone composite membrane and its effect on metal removal. J. Memb. Sci. 437, 90–98.

Shende, S., Joshi, K.A., Kulkarni, A.S., Charolkar, C., Shinde, V.S., Parihar, V.S., Kitture, R., Banerjee, K., Kamble, N., Bellare, J., Ghosh, S., 2018. *Platanus orientalis* leaf mediated rapid synthesis of catalytic gold and silver nanoparticles. J. Nanomed. Nanotechnol. 9, 2.

Shende, S., Joshi, K.A., Kulkarni, A.S., Shinde, V.S., Parihar, V.S., Kitture, R., Banerjee, K., Kamble, N., Bellare, J., Ghosh, S., 2017. *Litchi chinensis* peel: A novel source for synthesis of gold and silver nanocatalysts. Glob. J. Nanomedicine. 3(1), 555603.

Shinde, S.S., Joshi, K.A., Patil, S., Singh, S., Kitture, R., Bellare, J., Ghosh, S., 2018. Green synthesis of silver nanoparticles using *Gnidia glauca* and computational evaluation of synergistic potential with antimicrobial drugs. World J. Pharm. Res. 7(4), 156–171.

Xu, J., Zhang, L., Gao, X., Bie, H., Fu, Y., Gao, C., 2015. Constructing antimicrobial membrane surfaces with polycation: Copper (II) complex assembly for efficient seawater softening treatment. J. Memb. Sci. 491, 28–36.

Zhang, M., Field, R.W., Zhang, K., 2014. Biogenic silver nanocomposite polyethersulfone UF membranes with antifouling properties. J. Memb. Sci. 471, 274–284.

Zhang, Y., Wan, Y., Shi, Y., Pan, G., Yan, H., Xu, J., Guo, M., Qin, L., Liu, Y., 2016. Facile modification of thin-film composite nanofiltration membrane with silver nanoparticles for anti-biofouling. J. Polym. Res. 23(5), 105.

7 Membrane Reactors

*Moupriya Nag, Dibyajit Lahiri, Sayantani Garai,
Dipro Mukherjee, Ritwik Banerjee, Ankita Dey,
Smaranika Pattanaik, and Rina Rani Ray*

CONTENTS

7.1 INTRODUCTION

A reactor is a physical device or container in which biological or chemical reactions can be housed while maintaining a proper equilibrium and optimal environmental conditions that are desirable for the reactions. Reaction and separation are the most important parts of a biochemical or chemical reaction process in the industrial or commercial level and are mostly required to be carried out in separate units (Tsuru 2012).

According to the International Union of Pure and Applied Chemistry (IUPAC) definition (Gallucci et al. 2011), a membrane reactor (MR) is a physical device that can simultaneously perform a reaction and a membrane-based separation in the same physical device. Therefore,

DOI: 10.1201/9781003165019-7

membrane reactor technology refers to the combination of the two processes in a single unit, hence making the whole operation economic and the reactor management system compact and more efficient. This amalgamation of both chemical engineering and the process of membrane separation also helps shift the reaction equilibrium to desirable reactions. Membrane separation also allows the selective extraction of the products from the reaction mixture without hampering the catalyst concentration or keeping the enzyme concentration constant. The advantages of membrane separation techniques embedded in the reactor management system is multifold, and thus, a promising unit operation improvement can be achieved using MRs (Marcano and Tsotsis 2002; Seidel-Morgenstern 2010). Apart from being applied for separating purpose, membrane bioreactors (MBRs) are often used as alternative approaches to classical immobilization methods (catalytic MBRs). In such processes, membranes are used for the immobilization of biocatalysts such as enzymes or enzyme-producing microorganisms instead of having them suspended in the reaction mixtures. This improves the retention and stability of the catalysts as the membrane often acts as a matrix, supporting the catalyst molecules. The membranes used to support various immobilization techniques such as entrapment, physical adsorption, ionic binding, crosslinking, covalent binding, and so on.

MRs have gained popularity in various research and process technical aspects. Initially, the development of MRs emerged as a strategy to aid in the treatment of wastewater or municipal sewage in the late 1980s and 1990s. MRs, in the past, gained recognition for the primary purpose of wastewater treatment through ultrafiltration and microfiltration of toxic wastes and impure substances from the wastewater. Recent decades have witnessed substantial worldwide research and process development efforts centered on MR technology (Marcano and Tsotsis 2002). There have been numerous studies investigating the efficiency and improvement of MRs; their operating conditions, geometric, and hydrodynamic parameters; separation optimization; biochemical analysis for improvement of the immobilization stability; and more (Giorno et al. 2003). An equivalent amount of research has been carrying carried out to improve the existent technology even further through modeling, simulation, modification, polymer technology, operability analysis studies, and so on (Bishop and Lima 2020; Zhong et al. 2010; Nakajima et al. 1989).

MBRs are extensively used in industrial wastewater management and domestic or municipal sewage treatment. Apart from this primary area of application, MBRs are also used for the production of drugs or medicines, purification of isomers or enantiomers, production of nutritive such as vitamins and amino acids, synthesis and purification of important enzymes, and innumerable other commercial aspects. For example, multiphase MBRs have been reported to be used in the production of diltiazem, a drug used in angina and hypertension patients. MRs were reported to enhance the production rates of chiral intermediates by improving the biotransformation of the reactants through enhanced enzyme-substrate contact due to enzyme immobilization (Lopez and Matson 1997). In a similar study, it was used in the synthesis of lovastatin, a drug used for maintaining cholesterol levels (Yang et al. 1997). Furthermore, MRs have been used in the production and purification of analgesic compounds like kyotorphin (Belleville et al. 2001), anticancer drugs, ibuprofen esters (Long et al. 2005), and more. MRs have been reported to be used in enzymatic transformation, the production of oligosaccharides (Martin et al. 2001), malic acid synthesis (Giorno et al. 2001), and so on.

Despite the extensive application areas of MRs, the limited success and popularization of this technology is mainly because of the scale-up difficulties and cost of membranes. Membrane systems are comparatively costly and less robust. The rate-limiting factors and the lifetime of immobilized enzymes or the sheer stress factors on the membranes impose difficulties in scaling up of the technology. However, MBRs are still considered a boon in wastewater management operations because of their efficiency in purifying and removing chemical effluents and volatile components, which is extremely hard to achieve by other separation methods or by conventional chemical treatments.

7.2 CONCEPT OF REACTORS

Reactors or biochemical reactors are physical devices or vessels in which a biochemical reaction can be carried out in desired equilibrium conditions. Bioreactors are specifically manufactured devices or systems designed to support a biologically active reaction. Bioreactors involve the use of biotic organisms like bacteria, fungi, plant and animal cells, or biochemically active molecules or substances such as enzymes. A reactor management system or a bioreactor management system refers to the management of specifically optimized environments or environmental conditions within the reactor vessel so that the biochemical reactions can provide maximum output in terms of product. The reaction environment can be optimized through the maintenance of proper pH, temperature, ionic strength of the solution or reaction mixture, physical agitation of the reaction mixture, dissolved oxygen levels, and so on in order to achieve high yield of the bioprocess. Bioreactors or reactors can be classified into many types based on operation modes, process requirements, presence or absence of oxygen, reactor design, and the like. The main basis of classification, however, continues to be on the mode of operation of the reactors, wherein they can be classified into batch reactors, continuous reactors, and semi-batch or fed-batch reactors

7.2.1 BATCH REACTORS

Batch processes refer to a partially closed system in which the reactants are added initially, all at once, and are then processed in fixed batches or volumes until the conversion is achieved, and only then, at the end of the operation, the products are aseptically removed from the mixture through discharge tubes. In a batch process, only gaseous exchange occurs during the course of operation, which can be accompanied by the incorporation of antifoam and pH control agents.

A typical batch reactor is designed to contain a reactor vessel or a tank varying in size from less than 1 liter to more than 15,000 liters and is generally equipped with an agitator and a temperature management system. The reactors operate in a batch mode; that is, the batch reactor is a nonsteady, transient reactor that implies that the extent of product conversion is a function of time. The advantages of using batch reactors are the variability and versatility of the reactors, which allow a wide area of applications. Batch reactors are used in various industries, both in large and small-scale production, and are especially helpful in the study of growth and reaction kinetics. However, a major disadvantage of the batch system is the prolonged idle and operation times.

7.2.2 CONTINUOUS REACTORS

Continuous reactors, also known as continuous stirred tank reactors (CSTRs) comprise a tank having a constant volume that is being supplied continuously to the reactor tank. Unlike batch culture, reactants are constantly supplied to the system through specific inlet valves (continuous feed), and the product is constantly derived at a steady rate. CSTRs are equipped with influent and effluent ports for the inflow of reactants and the harvest of products, respectively (Chan et al. 2009). CSTRs are steady-state equipment, which means that extent of conversion is not a function of time. As the reactants and products follow a constant flow rate or diffusion, the volume of the tank is maintained at a constant level, and the extent of reaction conversion will be a function of the reaction volume. An ideal CSTR can be considered to employ homogeneous mixing, without any variations in temperature, concentration, fluid properties, and reaction rate.

There are two types of CSTR operation strategies: chemostat and turbidostat. In a chemostat, the reactants are added in excess and the reaction mixture is maintained at a constant volume by setting the inlet and outlet flow rates equal. The operating conditions for a chemostat are such that the reactor is needed to monitor and maintain a constant chemical composition. The turbidostat, on the other hand, is designed to maintain a constant cell concentration by maintaining the turbidity of reaction mixture through spectrophotometric monitoring.

The advantages of CSTRs are constant product formation, maintenance of reaction at exponential phase, and less idle time, among others. CSTRs can also be used in series with more than one bioreactor with different conditions in each one.

7.2.3 FED-BATCH CULTURE

A semi-batch or a fed-batch reactor is a semi-flow reactor wherein one or more reactants are fed continuously while the products are discharged in batch mode. It is, thus, a modification of a batch reactor. This procedure means that the concentration of one or more of the nutrients in the reaction medium can be altered by changing the feed rate during the run according to the feedback of control parameters such as dissolved oxygen (DO), pH, or respiratory quotient (RQ). This offers major operational flexibility and a better management system of reactions compared to pure batch systems.

Apart from operating modes, in terms of design and scale-associated factors, various types of reactors have been designed for specific purposes, reaction types, and in order to improve efficiency or yield. These are reactors designed by the improvisation of the type of packing material used, or catalyst immobilization techniques, which can increase the yield of products. The various types of highly efficient reactor systems are discussed as follows.

7.2.4 BUBBLE COLUMN BIOREACTOR

Bubble column reactors are a subset of pneumatic gas–liquid reactors that use an injection of compressed air from the base of the reactor vessel to mix the different phases of the reaction mixture. The compressed air, once injected into the liquid mixture, moves up through the reactants forming bubbles with a superficial gas velocity of approximately 0.03 to 1 m/s. Such fast-rising gas bubbles facilitate the mixing of the reaction mixture, thus providing a cheap and simple method of mixture agitation where other mechanical agitators or baffles are not used.

7.2.5 PACKED-BED REACTORS

Packed-bed reactor (PBR) refers a tubular reactor system packed with solid particles of packing materials inside which a catalyst is immobilized (Fogler 2006). This increases the nutrient exchange per unit volume of reaction mixture due to higher contact area, thereby increasing the conversion rate. The catalyst is packed in the column, and the nutrients are fed from either the bottom or the top of the reactor (Martinov et al. 2010). There are several types of packing materials available such as ceramic pieces, volcanic rocks, clay balls, polyethylenevinylacetate, and so on (Hadjiev et al. 2007).

7.2.6 FLUIDIZED-BED REACTORS

Fluidized-bed reactors (FBRs) are developed by the use of small carriers that results in the development of a bed within the column by the use of various types of influent flowing mass. The media particles remain distributed on the basis of the size of their gradient. Small-sized materials are used for this technology for the purpose of enhancement of the specific surface area. The separation of particles takes place when the force of gravity exceeds the driving force within the reactor system (Lewandowski and Boltz 2011).

7.2.7 AIRLIFT REACTORS

Airlift reactors are another type of gas–liquid reactor, which is a specialized version of the fluidized bed reactor. The principle of working of an airlift reactor is based on the working of a draught or draft tube. The main reactor vessel is divided into two parts which is interconnected by the means of a baffle or the tube. In one of the zones, the air or gas is pumped in through a sparger. The compartment is also called the riser compartment as the gas flows upward in this zone. In the other zone, the downward flow of the gas occurs, which aids in mixing the reaction mixture. Airlift reactors are highly efficient and widely used in wastewater treatment, single-cell protein production, and methanol production, among others.

7.2.8 MRs

As discussed, MRs use specific membranes for separation, purification, and catalyst immobilization purposes in batch, semi-batch, or continuous flow reactors. The amalgamation of the two technologies, that is, the membrane technology and the reactor technology, enables us to have the best of both worlds. A more comprehensive account of MRs is covered in the later parts of this chapter.

7.3 VARIOUS TYPES OF MEMBRANES ASSOCIATED WITH REACTORS

The membranes that are used in MR can be divided into various domains based on their geometrical shape, the nature of the membranes, and the separation criteria (Khulbe 2007). The main types of membranes are organic membranes, inorganic membranes, and their hybrids and their choice are dependent on parameters like productivity, membrane longevity, sustenance of optimal operating conditions – both mechanical and chemical – selectivity of separations, and, most important, the cost of the membrane. The substantial development and popularity of MRs, due to the discovery of new membranes, which increases its applicability, are reflected in numerous scientific journals and have grown manifold in the past few decades (McLeary et al. 2006).

First, the membranes can be divided into biological and synthetic ones. The *biological membranes* are more economical as they can be manufactured very easily but has limited usage given their low work range. The major drawbacks that hinder the large-scale usage of the biological membranes are that they can be operated only at a specific temperature range (i.e., ideally at room temperature and always below 100°C), lower pH tolerance and cleaning-up techniques. These biological membranes are prone to microbial degradation as well and thus are generally avoided (Xia 2003). The *synthetic membranes*, on the other hand, can be further subcategorized into organic and inorganic ones. A major example of an organic membrane is a polymeric membrane that can operate up to 300°C (Catalytica 1988). The organic membranes that are generally used in the industries are generally made from natural polymers, like cellulose, wool, and polyisoprene (rubber), and synthetic polymers, like polystyrene, polytetrafluoroethylene (Teflon), and polyamide. Inorganic membranes are most commonly used in various fields, owing to their wider pH tolerance, greater temperature range (>250°C), and resilience to degradation by chemicals.

In the perspective of the morphology or potentially layer structure, the inorganic layers can be even partitioned into metallic and porous membranes. Metallic membranes are generally supported or unsupported. Numerous benefits are provided by the supported dense membranes (SDMs) as compared to porous membranes (like ceramic). Specifically, numerous endeavors were committed to creating thick metallic layers stored on porous help (alumina, silica, carbon, and zeolite) for isolating hydrogen with a noncomplete selectivity yet bringing the expenses related down to the thick metallic layers (Lin 2001). The porous membranes are categorized by IUPAC according to the pore size. The ones with a pore diameter smaller than 20Å are known as microporous, those between 20Å and 500Å are called mesoporous, and above 500Å, they are called microporous.

7.3.1 POLYMERIC MEMBRANES

Even though all the polymeric substances can be used as a membrane in reactors, due to certain physical and chemical properties, the actual number of polymeric membranes used practically are limited. The precise properties from structural factors are the main guidelines kept in mind when choosing a particular polymer as a membrane and thus are not randomly chosen. In a scientific journal, Ozdemir et al. (2006) provide insight into the commercial applications of polymers as membranes in reactors. However, these kinds of membranes are generally not favored as in industries there is an involvement of high temperatures, which is not suited for polymers. The solution to this problem is the usage of inorganic membranes.

7.3.2 Inorganic Membranes

The main advantage of using inorganic membranes in reactors is mainly the temperature tolerability of the materials. Common inorganic membranes include ceramic, zeolite, metal oxides (like zirconia, alumina), silica, palladium, silver, and their respective alloys, which are operational at elevated temperatures above 250°C and up to 900°C. In some cases, the ceramic membranes are used at temperatures over 1000°C as reported by Van Veen in 1996. Another major advantage of these kinds of membranes is that they are resistant to both chemical and other degradations but pose disadvantages because of their high cost. Also, these membranes have high mechanical stability, resistance to sudden pressure changes, and solvent-resistant.

7.3.2.1 Palladium-Based Membranes

The main reactions in which palladium-based membranes are used are primarily dehydrogenation (Wood 1968; Itoh 1987, 1990), hydrogenation and hydrogen oxidation (1990; Zhao et al. 1990), and the reforming of steam. This is mainly due to the membrane's high selectivity to hydrogen. Palladium amalgams are regularly liked on the grounds that unadulterated palladium will, in general, become fragile after rehashed patterns of hydrogen absorption and desorption (Zaman et al. 1994). A large part of the early investigation of the utilization of palladium-based layers was been done in the previous Soviet Union (Gryaznov et al. 1987). Broad investigations have been made over the course of the years on the penetrability just as mechanical properties and longevity of these layers (Lewis et al. 1988). In a journal, a group of scientists used palladium–Ag membrane at an approximate temperature of 700°C and up to 40 bar pressure (Schmitz et al. 1988). Compared to the balance value, the conversion rate has increased by 25%. They also studied the penetration of H2 into palladium, palladium–silver, stoichiometric nickel–nickel–titanium alloys, copper–palladium-coated vanadium membranes, and double layers. The diaphragm is made of a vanadium-plated Pd–Ag alloy (atomic ratio 75/25). The permeability of the titanium and nickel membranes is lower; the performance of the vanadium-coated membranes is better than that of pure palladium, while the permeability of Pd–Ag membranes is higher (Zaman et al. 1994).

7.3.2.2 Ceramic and Glass Membranes

Inorganic layers were used as separators and reactors as a result of the porous ceramic and glass membrane method. Ceramic membranes are generally made of oxides of silica, titanium, and aluminum and have the advantage of being chemically inert and unchanged at elevated temperatures. Owing to these properties, the ceramic membranes in microfiltration and ultrafiltration are applicable in food, pharmaceutical, and other biotechnological approaches, where membranes are required to undergo multiple steam sterilization and cleaning with chemicals. These membranes are also used as gas separators in MRs. Even after being quite useful to industries using MRs, ceramic membranes have certain limitations that make using them disadvantageous. At higher temperatures, the sealing of the membranes in the reactor becomes difficult and sometimes, the membranes might crack as well. Some ceramic substances used as membranes like perovskite are quite unstable, which is another con for these kinds of substances. In recent years, composite alumina membranes have been widely used in MRs and separation applications. In a typical composite tubular membrane, the innermost layer is approximately of 5µm with pore diameter of about 40Å. The successive layers are thicker and have larger pores ranging from 2000 to 8000Å, which is supported by an approximately 0.1-cm-thick layer with a pore diameter ranging from 10 to 15µm (Zaman et al. 1994).

Glass membranes are usually made by a combination of heat treatment and chemical leaching, while ceramic membranes are made by slip casting. The details of the preparation method are carefully preserved; various reviews and patent information provide useful information (Bhabe 1991). The nature of the film is a lot of wards on the planning strategies and a nearby adherence to an unbending convention is important to acquire layers of reliable quality. The developed membrane also undergoes a series of tests like X-ray diffraction, scanning electron microscopy (SEM), TEM analysis, and others (Leenaars et al. 1984; Anderson et al. 1988; Gieselmann et al. 1988; Okubu et al. 1990; Larbot et al. 1988; Table 7.1).

7.3.2.3 Carbon Membranes

Carbon molecular sieve (CMS) membranes have been identified as promising candidates for gas separation, both in terms of separation performance and stability. A CMS membrane is a porous solid containing narrow pores approximately the same size as the molecules of the diffusion gas. Molecular sieves can effectively separate tiny size differences. These membranes can be divided into two categories namely, supported and unsupported ones, and are manufactured by the process of pyrolysis of thermosetting polymers like PFA (poly furfural alcohol), PVDC (poly vinylidene chloride), polyacrylonitrile, and phenol formaldehyde.

TABLE 7.1
Application of Porous Inorganic Membrane Reactors

Membrane type	Reaction	Operating condition	Reactor configuration	Reference
Glass, alumina, composite alumina	H_2S decomposition	873–1073 K	Inert membrane packed bed reactor (IMPBR)	Kameyama et al. 1981, 1983
	Ethane dehydrogenation	723–873 K Catalyst: Pt	Catalytic membrane reactor (CMR), packed bed catalytic membrane reactor (PBCMR)	
Composite alumina	Propane dehydrogenation	753–898 K Catalyst: Pt/γ- alumina		Ziaka et al. 1993
	n-butane dehydrogenation	673–773 K Catalyst: Pt/SiO_2		Zaspalis et al. 1991
	Cyclohexane dehydrogenation	470 K Catalyst: Pt	IMPBR	Okubu et al. 1991
	Steam reforming of methane	718–883 K Catalyst: Ni/alumina		Minet et al. 1992
	Ethylbenzene dehydrogenation	828–875 K Catalyst: Iron oxide		Anderson et al. 1990
	Methanol dehydrogenation	573–773 K Catalyst: γ- alumina/γ- alumina with silver	CMR, IMPBR	Zaspalis et al. 1991
Glass	Cyclohexane dehydrogenation	470–570 K Catalyst: Pt	CMR	Sun et al. 1988
		460 K Catalyst: Pd	IMPBR, PBCMR	Cannon et al. 1992
	Methanol dehydrogenation	573–673 K Catalyst: Ag	IMPBR	Zhao 1989
Composite titania and composite alumina	Reduction of nitrogen oxide with ammonia	573–623 K Catalyst: V_2O_5	CMR	Zaspalis et al. 1991

7.3.2.4 Zeolite Membranes

Zeolites are microporous crystalline aluminosilicates with uniform pore diameters. Zeolite is used as a catalyst or adsorbent in the form of micron or submicron crystallites encased in millimeter particles. However, these membranes have comparatively inferior gas fluxes than other inorganic membranes and the thermal effect, where these membranes show negative thermal expansion (where they shrink in higher temperatures), which are the main drawbacks of using these kinds of membranes. In the thermal effect, the zeolite shrinks at elevated temperatures, while the support continuously expands, resulting in thermal stress on the attachment of the membrane to the support and on the connection of the individual microcrystals within the zeolite layer (Cejka).

7.4 VARIOUS TYPES OF MRS

Over the past couple of years, membrane reactors are gaining immense popularity in various process industries. The most important sector in which MRs are extensively used is the hydrogen production industry. MRs are deployed to produce ultra-pure hydrogen gas in industries. However, MRs can also be used in other industries as well. Based on the requirement of the process, the reactors can be of various types. Here, we discuss some of the common MRs that have been used in recent years.

7.4.1 FLUIDIZED-BED MRS

The union of noncatalytic membranes (dense or porous) into an FBR combines the advantage of not only the separation through membrane but also the benefits from the fluidization regime. Unlike the packed-bed MRs and PBRs, which suffer from the same limitations like high-pressure fluctuations, complex heat removal and supply mechanisms, and low membrane surface area per volume of the reactor, the fluidized-bed membrane reactor (FBMR) has some major advantages for the users. There is no drop in pressure and no internal heat/mass transfer limitations. This is due to the deployment of minute particles in the reactor. Isothermal operation and the elasticity in membrane heat-transfer surface area prove beneficial as well. Compartmentalization reduces the axial-gas back mixing, which also improves the overall fluidization behavior, which is another advantage. However, the FBMR also has certain limitations like erosion of internal components and catalyst attrition. Difficulties in reactor construction and sealing at the wall are other obstacles to FBMR operation. The former limitation can become critical in the case of a highly selective thin-layer membrane being used in the fluid bed. The overall performance of the reactor deteriorates if the permselectivity is reduced as a result of internal erosion. Hence, membranes to be utilized in fluidized layer reactors ought to be ensured by disintegration, maybe by utilizing a permeable media between the film layer and the fluidized bed.

Various groups of scientists have studied the applications of the FBMRs for the production of pure hydrogen gas (Gallucci 2008b). For this situation, as examined in the initial segment of the survey, Pd-based films are embedded in fluidized bed reactors where the transforming of hydrocarbons happens. Then again, fluidized bed reactors have likewise been proposed for various applications. Deshmukh et al. (2005a, 2005b) have proposed a membrane-assisted FBMRs for oxidizing methanol partially. The gas-phase back mixing was studied by cold experiments initially, using tracer injection technique, and bubble to emulsion phase by ultrasonic experiments.

7.4.2 PEROVSKITE MRS

The first usage of perovskite membranes, which can be traced back to 1985, for studying the flux of oxygen through 1-cm disks shaped perovskite material. Even after decades of their first reported usage, perovskite membranes do not find much application in the present day mainly due to their limitations like difficulties in module sealing at elevated temperatures and poor membrane stability.

In recent years, perovskite has been used to oxidize NH_3 to NO for the production of HNO_3 acid (Sun et al. 2009). In reality, about 80% of ammonia is utilized for manure manufacture, and a major part is first changed over to nitric corrosive through high-temperature oxidation on platinum–rhodium combination catalyst. This response is notable for quite a long time and furthermore all around advanced as far as catalyst. Notwithstanding, still, some specialized issues must be confronted. Specifically, this activity is very expense serious additionally because of impetus misfortune as oxides, which is being studied by exploration of other catalysts like Cr_2O_3. Nitrous oxide emissions are the main concern in these industries, which must be captured as it is expensive. The separation of oxygen and the reaction in one unit is carried out by the application of a perovskite membrane reactor (Sun et al. 2009).

7.4.3 Catalytic MRs

A direct review of the main researchers of the catalytic membrane reactor (CMR) is quite complicated because some authors mistakenly refer to the CMR as a reactor in which the catalyst is filled into the reactor in some way. It is called a packed bed membrane reactor. CMR is a special type of reactor in which the membrane acts as both a separation layer and a catalyst. The membrane is either autocatalytic or made catalytic by the coating of a dense material on the surface or by casting the polymeric material and catalytic material. The experimental and theoretical research of the CMR is introduced. The Mendes group is very active in the modeling of polymer CMRs. They used fairly detailed models to model various reaction systems in polymer CMR. Both polymer and inorganic CMRs are used for experimental work. Porous polymeric membranes with enhanced flux with the casting machine was produced by Fritsch (2006). As mentioned earlier, the author used two different routes to produce catalytic membranes: a casting solution containing a catalyst and a catalyst material used to fill the pores. The membrane has been used to hydrogenate sunflower oil into edible oil. The proposed method is very interesting, because the author uses a high-throughput catalytic membrane to solve the problem of separating catalysts from edible oil (commonly used catalysts: expensive or toxic) and the problem of droplets.

7.4.4 Photocatalytic MRs

An interesting new system to consider is the photocatalytic membrane reactor system, in which photocatalysis is improved to some extent by membrane separation. There are mainly two ways to build a photocatalytic membrane. In the first method, a photocatalytic MR is a reactor in which a membrane is in contact with a reagent and in which light is emitted from an internal or external light source. Figure 7.1 shows the schematic representation of a typical integrated photocatalytic MR.

The second method of working with a photocatalytic membrane is to differentiate between the reaction and membrane separation in two different steps (Azrague 2005). Figure 7.2 depicts the typical photocatalytic reactor coupled with membrane separation system.

The film frequently fills in as a separator for the suspended impetus particles from the treated media. In other cases, the photocatalyst can be immersed in the middle of the membrane, which also acts as a carrier, or the membrane itself can be photocatalytic. The membrane can also be used as a separator for reaction products. Typical applications of photocatalytic membrane reactors are the photodegradation of water-based pollutants, the photoreaction of high-value products, and the photooxidation of pollutant vapors. The use of membranes as an external separation system reduces the problems of membrane filtration research. In this case, a commercially available membrane filtration system can usually be used without any problems.

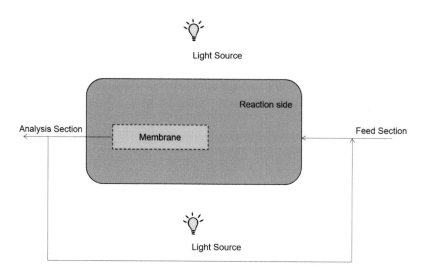

FIGURE 7.1 A typical integrated photocatalytic MR.

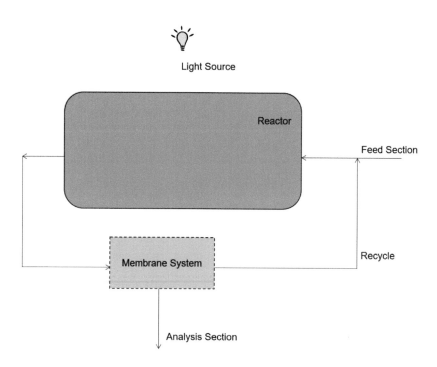

FIGURE 7.2 Typical photocatalytic reactor coupled with membrane separation system.

7.4.5 MBRs

An MBR, or the subject of MBRs, is an accumulation of concepts of membrane filtration and traditional biological wastewater treatment plants. The subject is almost technologically the same as that of a conventional wastewater treatment plant, but it is not applicable in the case of the separation of wastewater that has undergone treatment and activated sludge. In the process of installing an MBR, the process of separating the wastewater is not done by the process of sedimentation in a secondary clarification tank, but it is done by a membrane filtration process (Huang et al. 2003).

The primary batch of MBRs is composed of crossflow operational membranes that are usually arranged on the outside and installed in the external part of the activated sludge tank. The theory of crossflow, along with its association with high flow velocity, is rapidly used for preventing the construction of solids on the surface of the membrane, which is also known as cake-layer synthesis (Wang et al. 2003). The procedure of the operation of crossflow needs large amounts of energy for the formation of the sludge velocity on the surface of the membrane for the maintenance of both the needed pressure drop required for the process of permeation and the high crossflow velocity in case of the membrane cleaning (Mansell et al. 2003). Looking at the energy needs, the field of MBRs was seen as nonviable in the application of the corporation wastewater treatment process. Moreover, the requirement of the crossflow recirculation pump, along with the association of excessive shear and high pressure, was looked on at having a dangerous effect on the size of the floc and how stable it is inside the system. A significant creation of membranes was seen when research scholars tried experimenting with submerging the membranes inside an aeration tank. To achieve the proper permeation, the technology was made such that it utilizes a decreased amount of pressure in the opposite of pressure tubes, which is installed externally and is required for the high amount of over-pressure. This kind of submerged membrane filtration in a biological system is usually known a SMBR, or submerged membrane bioreactor (Holloway et al. 2003). The energy requirement was also decreased to a great extent. Decreased pressure is applied in the process of permeate extraction, and it is much lower than what is needed for the process of crossflow permeation. Moreover, an important portion of the crossflow technology is the recirculation pump is not present in the SMBR structure (E. Scholes et al. 200313; Table 7.2).

TABLE 7.2
A List of Large MBR Plants Undergoing Treatment in Municipal Wastewater That Have Been Commissioned during Approximately the Last 6 Years

MBR plant	MBR company	Location and country	Year of commissioning	Peak daily flow (m³/d)	New or upgrade
Henriksdal	GE WPT	Stockholm/Sweden	2018	864,000	Upgrade
Seine Aval	GE WPT	Acheres/France	2016	357,000	Upgrade
Canton	Ovivo	Ohaio/USA	2015	333,000	Upgrade
Water Affairs Integrative EPC	OW	Xingyi, Guizhou/China		307,000	New
Euclid	GE WPT	Ohaio/USA	2020	250,000	Upgrade
Yunnan	OW	Kunming/China	2013	250,000	Upgrade
Shunyi	GE WPT	Beijing/China	2016	234,000	New
Macau	GE WPT	Macau Special Administrative Region/China	2017	210,000	New
Fuzhou Yangli	Memstar/ United Enviro	Fujian/China	2015	200,000	New
Wuhan, Sanjiang	OW	Hubei Province/China	2015	200,000	Upgrade
Brussels Sud	GE WPT	Brussels/Belgium	2017	190,000	Upgrade
Macau	GE WPT	Macau/China	2014	189,000	New
Riverside	GE WPT	California/USA	2014	186,000	Upgrade
Brightwater	GE WPT	Washington/USA	2011	175,000	New
Visalia	GE WPT	California/USA	2014	171,000	Upgrade
Cox Creek	GE WPT	Maryland/USA	2016	170,000	Upgrade
Qinghe	OW/MRC	Beijing/China	2012	150,000	New

(Continued)

TABLE 7.2 *(Continued)*

A List of Large MBR Plants Undergoing Treatment in Municipal Wastewater That Have Been Commissioned during Approximately the Last 6 Years

MBR plant	MBR company	Location and country	Year of commissioning	Peak daily flow (m³/d)	New or upgrade
Jilin (Phase I)	OW	Jilin Province/China	2015	150,000	Upgrade
Jilin (Phase II)	OW	Jilin Province/China	2014	150,000	New
Yantai Taoziwan	OW	Shandong Province/China	2014	150,000	New
Nanjing Chengdong	OW	Jiangsu Province/China	2013	150,000	New
Carré de Reunion	KMS	Versailles/France	2015	144,000	Upgrade
Changsha 2nd	OW	Hunan Province/China	2014	140,000	New
North Las Vegas	GE WPT	Nevada/USA	2011	136,000	New
Assago	GE WPT	Milan/Italy	2016	125,000	New
Daxing Huangcun	OW	Beijing/China	2013	120,000	Upgrade
Jinyang	OW	Shanxi/China	2015	120,000	New
Daxing Huangcun	OW	Beijing/China	2013	120,000	Upgrade
SBGE	GE WPT	Brussels/Belgium	2018	120,000	Upgrade
Ballenger McKinney	GE WPT	Maryland/USA	2015	135,000	Upgrade
Yellow River	GE WPT	Georgia/USA	2011	114,000	Upgrade
Aquaviva	GE WPT	Cannes/France	2013	108,000	New
Urumqi Ganquanpu	OW	Xinjiang Uygur/China	2014	105,000	New
Busan	GE WPT	Busan/Korea	2012	102,000	New

7.4.5.1 Principles and Background of MBRs

Filtration is known as the process of separating two or more substances from a fluid stream. In traditional use, it is generally referred to the separation of insoluble components or solids from a stream of any liquid. Membrane filtration is simply a further extension of the application, including the separation of dissolved solids inside liquid streams, because the membrane-related methods in the water treatment are usually required for removing different types of substances that range from salts to microorganisms (Yusuf et al. 2003). Different types of membrane processes can be divided into different, similar types of subtypes. Three of these subtypes are the pore size, the molecular weight cutoff, and the pressure at which they are being operated. As the pore size is generally reduced or the molecular weight cutoff is gradually reduced, the pressure acting on the membrane for the purpose of separation of the water from different other materials usually rises.

The membrane processes controlled by pressure with the help of microfiltration, as well as reverse osmosis, are very specific procedures with respect to the corresponding pore sizes (Cote et al. 2003). The separation that takes place inside the MF, or microfiltration, can be useful in removing suspended compounds or particulates that range from 0.1 to 10 micrometers. Again, it is seen that UF, or the process of ultrafiltration, is generally used for recovering macromolecules ranging from size 0.01 to 0.1 micrometer. However, NF, or the process of nanofiltration, can be useful with removing the particulate of size ranging from 0.001 to 0.01 micrometer. RO, or reverse, osmosis membranes are responsible for the separation of materials smaller than 0.001 micrometer (Skouteris et al. 2003). The principal and working procedure of RO need very high pressure. It can be as high as 150 bar sometimes to overcome the osmotic pressure. However, the hydrodynamic pressure that is needed, including for the flow through MF and UF membranes, lies usually in the range of 0.1 to 10 bar pressure range (Wang et al. 2003; (Tables 7.3 and 7.4).

7.4.5.1.1 Biofilm-Based MRs

The membrane biofilm reactor (MBfR) is a high-tech innovation that can transfer H_2 to microorganisms efficiently, consistently, and safely. It forms a natural relationship between a membrane and

TABLE 7.3

Application of MBRs in Industrial Wastewater Treatment

Sl no.	Wastewater source	Country of application	Size of operation	Membrane configuration	Treatment efficiency
1.	Food industry	USA	Full scale 600 m³/d	Microfiltration	Effluent TSS 9 mg/l
2.	Paint industry	USA	Full scale 113 m³/d	Ultrafiltration external	COD removal 94 %
3.	Various sources	Germany	Pilot scale 0.2–24.6 m³/d	Ultrafiltration external	COD removal 97 %
4.	Cosmetic industry	France	Full scale	Ultrafiltration external	COD removal 98 %
5.	Tannery industry	Germany	Full scale 500–600 m³/d	Ultrafiltration external	COD removal 93 %
6.	Electrical industry	Germany	Full scale 10 m³/d	Ultrafiltration external	COD removal 97 %

TABLE 7.4

Applications of MBRs in Landfill Leachate and Sludge Digestion

Sl no.	Source of wastewater	Country of application	Treatment efficiency	Size of operation	Membrane configuration
1.	Landfill leachate	France	Not available	Full scale 50 m³/d	Ultrafiltration external
2.	Landfill leachate	Germany	COD removal 80%	Full scale 264 m³/d	Ultrafiltration external
3.	Landfill leachate	Germany	COD removal 90%	Full scale 250 m³/d	Ultrafiltration external
4.	Sludge digestion (anaerobic)	South Africa	Not available	Pilot scale 0.13 m³/d	Microfiltration external

a biofilm, inhabited commonly by a consortium of different bacterial species Microorganisms are immobilized on the membrane surface where the membrane concurrently serves to provide substrate transport from a liquid and/or gas. Here the membranes act as a dynamic substratum where the microbial biofilm can reduce oxyanions in the water. The concept of microbial MBRs is based on the integration of a bioreactor containing suspended biomass with an MF/UF process. The exopolysaccharide layer of the biofilm matrix also traps other exogenous substances, including nucleic acids, proteins, minerals, nutrients, and cell wall material, and protects cells against desiccation. A special type of MBfR is an aerated membrane biofilm reactor (MABR) in which the membrane serves additionally to provide oxygenation. Often the reactor outlines the utilization of microorganisms like *Geobacter* spp and *Rhodoferax*, and those are able to transport electrons directly to the anode by means of cytochromes present on the outer membrane and transform energy from a number of substrates to generate electrical energy.

7.4.5.2 System Configurations

MBRs are formed of two major components. The first is the biological component, which is formed because of the biodegradation of the waste components, and the membrane component, whose main function is for the physical separation of the water that has gone through treatment from the mixed liquor (Brindle et al. 2003). Membrane bioreactor systems can be differentiated into two main parts depending on their structure. The first part is usually known as the SMBR system. It is formed by the outer skin membranes that are present inside the bioreactor. The driving force through the membrane is received with the help of creating pressure on the bioreactor or creating some negative pressure on the permeate area (Zuthi et al. 2003). The process of cleaning the membrane is done with the help of permeate back pulsing in regular intervals and chemical backwashing on special occasions. A diffuser is normally kept straight below the membrane compartment for facilitating the scouring of the surface of the filtration. Various processes like mixing and aeration are also done with the help of this compartment. Anoxic or anaerobic compartments can be considered for enabling the simultaneous removal of biological nutrients.

The second compartment or structure is the external Membrane bioreactor. It contains the recirculation of the mixed liquor by a membrane compartment which is present just beyond the bioreactor. Both inner-skin and outer-skin membranes might be useful in this application (Naessens et al. 2003). The driven force is the pressure generated in high crossflow velocity along with the membrane surface. The structure of the recirculation and more and more resilient polymeric membranes in association with the reduced pressure requirement and increased permeate flux have created an acceleration all over the world for the commercial use of SMBRs (Naessens et al. 2003).

Different kinds and structures of membranes are usually used for Membrane bioreactor applications. These contain frame; hollow fiber; organic, which can include polyethylene, polysulfone, and others; metallic; rotary disk; inorganic or ceramic MF; tubular; plate; and UF membranes. The pore size in the membranes that are in use usually ranges from 0.01 to 0.4 micrometer. The fluxes received usually range from 0.05 to 10 micro-decimeters, and they largely depend on the membrane material and structure (Buisson et al. 2003). The normally occurring values for the inner-skin membranes usually range from 0.5 to 2.0 micro-decimeters, and for outer-skin membranes, the usually occurring membrane ranges from 0.2 to 0.6 micro-decimeter at 20 Co., and the trans-membranes pressure that is in application usually ranges from 20 to 500 kpa for inner-skin membranes and from −10 to −80 kpa for outer-skin membranes. The membranes that are usually applied in the MBR systems should also satisfy other requirements (Ciek et al. 2003). Different experiments and research work has been done on the selection of membrane components and structures and the application of different operating parameters; many types of useful research work, journals, papers, and the like that are of industrial relevance can be found (Drews et al. 2003).

7.5 CONCLUSION

MBR theory is almost the same to traditional wastewater treatment processes that occur in biology. However, in the process of separating the sludge that has undergone activation and wastewater that has undergone treatment. In the system of membrane bioreactors, this separation is performed by membrane filtration (Meng et al. 2003). However, in the traditional systems, it is performed as secondary clarification. Treating the MBR system gives us a greater degree of treating measurements in terms of solids that are in suspension and the removal of organic matter. Again, another process can be run in a nitrification or denitrification mode for the removal of nitrogenous compounds, which can be used in a combination with a coagulant in the process of removing phosphorus (Scholzy et al. 2003). The technology related to MBRs possesses the chance of having greater efficiency and capabilities and has a wide field of different applications. These includes solid waste digestion and municipal and industrial wastewater treatment. The large-scale systems are mainly used in operational purposes and in different parts of the world and also in substantial growth in the size and number of installations. This is assumed usually to be a viable method that can serve as an alternative to many of the wastewater treatment processes and their different difficulties, for example, water quality issues (Pillary et al. 2003). However, the MBR, a versatile technology, may be further developed for efficient, cost-effective and nontoxic way of decontamination of wastewater.

BIBLIOGRAPHY

Abashar, M.E.E., Elnashaie, S.S.E.H., Feeding of oxygen along the height of a circulating fast fluidized bed membrane reactor for efficient production of hydrogen, Chem. Eng. Res. Des. 85 (2007) 1529–1538.

Abass, O.K., Wu, X., Guo, Y., Zhang, K., Membrane bioreactor in China: A critical review, Int. J. Membr. Sci. Technol. 2 (2015) 29–47.

Abegglen, C., Joss, A., McArdell, C.S., Fink, G., Schlüsener, M.P., Ternes, T.A., Siegrist, H., The fate of selected micropollutants in a single-house MBR, Water Res. 43 (2009) 2036–2046.

Anderson, M.A., Gieselmann, M.J., Xu, Q., Titania and alumina ceramic membranes, J. Membrane Sci. 39 (1988) 243.

Anderson, M.A., Lechuga, F.T., Xu, Q., Hill, C.G., Catalytic ceramic membranes and membrane reactors, ACS Symp. Ser. 437 (1990) 198.

at IFAT Munich, Germany, 2008.

Azrague, K., Puech-Costes, E., Aimar, P., Maurette, M.T., Benoit-Marquie, F., Membrane photoreactor (MPR) for the mineralisation of organic pollutants from turbid effluents, J. Membrane Sci. 258 (2005) 71–77.

Basile, A., Gallucci, F., Membranes for membrane reactors (Preparation, Optimization and Selection) II, 10.1002/9780470977569(). doi:10.1002/9780470977569.

Belleville, M.P., Lozano, P., Iborra, J.L., Rios, G.M., Preparation of hybrid membranes for enzymatic reaction. Sep Purif Technol. 25 (2001) 229–233.

Bhabe, R.R., Inorganic Membranes: Synthesis, Characteristics and Applications, Reinhold, New York, 1991.

Bishop, B.A., Lima, F.V., Modeling, simulation, and operability analysis of a nonisothermal, countercurrent, polymer membrane reactor, Processes. 8 (2020) 78. https://doi.org/10.3390/pr8010078

Bouju, H., Buttiglieri, G., Malpei, F., Comparison of pharmaceutical substances removal between a full scale conventional activated sludge process and a MBR pilot plant, Proceedings of MBR-Network workshop, Berlin, 2009.

Brepols, C., Large Scale Membrane Bioreactors for Municipal Wastewater Treatment, IWA Publishing, London, UK, 2011.

Brindle, K., Stephenson, T., The application of membrane biological reactors for the treatment of wastewater, Biotechnology and Bioengineering. 49(6) (1996) 601–610.

Buisson, H., Cote, P., Praderie, M., Paillard, H., The use of immersed membranes for upgrading wastewater treatment plants, Water Science and Technology. 37(9) (1998) 89–95, 1998.

Cannon, K.C., Hacskaylo, J.J., Evaluation of palladium impregnation on the performance of a Vycor glass catalytic membrane reactor, J Membrane Sci. 65 (1992) 259.

Catalytica®, Catalytic membrane reactors: Concepts and applications, Catalytica Study N. 4187 MR, 1988.

Cejka, J., Van Corna, H., Corma, A., Schuth, F., Introduction to Zeolite Science and Practice, Studies in Surface Science and Catalysis, Elsevier.

Champagnie, A.M., Tsotsis, T.T., Minet, R.G., Webster, I.A., A high temperature catalytic membrane reactor for ethane dehydrogenation, Chem. Eng. Sci. 45 (1990) 2423.

Chan, Y.J., Chong, M.F., Law, C.L., et al., A review on anaerobic – aerobic treatment of industrial and municipal wastewater. Chem Eng J. 155 (2009) 1–18. https://doi.org/10.1016/j.cej.2009.06.041

Charcosset, C., Membrane processes in biotechnology: An overview. 24(5) (2006) 482–492. doi:10.1016/j.biotechadv.2006.03.002

Chin, S.S., Lim, T.M., Chiang, K., Fane, A.G., Factors affecting the performance of a low-pressure submerged membrane photocatalytic reactor, Chem. Eng. J. 130 (2007) 53–63.

Choo, K.-H., Tao, R., Kim, M.-J., Use of a photocatalytic membrane reactor for the removal of natural organic matter in water: Effect of photoinduced desorption and ferrihydrite adsorption, J. Membrane Sci. 322 (2008) 368–374.

Choubert, J.-M., Pomiès, M., Martin Ruel, S., Coquery, M., Influent concentrations and removal performances of metals through municipal wastewater treatment processes, Water Sci. Technol. 63 (2011) 1967–1973.

Ciek, N., A review of membrane bioreactor and their potential application in the treatment of agricultural wastewater, Canadian Biosystems Engineering. 45(6) (2003) 37–47.

Comparison of kinetic resolution between two racemic ibuprofen esters in an enzymic membrane reactor, Process Biochemistry. 40(7) (2005) 2417–2425. ISSN 1359–5113. https://doi.org/10.1016/j.procbio.2004.09.014

Comte, S., Guibaud, G., Baudu, M., Biosorption properties of extracellular polymeric substances (EPS) resulting from activated sludge according to their type: Soluble or bound, Proc. Biochem. 41 (2006) 815–823.

Cote, P., Buisson, H., Pound, C., Arakaki, G., Immersed membrane activated sludge for the reuse of municipal wastewater, Desalination. 113(2–3) (1997) 189196.

Deng, L., Guo, W., Ngo, H.H., Zhang, J., Liang, S., Xia, S., Zhang, Z., J, L., A comparison study on membrane fouling in a sponge-submerged membrane bioreactor and a conventional membrane bioreactor, Biores. Technol. 165 (2014) 69–74.

Deshmukh, S.A.R.K., Heinrich, S., Mörl, L., van Sint Annaland, M., Kuipers, J.A.M., Membrane assisted fluid-ized bed reactors: Potentials and hurdles, Chem. Eng. Sci. 62 (2007) 416–436.

Deshmukh, S.A.R.K., Laverman, J.A., Cents, A.H.G., Van Sint Annaland, M., Kuipers, J.A.M., Development of a membrane-assisted fluidized bed reactor. 1. Gas phase back-mixing and bubble-to-emulsion phase mass transfer using tracer injection and ultrasound experiments, Ind. Eng. Chem. Res. 44 (2005a) 5955–5965.

Deshmukh, S.A.R.K., Laverman, J.A., Van Sint Annaland, M., Kuipers, J.A.M., Development of a membrane-assisted fluidized bed reactor. 2. Experimental demonstration and modeling for the partial oxidation of methanol, Ind. Eng. Chem. Res., 44 (2005b) 5966–5976.

de Souza Figueiredo, K.C., Martins Salim, V.M., Borges, C.P., Synthesis and characterization of a catalytic membrane for pervaporation-assisted esterification reactors, Catal. Today. 133–135 (2008) 809–814.

DeWever, H., Weiss, S., eemtsma, T., Vereecken, J., Muller, J., Knepper, T., Rorden, O., Gonzalez, S., Barcelò, D., Hernando, M.D., Comparison of sulfonated and other micropollutants removal in membrane bioreactor and conventional wastewater treatment, Water Res. 41 (2007) 935–945.

Di Fabio, S., Lampis, S., Zanetti, L., Cecchi, F., Fatone, F., Role and characteristics of problematic biofilms within the removal and mobility of trace metals in a pilotscale membrane bioreactor, Process Biochem. 48 (2013) 1757–1766.

Di Fabio, S., Malamis, S., Katsou, E., Vecchiato, G., Cecchi, F., Fatone, F., Are centralized MBRs coping with the current transition of large petrochemical areas?

Di Fabio, S., Membrane Bioreactors for Advanced Treatment of Wastewaters from a Large Petrochemical Industrial Area, University of Verona, Verona, Italy, 2011.

Drews, A., Membrane fouling in membrane bioreactors – characterisation, contradictions, cause and cures, J. Membr. Sci. 363 (2010) 1–28.

Fatone, F., Eusebi, A.L., Pavan, P., Battistoni, P., Exploring the potential of membrane bioreactors to enhance metals removal from wastewater: Pilot experiences, Water Sci. Technol. 57 (2008) 505–511.

Fogler, H.S., Elements of Chemical Reaction Engineering, Prentice Hall, Upper Saddle River, 2006.

Fritsch, D., Bengtson, G., Development of catalytically reactive porous membranes for the selective hydrogenation of sunflower oil, Catal. Today. 118 (2006) 121–127.

Fuertes, A.B., Centeno, T.A., Preparation of supported asymmetric carbon molecular sieve membranes, J. Membrane Sci. 144 (1998) 105–111.

Gabarrón, S., Dalmau, M., Porro, J., Rodriguez-Roda, I., Comas, J., Optimization of full-scale membrane bioreactors for wastewater treatment through a model-based approach, Chem. Eng. J. 267 (2015) 34–42.

Gallucci, F., Basile, A., Hai, F.I., Introduction: A review of membrane reactors, in: A. Basile, F. Gallucci (Eds.), Membranes for Membrane Reactors: Preparation, Optimization and Selection (pp. 1–61), John Wiley & sons, United Kingdom, 2011.

Gallucci, F., Van Sint Annaland, M., Kuipers, J.A.M., Autothermal reforming of methane with integrated CO2 capture in a novel fluidized bed membrane reactor. Part 1: Experimental demonstration, Topics in Catal. 51 (2008a) 133–145.

Gallucci, F., Van Sint Annaland, M., Kuipers, J.A.M., Autothermal reforming of methane with integrated CO2 capture in a novel fluidized bed membrane reactor. Part 2: Comparison of reactor configurations, Topics in Catal. 51 (2008b) 146–157.

Gieselmann, M.J., Anderson, M.A., Moosemiller, M.D., Hill, CC., Physico-chemical properties of supported and unsupported (Y-A & O3 and TiOz ceramic membranes, Sep. Sci. Technol. 23 (1988) 1695.

Giorno, L., De Bartolo, L., Drioli, E., Membrane bioreactors for biotechnology and medical applications, in: D. Bhattacharyya, D.A. Butterfield (Eds.), New Insights into Membrane Science and Technology: Polymeric and Biofunctional Membranes. Elsevier, 2003. Chap 9.

Giorno, L., Drioli, E., Carvoli, G., Cassano, A., Donato, L., Study of an enzyme membrane reactor with immobilized fumarase for production of L-malic acid, Biotechnol. Bioeng. 72 (2001) 77–84. https://doi.org/10.1002/1097-0290(20010105)72:1<77::AID-BIT11>3.0.CO;2-L

Grant, S.B., Ambrose, R.F., Deletic, A., Brown, R., Jiang, S.C., Rosso, D., Cooper, W.J., Marusic, I., Saphores, J.D., Feldman, D.L., Hamilton, A.J., Fletcher, T.D., Cook, P.L.M., Stewardson, M., Sanders, B.F., Levin, L.A., Taking the "waste" out of, Science. 337 (2012) 681–686.

Gryaznov, V.M., Gulyanov, S.G., Serov, Yu, M., Yagodovskii, V.D., The hydrogenation of COz on a nickel-coated palladium-ruthenium membrane catalyst, Russ. J. Phys. Chem., 55 (1981) 730.

Gryaznov, V.M., Mischenko, A.P., Sarylova, M.E., Catalyst for cyclization of pentadiene-1,3 into cyclopentere andcyclopentane, GB Pat. 2 (1987) 187, 758A.

Gryaznov, V.M., Smirnov, V.S., Slinko, M., Binary palladium alloys as selective membrane catalyst, in: G.C. Bond, P.B. Wells, F.C. Tompkins (Eds.), Proc. 6th Int. Congr. Catalysis, Vol. 2, The Chemical Society, London, 894.

Hadjiev, D., Dimitrov, D., Martinov, M., Sire, O., Enhancement of the bio lm formation on polymeric supports by surface conditioning, Enzyme and Microbial Technology. 40(4) (2007) 840–848.

He, S.-B., Xue, G., Kong, H.-N., Zeolite powder addition to improve the performance of submerged gravitation-filtration membrane bioreactor, J. Environ. Sci. 18 (2006) 242–247.

Holloway, R.W., Achilli, A., Cath, T.Y., The osmotic membrane bioreactor: A critical review, Environ. Sci.: Water Res. Technol. 1 (2015) 581–605.

Huang, L., Lee, D.-J., Membrane bioreactor: a mini review on recent R & D works, Bioresour. Technol. (2015).

Huang, X., Meng, Y., Liang, P., Qian, Y., Operational conditions of a membrane filtration reactor coupled with photocatalytic oxidation, Sep. & Pur. Techn. 55 (2007) 165–172.

Huisjes, E., Colombel, K., Lesjean, B., The European MBR market: Specificities and future trends, Final MBR-Network workshop Salient outcomes of the European projects on MBR technology, Berlin, Germany, 2009.

Itoh, N., A membrane reactor using palladium, AIChE J. 33 (1987) 1576.

Itoh, N., Simulation of a bifunctional palladium membrane reactor, J. Chem. Eng. Jpn. 23 (1990) 81.

Itoh, N., Govind, R., Development of a novel oxidative palladium membrane reactor, AIChE Symp. Ser. 268 (1989) 10.

Itoh, N., Shindo, Y., Harya, K., Ideal flow models for palladium membrane reactors, J. Chem. Eng. Jpn. 23 (1990) 420.

Itoh, N., Xu, W., Haraya, K., Basic experimental study on palladium membrane reactors. J. Membrane Sci. 66 (1992) 149.

Jegatheesan, V., Pramanik, B.K., Chen, J., Navaratna, D., Chang, C.Y., Shu, L., Treatment of textile wastewater with membrane bioreactor: A critical review, Bioresour. Technol., 204 (2016) 202–212.

Joss, A., Zabczynski, S., Göbel, A., Hoffmann, B., Löffler, D., McArdell, C.S., Ternes, T.A., Thomsen, A., Siegrist, H., Biological degradation of pharmaceuticals in municipal wastewater treatment: proposing a classification scheme, Water Res. 40 (2006) 1686–1696.

Judd, S., The MBR Book: Principles and Applications of Membrane Bioreactors in Water and Wastewater Treatment, Elsevier, Oxford, 2006.

Kadic, E., Heindel, T.J., An introduction to bioreactor hydrodynamics and gas-liquid mass transfer (Kadic/an introduction to bioreactor hydrodynamics and gas-liquid mass transfer), Bubble Column Bioreactors. (2014) 10.1002/9781118869703(), 124–167. doi:10.1002/9781118869703.ch7

Kameyama, T., Dokiya, M., Fujishige, M., Yokokawa, H., Fukuda, K., Possibility for effective production of hydrogen from hydrogen sulfide by means of a porous Vycor glass membrane, Ind. Eng. Chem. Fundam. 20 (1981) 97.

Kameyama, T., Dokiya, M., Fujishige, M., Yokokawa, H., Fukuda, K., Production of hydrogen from hydrogen sulfide by means of selective diffusion membrane, Int. J. Hydrogen Energy. 8 (1983) 5.

Katsou, E., Malamis, S., Cecchi, F., Fatone, F., Fate and removal of trace metals from urban wastewater by membrane bioreactors: pilot and full-scale experiences (Chapter 11 membrane technologies for water treatment: removal of toxic trace elements), in: A. Figoli, J. Hoinkis, J. Bundschuh (Eds.), Sustainable Water Developments, CRC Press, 6000 Broken Sound Parkway NW, Suite 300, Boca Raton, FL 33487–2742, 2016.

Katsou, E., Malamis, S., Mamais, D., Bolzonella, D., Fatone, F., Occurrence, fate and removal of PAHs and VOCs in WWTPs using activated sludge processes and membrane bioreactors: Results from Italy and Greece, in: A. Forsgren (Ed.), Wastewater Treatment: Occurrence and Fate of Polycyclic Aromatic Hydrocarbons (PAHs), CRC Press, Boca Raton, FL, 2015.

Khulbe, K.C., Feng, C.Y., Matsuura, T., Synthetic Polymeric Membranes, Characterization by Atomic Force Microscopy, Springer, 2007.

Kraemer, J.T., Menniti, A.L., Erdal, Z.K., Constantine, T.A., Johnson, B.R., Daigger, G.T., Crawford, G.V., Crawford practitioner's perspective on the application and research needs of membrane bioreactors for municipal wastewater treatment, Bioresour. Technol. 122 (2012) 2–10.

Krzeminski, P., van der Graaf, J.H.J.M., van Lier, J.B., Specific energy consumption of membrane bioreactor (MBR) for sewage treatment, Water Sci. Technol. 65 (2012) 380–394.

KYA, Determination of measures, conditions and procedures for the reuse of treated wastewater and other provisions, in: KYA (Ed.) FEK 354/B'/8.3.2011, Greece, 2011.

Larbot, A., Fabre, J.P., Guizard, C., Cot, L., Inorganic membranes obtained by sol-gel techniques, J. Membrane Sci. 39 (1988) 203.

Lee, D.S., Jeon, C.O., Park, J.M., Biological nitrogen removal with enhanced phosphate uptake in a sequencing batch reactor using single sludge system, Water Res. 35 (2001) 3968–3976.

Leenaars, A.F.M., Burggraaf, A.J., The preparation and characterization of alumina membranes with ultra-fine pores, J. Colloid Interface Sci. 105 (1985) 27.

Leenaars, A.F.M., Keizer, K., Burggraaf, A.J., The preparation and characterization of alumina membranes with ultra-fine pores, J. Mater. Sci. 19 (1984) 1077.

Lesjean, B., Gnirss, R., Vocks, M., Luedicke, C., Does MBR represent a viable technology for advanced nutrients removal in wastewater treatment of small communities? in: EWA/JSWA/WEF – Proceedings of the 3rd Joint Specialty Conference Sustainable Water Management in response to 21st century pressures.

Lesjean, B., Tazi-Pain, A., Thaure, D., Moeslang, H., Buisson, H., Ten persistent myths and the realities of membrane bioreactor technology for municipal applications, Water Sci. Technol. 63 (2011) 32–39.

Lewandowski, Z., Boltz, J., Biofilms in water and wastewater treatment, in: Treatise on Water Science, Academic, Oxford, 2011, pp. 529–570.

Lewis, F.A., Kandasami, K. Baranowski, B., The uphill diffusion of hydrogen-strain-gradient induced effects in palladium alloy membranes, Platinum Met. Rev. 32 (1988) 22.

Li, C., Cabassud, C., Guigui, C., Evaluation of membrane bioreactor on removal of pharmaceutical micropollutants: a review, Desalination Water Treat. 55 (2015) 845–858.

Lin, Y.S., Microporous and dense inorganic membranes: Current status and prospective, Sep. Purif. Tech. 25 (2001) 39–55.

Lopez, J.L., Matson, S.L., A multiphase/extractive enzyme membrane reactor for production of diltiazem chiral intermediate, Journal of Membrane Science. 125(1) (1997) 189–211. ISSN 0376–7388. https://doi.org/10.1016/S0376-7388(96)00292-X.

Luo, Y., Guo, W., Ngo, H.H., Nghiem, L.D., Hai, F.I., Zhang, J., Liang, S., Wang, X.C., A review on the occurrence of micropollutants in the aquatic environment and their fate and removal during wastewater treatment, Sci. Total Environ. (2014) 473–474.

Mahecha-Botero, A., Boyd, T., Gulamhusein, A., Comyn, N., Lim, C.J., Grace, J.R., Shirasaki, Y., Yasuda, I., Pure hydrogen generation in a fluidized-bed membrane reactor: Experimental findings, Chem. Eng. Sci. 63 (2008) 2752–2762.

Malamis, S., Andreadakis, A., Mamais, D., Noutsopoulos, C., Can strict water reuse standards be the drive for the wider implementation of MBR technology?, Desalination Water Treat. 53 (2015) 3303–3308.

Mansell, B., Schroeder, E., Biological denitrification in a continuous flow membrane reactor, Water Science and Technology, 38(1) (1998) 9–14.

Marcano, J.G.S., Tsotsis, T.T. Catalytic Membranes and Membrane Reactors, Wiley-VCH, Weinheim, 2002.

Martinov, M., Hadjiev, D., Vlaev, S., Gas: Liquid dispersion in a "brous "xed bed bio"lm reactor at growth and non-growth conditions, Process Biochemistry. 45(7) (2010) 1023–1029.

McLeary, E.E., Jansen, J.C., Kapteijn, F., Zeolite based films, membranes and membrane reactors: Progress and prospects, Microp. & Mes. Mat. 90 (2006) 198–220.

Meng, F., Chae, S.-R., Drews, A., Kraume, M., Shin, H.-S., Yang, F., Recent advances in membrane bioreactors (MBRs): Membrane fouling and membrane material, Water Res. 43 (2009) 1489–1512.

Meng, F., Chae, S.-R., Shin, H.-S., Yang, F., Zhou, Z., Recent advances in membrane bioreactors: Configuration development, pollutant elimination, and sludge reduction, Environ. Eng. Sci. 29 (2011) 139–160.

Metcalf, E. (ed.), Wastewater Engineering: Treatment and Reuse, 4th edition, McGraw-Hill Companies, Inc., New York, 2003.

Minet, R.G., Vasileindis, S.P., Tsotsis, T.T., Experimental studies of a ceramic membrane reactor for the steam/methane reaction at moderate temperatures (400–7OO"C), ACS Symp. Ser. 37 (1992) 245.

Molinari, R., Caruso, A., Poerio, T., Direct benzene conversion to phenol in a hybrid photocatalytic membrane reactor, Catal. Today. 144 (2009) 81–86.

Mozia, S., Morawski, A.W., Toyoda, M., Inagaki, M., Effectiveness of photodecomposition of an azo dye on a novel anatase-phase TiO_2 and two commercial photocatalysts in a photocatalytic membrane reactor (PMR), Sep. & Pur. Tech. 63 (2008) 386–391.

Mozia, S., Morawski, A.W., Toyoda, M., Tsumura, T., Effect of process parameters on photodegradation of Acid Yellow 36 in a hybrid photocatalysis: Membrane distillation system, Chem. Eng. J. 150 (2009) 152–159.

Mutamim, N.S.A., Noor, Z.Z., Hassan, M.A.A., Olsson, G., Application of membrane bioreactor technology in treating high strength industrial wastewater: A performance review, Desalination. 305 (2012) 1–11.

Naessens, W., Maere, T., Nopens, I., Critical review of membrane bioreactor models: Part 1: biokinetic and filtration models, Bioresour. Technol. 122 (2012) 95–106.

Naessens, W., Maere, T., Ratkovich, N., Vedantam, S., Nopens, I., Critical review of membrane bioreactor models – Part 2: Hydrodynamic and integrated models, Bioresour. Technol. 122 (2012) 107–118.

Nagamoto, H., Inoue, H., Analysis of mechanism of ethylene hydrogenation by hydrogen permeating palladium membrane, J. Chem. Eng. Jpn. 14 (1981) 377.

Nagamoto, H., Inoue, H., A reactor with catalytic membrane permeated by hydrogen, Chem. Eng. Commun. 34 (1985) 315.

Nagamoto, H., Moue, H., The hydrogenation of 1,3- butadiene over a palladium membrane, Bull. Chem. Sot. Jpn. 59 (1986) 3935.

Nakajima, M., Watanabe, A., Jimbo, N., Nishizawa, K., Nakao, S., Forcedflow bioreactor for sucrose inversion using ceramic membrane activated by silanization, Biotechnol Bioeng. 33 (1989) 856.

Ng, H.Y., Tan, T.W., Ong, S.L., Membrane fouling of submerged membrane bioreactors: Impact of mean cell residence time and the contributing factors, Environ. Sci. Technol. 40 (2006) 2706–2713.

Nguyen, L.N., Hai, F.I., Kang, J.G., Price, W.E., Nghiem, L.D., Removal of trace organic contaminants by a membrane bioreactor-granular activated carbon (mbrgac) system, Bioresour. Technol. 113 (2012) 169–173.

Novotny, V., Water and energy in the cities of the future: Achieving net zero carbon and pollution emissions footprint, Proceedings of the IWA World Water Congress and Exhibition, Montréal, Canada, 2010.

Okubu, T., Haruta, K., Kusakabe, K., Morooka, S., Anzai, H., Akiyama, S., Equilibrium shift of dehydrogenation at short space-time with hollow fiber ceramic membrane, Ind. Eng. Chem. Res. 30 (1991).

Okubu, T., Watanabe, M., Kusakabe, K., Morooka, S., Preparation ofγ-alumina thin membrane by solgel processing and its characterization by gas permeation, J. Mater. Sci. 25 (1990) 4822.

Olano-Martin, E., Mountzouris, K., Gibson, G., Rastall, R., Continuous production of pectic oligosaccharides in an enzyme membrane reactor, Journal of Food Science. 66 (2001) 966–971. https://doi.org/10.1111/j.1365-2621.2001.tb08220.x

Ozdemir, S.S., Buonomenna, M.G., Drioli, E., Catalytic polymeric membranes: Next term Preparation and application, Appl. Catal. A: Gen. 307 (2006) 167–183.

Pillary, V.L., Townsed, B., Buckley, C.A., Improving the performance of anaerobic digesters at wastewater treatment works: The coupled cross-flow microfiltration/digester process. Water Science and Technology. 30(12) (1994) 329337.

Prokopiev, S.I., Aristov, Y.I., Parmon, V.N., Girodano, N., Intensification of hydrogen production via methane reforming and the optimization of HZ:CO ratio in a catalytic reactor with a hydrogen permeable wall, Int. J. Hydrogen Energy. 17 (1992) 275.

Radjenovic, J., Petrovic, M., Barceló, D., Fate and distribution of pharmaceuticals in wastewater and sewage sludge of the conventional activated sludge (CAS) and advanced membrane bioreactor (MBR) treatment, Water Res. 43 (2009) 831–841.

Rahimpour, M.R., Enhancement of hydrogen production in a novel fluidized-bed membrane reactor for naphtha reforming, Int. J. Hydrogen En. 34 (2009a) 2235–2251.

Remy, M., van der Marel, P., Zwijnenburg, A., Rulkens, W., Temmink, H., Low dose powdered activated carbon addition at high sludge retention times to reduce fouling in membrane bioreactors, Water Res. 43 (2009) 345–350.

Rezaei, M., Mehrnia, M.R., The influence of zeolite (clinoptilolite) on the performance of a hybrid membrane bioreactor, Biores. Technol. 158 (2014) 25–31.

Santos, A., Ma, W., Judd, S.J., Membrane bioreactors: two decades of research and implementation, Desalination. 273 (2011) 148–154.

Schmitz, J., Lucke, L., Herzog, F., Glaubitz, D., Permeation membranes for the production of hydrogen at high temperatures, in: T.N. Veziroglu, A.N. Protsenko (Eds.), Hydrogen Energy Progress VII, Vol. 2, Int. Assoc. Hydrogen Energy, Pergamon, Oxford, 1988, pp. 8, 19–830.

Scholes, E., Verheyen, V., Brook-Carter, P., A review of practical tools for rapid monitoring of membrane bioreactors, Water Res. 102 (2016) 252–262.

Scholzy, W., Fuchs, W., Treatment of oil contaminated wastewater in a membrane bioreactor, Water Research. 34(14) (2000) 3621–3629.

Seidel-Morgenstern, A., Membrane Reactors, Wiley-WCH, Weinheim, 2010.

Skouteris, G., Saroj, D., Melidis, P., Hai, F.I., Ouki, S., The effect of activated carbon addition on membrane bioreactor processes for wastewater treatment and reclamation – a critical review, Bioresour. Technol. 185 (2015) 399–410.

Sun, F., Sun, B., Hu, J., He, Y., Wu, W., Organics and nitrogen removal from textile auxiliaries wastewater with A2O-MBR in a pilot-scale, J. Hazard. Mater. 286 (2015) 416–424.

Sun, S., Rebeilleau-Dassonneville, M., Zhu, X., Chu, W., Yang, W., Ammonia oxidation in Ba0.5Sr0.5Co0.8Fe0.2O3-d membrane reactor, Catal. Today, in press, 2009.

Sun, Y.M., Khang, S.J., Catalytic membrane for simultaneous chemical reaction and separation applied to a dehydrogenation reaction, Ind. Eng. Chem. Res. 27 (1988) 1136.

Tai, C., Snider-Nevin, J., Dragasevich, J., Kempson, J., Five years operation of a decentralized membrane bioreactor package plant treating domestic wastewater, Water Pract. Technol. 9 (2014) 206–214.

Téllez, C., Menédez, M., Santamarı´a, J., Kinetic study of oxidative dehydrogenation of butane on V/MgO catalysts, J. Catalysis. 183 (1999) 210221.

Tsuru, T., Membrane reactors. AccessScience, 2012. Retrieved July 18, 2021, from https://doi.org/10.1036/1097-8542.YB120346

Tsuru, T., Kan-no, T., Yoshioka, T., Asaeda, M., A photocatalytic membrane reactor for VOC decomposition using Pt-modified titanium oxide porous membranes, J. Membrane Sci. 280 (2006) 156–162.

van Haandel, A.C., van der Lubbe, J., Design and optimisation of activated sludge systems, in: IWA (Ed.), Handbook of Biological Wastewater Treatment, 2012.

Van Veen, H.M., Bracht, M., Hamoen, E., Alderliesten, P.T., Feasibility of the application of porous inorganic gas separation membranes in some large-scale chemical processes, in: A.J. Burggraaf, L. Cot (Eds.), Fundamentals of Inorganic Membrane Science and Technology (vol. 14, pp. 641–681), Elsevier, 1996.

Visvanathan, C., Benamin, R., Parameshwaran, K., Membrane separation bioreactors for wastewater treatment, Critical Reviews in Environmental Science and Technology. 30(1) (2000) 1–48.

Wang, X., Chang, V.W.C., Tang, C.Y., Osmotic membrane bioreactor (OMBR) technology for wastewater treatment and reclamation: Advances, challenges, and prospects for the future, J. Membr. Sci. 504 (2016) 113–132.

Wang, Z., Ma, J., Tang, C.Y., Kimura, K., Wang, Q., Han, X., Membrane cleaning in membrane bioreactors: A review, J. Membr. Sci. 468 (2014) 276–307.

Weiss, S., Reemtsma, T., Membrane bioreactors for municipal wastewater treatment: A viable option to reduce the amount of polar pollutants discharged into surface waters?, Water Res. 42 (2008) 3837–3847.

Wood, B.J., Dehydrogenation of cyclohexane on a hydrogen-porous membrane, J. Catal., 11 (1968) 30.

Wu, J., Chen, F., Huang, X., Geng, W., Wen, X., Using inorganic coagulants to control membrane fouling in a submerged membrane bioreactor, Desalination. 197 (2006) 124–136.

Xia, Y., Lu, Y., Kamata, K., Gates, B., Yin, Y., Macroporous Materials Containing Threedimensionally Periodic Structures, Chemistry of Nanostructured Materials, P. Yang (Ed.), World Scientific, 2003, pp. 69–100.

Young, T., Smoot, S., Peeters, J., Côté, P., Cost-effectiveness of membrane bioreactors treatment system for low-level phosphorus reduction from municipal wastewater, Water Pract. Technol. 9 (2014) 316–323.

Yusuf, Z., Abdul Wahab, N., Sahlan, S., Fouling control strategy for submerged membrane bioreactor filtration processes using aeration airflow, backwash, and relaxation: A review, Desalin. Water Treat. 57 (2016) 17683–17695.

Zaman, J., Chakma, A., Inorganic Membrane Reactors. 92(1) (1994) 1–28. doi:10.1016/0376-7388(94)80010-3

Zaspalis, V.T., van Praag, W., Keizer, K., van Ommen, J.G., Ross, J.R.H., Burggraaf, A.J., Reactions of methanol over alumina catalytically active membranes modified by silver, Appl. Catal. 74 (1991) 235.

Zaspalis, V.T., van Praag, W., Keizer, K., van Ommen, J.G., Ross, J.R.H., Burggraaf, A.J., Reactor studies using alumina separation membranes for the dehydrogenation of methanol and n-butane, Appl. Catal., 74 (1991) 223.

Zaspalis, V.T., van Praag, W., Keizer, K., van Ommen, J.G., Ross, J.R.H., Burggraaf, A.J., Reactor studies using vanadia-modified titania and alumina catalytically active membranes for the reduction of nitrogen oxide with ammonia, Appl. Catal. 74 (1991) 249.

Zhao, R., Itoh, N., Govind, R., Experimental investigation of a novel oxidative membrane reactor for dehydrogenation reactions, Paper presented at the ACS Symp. New Catalytic Materials and Techniques, Miami, FL, September 10–15, 1989.

Zhao, R., Itoh, N., Govind, R., Novel oxidative membrane reactor for dehydrogenation reactions, ACS Symp. Ser. 437 (1990) 216.

Zheng, J.M., Sousa, J.M., Mendes, D., Madeira, L.M., Mendes, A., Theoretical analysis of conversion enhancement in isothermal polymeric catalytic membrane reactors, Catal. Today. 118 (2006) 228–236.

Zhong, T., Seok-Jhin, K., Reddy, G., Dong, J., Smirniotis, P., Modified zeolite membrane reactor for high temperature water gas shift reaction, Journal of Membrane Science. 354(1–2) (2010) 114–122. ISSN 0376–7388. https://doi.org/10.1016/j.memsci.2010.02.057.

Zhu, Y., Bioprocessing for value-added products from renewable resources. Immobilized Cell Fermentation for Production of Chemicals and Fuels. (2007) 373–396. doi:10.1016/B978-044452114-9/50015-3

Ziaka, Z.D., Minet, R.G., Tsotsis, T.T., Propane dehydrogenation in a packed bed membrane reactor, AlChE J. 39 (1993) 526.

Zuthi, M.F.R., Ngo, H.H., Guo, W.S., Zhang, J., Liang, S., A review towards finding a simplified approach for modelling the kinetics of the soluble microbial products (SMP) in an integrated mathematical model of membrane bioreactor (MBR), Int. Biodeterior. Biodegrad. 85 (2013) 466–473.

8 Extended Investigation Processes in Advanced Wastewater Treatment for Water Reuse

*Sanchita Patwardhan, Sachin Palekar,
Nilesh S. Wagh, and Jaya Lakkakula*

CONTENTS

8.1   INTRODUCTION

The innovation of entrenched wastewater treatment processes is a growing concern among environmentalists and biotechnologists regarding shrinking the exploitation of freshwater resources caused by humans in the interest of modernization [1]. Immense research has been conducted over the years to develop an optimum wastewater treatment technique to prevail over the voluminous contamination generated by humans *via* intensifying industrialization and the water contamination caused by it [2]. It has been indicated that 10 million tons of waste are discharged into the water bodies *via* several industrial and agricultural activities worldwide, out of which less than 10% of the waste is treated, whereas approximately 90% of the waste is released without any pretreatment, which is causing constant loss in biodiversity and leading to water scarcity. A sustainable shift toward wastewater management is mandatory, including wiser wastewater management systems and innovative technologies for conserving water bodies [3].

The use of membrane technology for water decontamination is a traditional method that has been used since the 18th century [4]. Henceforth, numerous experiments are performed, and large numbers of membrane techniques are discovered for removal of diverse organic and inorganic contaminants released into water bodies. Recently, microfiltration (MF), ultrafiltration (UF), nanofiltration and various osmosis techniques are used alone or in combination with other methods like flocculation, oxidation, electrocoagulation, and others [5]. Extensive industrial and agricultural activities lead to the generation of large amounts of lethal contaminants in the water bodies, like pharmaceutical pollutants, herbicides, pesticides, organic and inorganic dyes, and fatal microorganisms, which are dangerous for aquatic life as well as human health. Today, the variety of membranes

DOI: 10.1201/9781003165019-8

are engineered and altered for the selective and complete removal of contaminants from a given wastewater sample. Improvements in the filtration performance of membranes are performed on a large scale to generate potable water before making the water available to the public [6]. Membrane techniques are also applied for desalination, irrigation, municipal wastewater management, and sewage treatment. Membrane techniques act as a bridge between the economical and sustainable processes of wastewater management and are one of the favored options in recent times since wastewater treatment has become a capital challenge for the world.

This chapter discusses different types of membrane technologies used worldwide today for wastewater treatment, as summarized in Table 8.1. It highlights the development of new membrane technologies that can replace the conventional membrane processes in use. It also reviews the applicability of the membrane with its filtration efficiency and ease of use.

8.2 MF

MF is a well-established and extensively practiced filtration process that is used to remove several organic and inorganic pollutants from wastewater. In the MF process, contaminated water containing suspended solids is passed through the membrane of the pore size ranging from 0.1 to 10 μm. The use of MF membranes in wastewater treatment was primarily suggested by Frick by introducing the cellulose nitrate membrane in 1855. Furthermore, this process has been applied in many small-scale industries for wastewater treatment. Due to remarkable research and advancements in technology, scientists have introduced several modified MF membranes that have higher pollutant rejection potentials and numerous applications in various industries. MF membranes show extensive relevance for the removal of different dyes, heavy metal ions, herbicides, microbes, and more. An MF membrane impregnated with a mixture of nano clay, polyethylene glycol (PEG), and zeolite was constructed using a dry-pressing technique [7]. Field emission scanning electron microscopy (FE-SEM) was used to determine the physiological characteristics of the membrane. This membrane was applied for the efficient removal of cationic dyes from textile industrial wastewater. The membrane exhibited excellent stability as it degraded only 0.8% and 0.5% by weight in acidic and alkaline solutions, respectively. The negatively charged membrane containing 30% zeolite, rejected approximately 95.55% of crystal violet and about 90.23% of methylene blue at its optimum environmental conditions. However, a significant decrease in the filtration rate of the membrane was observed after three filtration cycles. Furthermore, the membrane was heated at 300°C for the complete elimination of the cationic dyes absorbed on the membrane. The membrane was then used for the next filtration cycle [7].

The use of a polyvinylidene fluoride (PVDF) MF membrane fabricated with cellulose nanofibers and Meldrum's acid is an eco-friendly and economically useful approach to wastewater management. Morphological characteristics of the membrane were determined using scanning electron microscopy (SEM) and transmission electron microscopy (TEM). Whereas, chemical aspects of the membrane were evaluated with X-ray diffraction (XRD) analysis and Fourier transform infrared spectroscopy (FTIR). For the experiment, a modified membrane with a total filtration area of 9.4 cm² and a pore size of 50–100 nm was subjected to a filtration process at 200 m-Hg feed pressure. This membrane exhibits great potential in water purification by rejecting about 99% crystal violet dye and nanoparticles like iron (III) oxide. Due to its high stability and high pollutant absorbance, it can be used in large-scale industries [8].

A polypropylene microfiltration membrane having an average pore diameter of >0.2 μm was studied for efficient removal of anionic dyes like Direct Red 2 and salt from industrial wastewater. A large pilot-scale setup was constructed, which was composed of a filtration membrane that was 30 mm thick and 700 mm long. The temperature and pressure of the feed solution were maintained using a temperature cutoff and a pressure gauge. The MF process was also carried out at different feed pH, and it was observed that at pH 7.5 dye removal was maximum. The membrane was easily cleaned using glycine and various salts. The experiment illustrates the use of this MF membrane on commercial scale due to its reusability, cost-effectiveness, and high filtration rate per unit area (~100%) [9].

An MF membrane made from a blend of polyurethane and cellulose acetate was manufactured to remove carcinogenic dye, namely, Direct Blue, from industrial wastewater. The MF membrane used in the experiment had a pore diameter of 0.86 μm and a thickness equal to 175 μm. The wastewater containing Direct Blue was mixed with cationic surfactant, that is, cethylpyridinium chloride (1:4 and 1:8) for complete removal of the dye from the wastewater. The formation of micelles on the MF membrane due to interaction between the cethylpyridinium chloride and dye led to 100% removal of Direct Blue. The membrane, unaided, is sufficient for complete purification of dyes from contaminated water and thus proves to be an economical method for wastewater treatment on a commercial scale [10].

An MF membrane fabricated with polyethylene glycol and tannic acid was introduced for the successful removal of rhodamine B from polluted water. Polyethylene glycol and tannic acid were mixed, which led to the formation of a suspension *via* the development of hydrogen bonds. Furthermore, the suspension was passed through a polyethersulfone (PES) membrane under vacuum. The morphological characteristics of the membrane was determined *via* SEM and FE-SEM. Polyethylene glycol allowed the successful interaction of Rhodamine B and the tannic acid present on the membrane whereas tannic acid enhanced the absorption rate of the pollutant. The membrane was able to retain approximately 98.9% of Rhodamine from water at high flux. The study demonstrates the applicability of the membrane on the commercial scale due to its rapid and high pollutant retention rate [11].

A microfiltration membrane made from polyamide was tailored using multiple layers containing chitosan (CHI) and polystyrene sulphonate (PSS). The membrane exhibited an excellent potential in the retention of lethal herbicides like atrazine (ATZ) from contaminated water with a high flux of 1.89 m^3/m^2day. The experimental studies suggested that as the number of CHI/PSS bilayer increases, ATZ retention rate also increases. Therefore, MF membranes fabricated with a 9 CHI/PSS bilayer showed a maximum absorbance of ATZ on the membrane, that is, 92.23%. It was observed that the filtration efficiency of the membrane was increased slightly by adding some ions (calcium), surfactant (SDS), and humic acid. Also, the filtration performance of the membrane was increased by 5% in presence of salt (NaCl). The higher filtration of ATZ herbicide by this MF membrane shows great potential to meet the requirement of industrial utilization [12].

Industrial textile wastewater was purified using an engineered ceramic MF membrane. The MF membrane was customized with natural magnesite using a couple of techniques, including uniaxial pressing and sintering process. A bench-scale dead-end MF experiment was carried out at room temperature, and the efficiency of the membrane was constantly determined for its turbidity and COD extraction rate. The characteristics of this eco-friendly and cost-effective membrane were evaluated using SEM. The mechanical strength of the membrane was maintained using a higher sintering temperature to offer a higher tensile strength to the membrane during the dead-end MF process. The membrane, with an average pore size of 1.12 μm, efficiently removed about 99.9% of turbidity and 69.7% of chemical oxygen demand (COD) from contaminated textile water. The use of a natural magnesite–decorated MF membrane is one of the cost-effective and green methods for purification of water [13].

A bilayered MF membrane made from polyacrylonitrile (PAN)/polyethylene terephthalate (PET) was drafted with cellulose nanofibers. The multifunctional membrane concurrently absorbed bacteria like *E. coli*, bacterial viruses like MS2, and heavy metal ions, including hexavalent chromium ions and lead ions. The physical properties of the membrane were evaluated using SEM. SEM results indicated that the PAN nanofibers had an average diameter of 200 ± 30 nm and membrane had pore size of about 0.66 μm. Due to its high permeability and large charge density, the membrane absorbed approximately 100% of bacteria, viruses, and Cr (VI) and Pb (II) from the given sample of water. However, the retention rate of these pollutants from the water sample is highly related to the pH of the solution/water sample used. It was observed that the retention rate of Cr (VI) was maximum at pH 4, whereas lead (II) was absorbed utmost at pH6. On the other hand, the adsorption of MS2 bacterial viruses was maximum at neutral pH due to the maximum electrostatic interaction between negatively charged surface of the pathogen and the positively charged membrane. The membrane exhibited great potential for the decontamination of drinking water on a large scale [14].

Crossflow MF of emulsified oil wastewater was performed using an MF membrane made up of PVDF. The PVDF membrane was placed in a filtration unit on the alumina support, and a centrifugal pump was used to circulate the feed solution into the filtration unit at room temperature. The membrane had an average pore diameter of 0.22 μm, and a filtration area of 19 cm² was employed for the filtration process. The maximum oil retention was observed at the isoelectric point of emulsified oil wastewater when the crossflow velocity was kept 1 m/s at unit bar pressure. However, cake formation on the membrane due to the deposition of oil droplets remains a disadvantage for its use on a large scale. The membrane was easily cleaned using 0.1% sodium dodecyl sulfate, but a significant reduction in oil droplets rejection rate was seen on the subsequent decrease in pH (acidic pH) [15].

A carbonized, coal-based MF membrane was employed for oil retention from synthetic oily polluted water. Commercially available ningxia coal was pulverized and then molded using a hydraulic extruder. The tube-shaped molded membrane was carbonized at very high temperatures for an hour to get the final carbonized MF membrane. The membranes were evaluated for their morphological characteristics via SEM. The total filtration area of the membrane was 0.0275 m². The membrane of different pore sizes (0.6, 1.0, and 1.4 μm) was checked for its oil extraction efficiency at different feed concentrations and atmospheric conditions. Among different membranes, the carbonized membrane with pore size of 0.1 μm successfully absorbed approximately 97% of the oil from the wastewater at 0.10 MPa transmembrane pressure. The membrane can meet the demands of large-scale industries due to its low cost and high oil rejection efficiency [16].

MF membranes such as PVDF and cellulose-ester membranes were employed for algal removal from the given sample of water *via* cross filtration process as observed in Figure 8.1. Contaminated water containing green algae was pre-ozonized to elevate the viability of the algal cells. Furthermore, this contaminated water was passed through the membrane at high crossflow velocity. The filtration

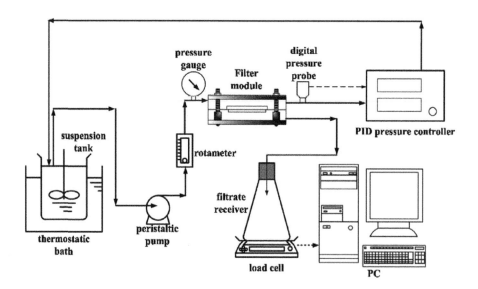

FIGURE 8.1 A schematic representation of the MF process equipment. Reprinted with permission from M.T. Hung and J.C. Liu, "Microfiltration for separation of green algae from water," *Colloids Surf B Biointerfaces*, 51 (2) 157–164, 2006. doi:10.1016/j.colsurfb.2006.07.003

membrane was placed in the tubular filtration module with a filtration area of 4 cm². The contaminated water containing an algal concentration of 13.9 mg/l was passed through the membranes with pore size of 0.22 μm. A transmembrane pressure of 60 kPa showed higher filtration efficiency with high fouling resistance due to the reduction in biomass loading onto the membrane. Out of the two MF membranes used, the PVDF membrane showed higher algal absorbance potential [17].

The electro-MF process was carried out using MF membranes made up of PES to remove humic substances from wastewater. In the experiment, the membrane was placed in between the anode (platinum) and the cathode (titanium), and the active filtration area was of 14.5 cm². The wastewater was passed through the MF membrane to remove humic substances due to the predictable electrochemical reactions. The MF process was carried out in both the presence and absence of the electric field. Due to the presence of high voltage across the membrane, a high rejection rate of some humic substances was observed. The absorbance of total organic carbon, trihalomethane formation potential (THMFP), and ultraviolet (UV) absorbance were also increased by about 50%. The use of electro-MF for removing organic contaminants from the wastewater can be beneficial due to its high decontamination potential and ease of the process [18].

A tube-shaped microfiltration membrane made up of ceramic material was employed for the removal of perfluorinated compounds (PFCs) including perfluorooctanoic acid (PFOA) and perfluorooctane sulfonate (PFOS). The microfiltration membrane with an average pore size of 0.1 μm was placed in between the cathode and an anode rod in a tubular support. An electro-MF process was carried out by passing the wastewater containing PFCs through an MF membrane with an active filtration area of 20.7 cm². The membrane successfully ejected approximately 70% of PFOA and PFOS from the given sample of water. The MF membrane also absorbed perfluorodecanoic acid, perfluorohexane sulfonate, and perfluorohexanoic acid and reduced dissolved organic carbon by about 80% from the industrial wastewater. The MF membrane was easy to clean using sodium hydroxide and methanol and was recycled almost by 100% without any deterioration in the filtration performance of the membrane [19].

The crossflow electro-MF process was carried out using an MF membrane made up of acrylic polymers. The polluted water was parallelly passed through the membrane of pore size equal to 0.2 μm, which was placed in between the two electrodes for the efficient absorption of chromium hydroxide on the filtration membrane. The surface charges were customized using a dispersant for increasing the rate of filtration by the membrane and for developing a resistance against fouling. The direct current electric field was generated in between the two electrodes with an active filtration area of 80 cm². A significant increase of 30% in the filtration rate was observed after applying voltage across the membrane. Crossflow electro-MF is used in many commercial industries for wastewater treatment for the removal of many other heavy metal ions [20].

Crossflow MF was performed using a cellulose acetate membrane having an average pore size of 0.2 μm. The cellulose acetate MF membrane with an effective filtration area of 28 cm² was placed in a filtration unit, and optimum environmental conditions were maintained using operating valves. The industrial wastewater was mixed with an absorber, that is, red mud for the efficient removal of phosphate ions. A dead-end MF process was carried out at room temperature and constant transmembrane pressure. Then the wastewater was passed through the MF membrane at slightly acidic conditions (pH 5.2) and at constant flux rate. The membrane successfully ejected about 100% of the phosphate ions from the wastewater. The membrane shows its high filtration potential with constant stability and fouling resistance, which can replace many other commercially available MF membranes for heavy metal ion removal [21].

A ceramic microfiltration membrane infused with the cathode and an anode was used for the degeneration of p-chloroaniline (PCA) from polluted industrial water. The method used in combination of low-pressure filtration and electrochemical oxidation for degrading PCA from wastewater. The physical and chemical properties of the membrane were evaluated using SEM and X-ray electron photoelectron spectroscopy, respectively. At the neutral pH, when the voltage of 2 V was applied across the membrane, it rejected approximately 87.1% of PCA by electrochemically oxidizing it and

about 45.2% of total organic carbon from industrial wastewater. Furthermore, the PCA degradation mechanism was evaluated using flow-through mode as well as flow-by mode. It was observed
that degradation efficiency for PCA was 3.6 times more using flow-through mode of filtration. This
experiment demonstrates the large-scale applicability of the electrochemical MF process on an
industrial scale due to its diverse advantages, including stability and ease of the process [22].

8.3 NANOFILTRATION

Nanofiltration is comparatively a modern membrane filtration process, which is extensively used
on the industrial scale to eliminate particulate material from the porous membrane. In the process
of nanofiltration, wastewater from various sources is passed through membrane filters of the pore
size of about 0.001 micron to get potable water. Today, the use of nanofiltration has increased on a
large scale due to its potential to remove a large range of pollutants from industrial wastewater at a
low cost.

A nanofiltration membrane made up of ceramic material was used to remove organic pollutants, including bisphenol A (BPA) and COD, from the biologically treated water. The membrane
with an effective filtration area of 0.1 m^2 was installed into the filtration system with a pressure
gauge, feed tank, and flowmeter. The biologically treated wastewater was prefiltered through an
MF membrane to remove undissolved solids from the given water. It was observed that when the
concentration of BPA in the water sample was kept low, the nanofiltration membrane was able to
absorb approximately 100% of BPA from the given water sample. However, when the concentration
of BPA was increased, significant reduction in the performance of the membrane was observed. It
removed approximately 61–75% of BPA from the water sample containing higher concentration of
BPA. Simultaneously, the membrane absorbed 40–60% COD and other particulate matter from the
biologically treated water [23].

Three nanofiltration membranes, namely, NF-97, NF-99, and DSS-HR98PP, composed of polyamides (PA) were used in the experiment for successfully eliminating phenol and phenolic compounds from the given sample of water. In the experiment, an INDEVEN flat membrane test module
was used to demonstrate the efficiency of the three nanofiltration membrane used in the experiment.

The water was tested using a spectrophotometer for the phenol concentration after passing it
through the nanofiltration membranes. Among the three membranes, DSS-HR98PP NF membrane
showed a maximum rejection rate of the phenol (~80%). The rejection rate of the DSS-HR98PP NF
membrane was not affected because of the change in the pH of the polluted water used. On the other
hand, a significant reduction in the performance of the membrane was observed on the decrease
in pH level. Due to its high filtration performance and stability, the HR98PP NF membrane can be
applied on an industrial scale to remove phenol from the water [24].

Five different commercial nanofiltration membranes, including NF270, HL, DL, DK, and LF10,
were examined for their filtration performance against modified agricultural polluted water. The
agricultural wastewater was mixed with synthetic carboxylic acids, including acetic acid and butyric
acid, which were determined using headspace gas chromatography. Out of the five hydrophobic NF
membranes, it was observed that LF10 demonstrated the highest efficiency of acetic acid (72.2%)
and butyric acid (~70%) absorption at a slightly alkaline pH. The NF270 membrane absorbed 52.6%
of acetic acid and 69.7% of butyric acid from the given sample of water. Furthermore, the agricultural wastewater treated with carboxylic acids was mixed with salts like calcium chloride and
calcium carbonate. It was observed that the addition of these salts in particular concentrations to the
agricultural water increased the carboxylic acid rejection rate. The study suggests that the LF10 and
NF270 membranes can be used in many industries for the successful retention of carboxylic acid
on a large scale [25].

A highly hydrophilic nanofiltration membrane with a negatively charged surface was used for
the elimination of different antibiotics and hormones from synthetic wastewater. The water sample
for the experiment was prepared by mixing humic acid and salts like NaCl and CaCl$_2$ in defined

amount of antibiotics and hormones. The water sample was passed through the membrane absent any transmembrane pressure. The membrane with active filtration area of 14.6 cm², successfully rejected 80% of tetracycline and 50% of doxycycline from the given sample of water. On the other hand, the adsorption of hormones and sulfanamides was comparatively less than the antibiotics [26].

A commercially available thin-film composite nanofiltration membrane (NF-300) was used to remove arsenic from the drinking water (Figure 8.2). The water sample was prepared using tap water and arsenic salts, including sodium arsenate (NaH_3AsO_4), sodium arsenate heptahydrate (Na_2HAsO_4), and sodium sulfate (Na_2SO_4). The membrane successfully rejected 99.80% of arsenate ions, as well as other contaminants, from the given water sample. NF membrane was also successful in reducing the turbidity of water sample. The concentration of the arsenic absorbed was determined using an atomic absorption spectrometer. The sample was passed through the membrane three times, and it was observed that the performance of the membrane drops by 3% after every cycle. Due to its low cost and easy operation method, the membrane can be used commercially for purifying drinking water on a large scale [27].

The interfacial polymerization method was used to prepare a polyamide nanofiltration membrane impregnated with silica nanoparticles for the process of desalination from contaminated oily water. Atomic force microscopy was used to determine the membrane structure. The images resulting from the atomic force microscopy showed that the membrane was rough, highly porous, and hydrophilic in nature. The quantitative analysis of the silica particles on the NF membrane was carried out using the X-ray photoelectron spectroscopy (XPS) analysis technique. The incorporation of silica nanoparticles on the NF membrane enhances the roughness and hydrophilic nature of the membrane, which, in turn, increases the salt adsorption rate. The membrane was able to retain about 50% of the salts from the oily wastewater. The result of the study illustrated that the membrane showed different adsorption rates for the different salts. The adsorption rate was maximum for sodium sulfate and minimum for sodium chloride ($Na_2SO_4 > MgSO_4 > MgCl_2 > NaCl$) [28].

A thin-film composite nanofiltration membrane was constructed to eliminate different salts from the water. The membrane was prepared by interfacial polymerization of piperazine (PIP) and 3,3_,5,5_-biphenyl tetraacyl chloride (mm-BTEC) monomer on polyacrylonitrile (PAN) membrane. Scanning electron microscopy and X-ray photoelectron spectroscopy were used to determine the physical and chemical morphology of the membrane respectively. The membrane exhibited high

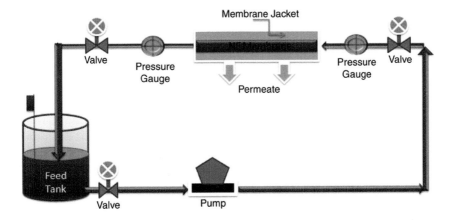

FIGURE 8.2 Illustration of the crossflow nanofiltration process to remove arsenate ions. Reprinted with permission from R.S. Harisha, K. M. Hosamani, R.S. Keri, S.K. Nataraj and T.M. Aminabhavi, "Arsenic removal from drinking water using thin film composite nanofiltration membrane," *Desalination*, vol. 252, no.1–3, pp. 75–80, 2010. doi:10.1016/j.desal.2009.10.022

salt rejection rate for all the salts $CaCl_2 > MgCl_2 > NaCl > Na_2SO_4$. However, the rejection rate was highest for calcium chloride (95.1%) salt under the pressure of 0.4 mPa [29].

Two nanofiltration membranes, NF1 and NF2, made up of PES were examined for their filtration efficiency at different transmembrane pressures (0.3 and 0.7 bars). Wastewater containing different pharmaceutical pollutants and salts was passed through the respective membrane at its optimum environmental conditions. Nanofiltration membranes showed greater adsorbance of bivalent ions compared to the monovalent ions. The NF1 membrane rejected 5–15% whereas NF2 rejected 25–35% of sodium chloride and removed 60% of naproxen and diclofenac from the wastewater. However, the rejection rate of carbamazepine and other pharmaceutical products was very low. As a result, further treatment of the water is necessary before it is used commercially [30].

A commercially available nanofiltration membrane was employed to eliminate chloride and sulfate ions from the polluted textile industrial water. The industrial wastewater containing chloride and sulfate ions was mixed with some organic solvents, including diisopropylamine (DIIPA), isopropylamine (IPA), and ethylamine (EA) for solventing out the desired ions in the form of precipitate. When a slightly alkaline wastewater sample (pH = 8) was passed through the NF membrane, 98% of the solvent was recovered. The precipitate obtained was then analyzed using FE-SEM, FTIR spectroscopy, and XRD analysis which indicated the presence of inorganic sulphates in range of 1000 to 1110 cm^{-1}. The membrane successfully removed 99.82% and 77.50% of the sulfate and chloride ions, respectively. An NF membrane can act as a best candidate for water purification in textile industries [31].

The interfacial polymerization method was operated on PES membrane to construct nanofiltration membranes impregnated with silica or polypiperazine. The surface morphology of the membrane was studied using SEM and atomic force microscopy. The chemical characteristics of the silica/polypiperazine-amide nanofiltration membrane were determined using attenuated total reflectance infrared. A thin-film composite polyamide NF membrane decorated with piperazine showed higher rejection of salts. The membrane with active filtration area of 75 cm^2 eliminated 97.3% of Na_2SO_4, 91.1% of $MgSO_4$, 50.7% of $MgCl_2$ and 50.7% of NaCl. It also removed color up to 99% from the given water sample at neutral pH [32].

Three commercial nanofiltration membranes namely, TFC-SR2, NF270, and MPS-34 were applied to remove pharmaceutical products, such as acetaminophen, caffeine, diazepam, diclofenac, ibuprofen, naproxen, sulfamethoxazole, triclosan, and trimethoprim from the wastewater. MPS-34 and NF270 nanofiltration membranes exhibited excellent potential by rejecting about 90% of the pharmaceutical compounds under specific environmental conditions. On the other hand, TFC-SR2 membrane showed low filtration potential by removing 60% of the pharmaceutical compounds. Removing these compounds by membranes highly depends on the pH of the solution used. The membranes were studied for its antifouling property using Hermia's model. The experimental study successfully demonstrated the use of NF membrane for tertiary treatment of wastewater containing pharmaceutical products [33].

An ultra-thin nanofiltration membrane was impregnated with reduced graphene oxide *via* a vacuum-assisted assembly method. The morphological and chemical properties of the membrane were determined using SEM, atomic force microscopy, and TEM. The membrane was compactly packed with a uniform thickness and high porosity. The membrane removed approximately 20–60% of four different salt ions in the order $(Na_2SO_4) > (NaCl) > (MgSO_4) > (MgCl_2)$. The membrane also showed an excellent potential for dye rejection by removing 99.8% of methylene blue and 99.9% of direct red from the given sample of water. The NF membrane is one of the best candidates for filtration processes on an industrial scale due to its low cost and high dye rejection rate [34].

8.4   UF

The process of UF uses hydrostatic force to filter micropollutants such as bacteria, viruses, and endotoxins, as well as dissolved solutes, from wastewater or primary treated water. In the process of UF, the type of UF membrane used is based on the type and size of the pollutant to be filtered. However, this process of water decontamination employs a UF membrane with an average pore size of 0.005–0.1 μm. UF allows the recycling of water with a stable filtration rate with a low cost due to its small UF plant and low maintenance compared to other membrane processes. Various UF membranes made up of different materials like cellulose, polyvinylidene fluoride, and PES were tested by M. Bielska et al. for their efficiency in removing methylene blue from synthetic micellar solutions (sodium dodecylsulfate [SDS] and oxyethylated coconut fatty acid methyl esters [OMC-10] and their binary solutions). Methylene blue dissolved in deionized (DI) water and various concentrations of micellar solutions were used as contaminants in the experiment. The filtration process was studied by adding each micellar solution to the phenol solution. When SDS was mixed with OMC-10 in the ratio 4:1, a significant reduction in the critical micelle concentration was observed. Furthermore, when the mixture was purified through the cellulose membrane it retained 93–94% of phenol from the mixture. The method can be used in the large-scale filtration of phenol due to its high phenol retention rate [35].

A polymeric UF membrane has also been employed for the removal of priority pollutants, namely, 4-chlorophenol and other colloids, from the synthetic polluted water. Crossflow UF was performed for different concentrations of pollutant at varied applied pressures. The initial and final concentration of 4-chlorophenol was determined by direct spectrometry at 257 nm. It was observed that when the experiment was performed at a low pressure and a low concentration of 4-chlorophenol, the UF membrane successfully ejected 90% of 4-chlorophenol from the synthetic wastewater. Simultaneously, the membrane also exhibited complete removal of colloidal matter from the water. The membrane was washable and recyclable and exhibited excellent antifouling potential. The use of this membrane on large scale is an economical method for water purification process [36].

A PVDF UF membrane impregnated with vermiculite nanoparticles *via* the phase-inversion technique has also been explored. The membrane was used to eliminate organic pollutants, including humic acid and certain dyes, from the contaminated water. The physiological characteristics of the membrane were determined using various methods such as SEM and FTIR spectroscopy. The results indicated that hydrophilic membrane has a fingerlike porous and asymmetric structures. Vermiculite nanoparticles incorporated in the membrane had a particle size of 86.2 nm and 90.1 nm. The wastewater containing organic pollutants was passed through the membrane and the efficiency of the membrane for the removal of humic acid, and different dyes were detected using UV/visible spectrophotometer. Furthermore, the filtration performance of the PVDF membrane incorporated with vermiculite was compared to various PVDF membranes impregnated with different nanoparticles, including aluminum oxide, silicon oxide, and copper oxide. It was concluded that the vermiculite PVDF membrane showed a maximum retention rate for humic acid (94.5%) followed by Al_2O_3 (91.7) > SiO_2 (89) > CuO (88.3%). The vermiculite PVDF membrane also showed high rejection rate for different dyes like methylene blue, Congo red, malachite green oxalate, and safranin O. The membrane can be used for water purification purposes in many industries in which organic pollutants are produced on a large scale [37].

A flat sheet, micellar-enhanced ultrafiltration membrane made up of PES was evaluated for its ability to eliminate phenol from contaminated water. A comparative study was conducted, in which various concentrations of varied anionic (sodium dodecyl sulfate, SDS), cationic (Gemini surfactant -N1-dodecyl-N1,N1,N2,N2-tetramethyl-N2-octylethane-1,2-diaminium bromide [CG] and conventional cationic surfactant – dodecyl trimethyl ammonium bromide, DTAB), and nonionic surfactant ((dodecyloxy) polyethoxyethanol, Brij35) were added to contaminated water containing phenol in order to enhance the filtration performance of the membrane. The surface morphology of the membrane was examined using SEM and ATR-FTIR. The molecular weight cutoff for the membrane was 10 kDa, and the active filtration area was 0.06 m^2. The result of the experiment proved that when the cationic Gemini surfactant was added to phenolic wastewater, maximum retention of phenol (95.8%) was observed with high permeate flux. The efficiency of the membrane to eliminate phenol by solubilizing the phenol was in the order CG > DTAB > SDS > Brij35 [38].

A commercially available Ultrafiltration membrane has also been studied to remove heavy metal ions like Cd^{2+} and Zn^{2+} from wastewater using some macroligands as complexing agents. Synthetic wastewater was mixed with different macroligands like dextrin, polyethylene glycol, and diethyl-aminoethyl cellulose before subjecting it to the UF membrane. The best filtration performance by the membrane was achieved at neutral pH and 300 kPa pressure. When diethylaminoethyl cellulose was added to the wastewater, the ion retention rate was maximum, that is, 95% for Cd^{2+} and 99% for Zn^{2+}, due to the interaction of ions with diethylaminoethyl cellulose. On the other hand, lowest rejection of ions was observed when dextrin was added as a complexing agent due to its low molecular weight. Furthermore, a detailed study of some operational parameters can support the use of this UF process at commercial levels [39].

A UF membrane was engineered by adding a mixture of polysulfone, 4,4-sulfoxylphenol, and polystyrene into N,N-dimethylformamide (DMF) with constant stirring for 45 minutes. Later, the mixture was spread on a glass plate and merged into a water bath. Furthermore, the membrane was studied for its morphological characteristics using SEM, nuclear magnetic resonance, and FTIR. During the filtration process, the effluent, containing vat dyes including indigo and black sulfur dye, was passed through the customized UF membrane. At a slightly alkaline pH, the membrane reduced the concentration of indigo and sulfur dye by 87.24% and 64.04%, respectively. The membrane can act as a superior candidate for decolorizing textile effluents [40].

A hydrophilic polyacrylonitrile UF membrane was employed for the removal of arsenic (As (III) and As(V)) from two sources of groundwater. The UF membrane used in the experiment had a pore size of 0.45 μm with an active filtration area of 14.7 cm². When the groundwater sample was passed through the membrane in the absence of voltage, only 1–14% of arsenic rejection was observed, whereas in the presence of the voltage (25 V), arsenic rejection rate was over 79% due to the interaction of arsenic with organic matters in the groundwater on application of voltage. The total arsenic concentration of the wastewater was determined using OI analytical. The method of electro-UF can be used on large scale due to its advantages over conventional ultrafiltration method and high efficiency in water purification process [41].

The process of complexation-UF was applied to the decontamination of a water sample containing heavy metal ions. The UF membrane made from PES of a molecular weight cutoff equal to 10,000 Da was used in the experiment. The contaminated water for the study was prepared by mixing a polymeric carboxy methyl cellulose complexing agent with a heavy metal ion solution. When the contaminated water was passed through the UF membrane at a constant flow velocity of 7.5 L/h, 97.6% of Cu (II), 99.5% of Cr (III), and 99.1% of Ni (II) ions were rejected by the membrane due to the formation of complexes with carboxy methyl cellulose. The membrane was recycled by washing it with warm water and further treating it with a water solution containing sodium dithionite, sodium hydroxide, and citric acid. The experimental study shows the potential application of this process at an industrial scale by illustrating its advantages such as high efficiency, low cost, lower energy consumption, and fast reaction [42]. Crossflow UF was employed for the removal of phenyl-urea herbicides, including linuron, chlortoluron, diuron, and isoproturon. Different UF membranes, namely, thin-film composite polyamide membrane (GK) and two PES membranes of molecular weight 5000 and 20,000 Da, were used in the experiment. The amount of herbicide in the polluted water was determined using OI analytical. The filtration performance of the UF membrane was evaluated, and the adsorption of different herbicides was observed to be in the sequence: linuron > diuron > chlortoluron > isoproturon. A PES UF membrane with molecular weight cutoff of 5000 Da showed a maximum retention of linuron (~90%) and other herbicides. However, membrane fouling acts as a barrier for the long-term use of the membrane in the filtration process [43].

A UF membrane made up of cellulose acetate was applied for the removal of bacteriophages like MS2 and φX174 from the given sample of polluted water. A bench-scale UF process was used in the experiment. The bacteriophage rejection efficiency of the membrane was tested against different experimental conditions, including varied pH, temperature, and so on. It was observed that retention of the bacteriophages by the membrane was highly related to the pH and isoelectric point of the wastewater used. The result of the study indicated that, at a low isoelectric point, MS2 was adsorbed on a large scale, whereas at high pH and increasing isoelectric point, adsorption of φX174 was maximum. Further research in the experimental parameters for maximum retention of MS2 and φX174 during the UF process can replace other commercial techniques used currently for bacteriophage retention [44].

8.5 OSMOSIS

Osmosis is generally defined as the process through which the displacement of a solvent occurs from a high concentration gradient to a low concentration gradient through a porous membrane. The process of osmosis is a pressure-driven procedure to generate potable drinking water. Osmosis utilizes the natural phenomenon of osmosis to separate contaminants from wastewater based on its concentration gradient.

8.5.1 FORWARD OSMOSIS

The process of forward osmosis (FO) exploits natural osmosis process for the movement of purified water through a porous FO membrane while rejecting dissolved, as well as undissolved, pollutants on the other side of the membrane. Large numbers of FO membranes are engineered to improve filtration efficiency and overcome the disadvantages of other membrane processes like fouling and reverse flux.

An FO membrane was engineered and checked for its efficiency in desalination of water. The beads of polysulfone (PSF) was immersed in a mixture of 1-methyl-2-pyrrolidinone (NMP) and DMF and subsequently, it was cast on nonwoven polyester sheet with the help of a casting knife. Later, the membrane was rinsed with 1,3-phenylenediamine (MPD) and 1,3,5-benzenetricarbonyl trichloride (TMC), resulting in the formation of thin composite polyamide membranous sheet on casted PSF membrane. The membrane was then rinsed with a couple of aqueous solutions, such as sodium hypochlorite and sodium bisulfite. Finally, the membrane was washed with DI water and used for the filtration process. The surface morphology of the membrane was determined *via* SEM. The results indicated that the membrane possessed an effective filtration area of 20.02 cm^2. The average thickness of the membrane was observed to be 95.9 $\pm$12.6 μm. A crossflow FO method was carried out with feed flux <18 Lm^2·h^{-1} and a 1.5 molar concentration of sodium chloride in wastewater. It was observed that the FO membrane successfully ejected 97% of sodium chloride from the wastewater. The membrane exhibits its potential in large-scale application for water desalination processes [45].

A robust nanocomposite polysulfone carbon nanotubes coated with a thin polyamide membranous layer *via* phrase-inversion method was introduced to remove NaCl from the given sample of water. The membrane was characterized for its various parameters like surface morphology, mechanical strength, feed contact angle *via* FE-SEM, dynamic mechanical thermal analyses, and a contact angle meter, respectively. The membrane has an effective filtration area of 9.2 cm^2 with a pore size of 0.1 μm. The porous and fingerlike structures of the membrane allowed maximum retention of salts from the wastewater. Furthermore, the filtration performance of the membrane was compared with the commercially available FO membranes. The results suggested that FO membrane successfully separated <90% NaCl from the water sample. The membrane at its optimum environmental conditions removed approximately 97% of NaCl, which is much higher than many other commercially available FO membranes [46].

Two commercially available FO membranes, namely, thin-film composite (TFC) polyamide membrane and cellulose triacetate FO membrane were evaluated for their filtration potential against various pharmaceutical contaminants. Crossflow FO was carried out with FO membranes of active filtration area equal to 42 cm^2. The rejection of these pharmaceutical compounds is highly dependent on the hydrophobic interaction between these compounds and the FO membrane at a changing pH. Under alkaline conditions, a cellulose triacetate membrane coated with an FO membrane showed a maximum retention for all four pharmaceutical contaminants in wastewater in the sequence diclofenac (99%) > carbamazepine (95%) > ibuprofen (93%) ≈ naproxen (93%). However, a TFC FO membrane was stable and showed high retention of pharmaceutical compounds over large range of pH (3–8). The detailed research on the parameters affecting varied environmental conditions required for optimum retention of these contaminants by the membrane can lead to the generation of one of the best candidates for the FO filtration process [47].

A commercially available FO membrane was polymerized using polyoxadiazole-co-hydrazide (PODH) and polytriazole-co-oxadiazole-co-hydrazide and was used for the filtration of Congo red dye from the wastewater. The physical properties of the membrane were determined using FE-SEM and atom force microscopy. The membrane was symmetrical with an active filtration area of 10 cm^2. Due to its highly dense and negatively charged surface, the polymerized FO membrane retained a high concentration of Congo red dye. An admirable bactericidal, as well as bacteriostatic, property was demonstrated by the membrane due to the presence of cytotoxic oxadiazole and triazole moieties. An extremely low viability of the bacteria used (*E. coli* and *S. aureus*) was observed after filtration due to the hydrophobic nature and smooth plane of the membrane. The membrane was easily regenerated by processing it with alcohol. It also showed excellent antifouling properties with superior tensile strength [48].

Flat-sheet FO membranes were employed for the removal of certain hydrophilic-neutral, hydrophobic-neutral, and ionic micropollutants from the wastewater (Figure 8.3). The physical properties of the commercial RO membrane were observed using SEM. The membrane had an active filtration area of 60 cm^2 and showed a porous nature. After the filtration process, the surface of the membrane was studied using confocal laser scanning microscopy. It was observed that the hydrophilic ionic contaminants were absorbed maximum after fouling the membrane with negatively charged foulant (96–99%). On the other hand, the retention rate of hydrophilic neutral microcontaminants was reduced by 5% on fouling of the membrane (44–95%), whereas the hydrophobic micropollutants were efficiently absorbed after fouling of the membrane (48–92%). When the wastewater containing micropollutants was filtered through both FO and RO membranes, 96% of the micropollutants were rejected by the membranes. Due to its high rejection rate of hydrophilic micropollutants, the fouled FO membrane can be applied at a large scale for water decontamination processes [49].

A commercial polyamide FO membrane was coated with a macrovoid-free Matrimid substrate using a phase-inversion technique. The wastewater containing heavy metal salts (Na$_2$Cr$_2$O$_7$, Na$_2$HAsO$_4$, Pb (NO$_3$)$_2$, CdCl$_2$, CuSO$_4$, Hg(NO$_3$)$_2$) was mixed with a draw solute (Na$_4$[Co(C$_6$H$_4$O$_7$)$_2$]). A lab-scale FO system with an active filtration area of 4 cm^2 was used for the experiment. When the water sample was passed through the FO membrane at 60°C, 11 LMH flux, and unit molar

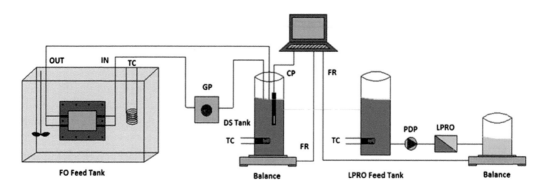

FIGURE 8.3 Schematics of forward osmosis system used to remove micropollutants. Reprinted with permission from R.V. Linares, V. Yangali-Quintanilla, Z. Li and G. Amy, "Rejection of micropollutants by clean and fouled forward osmosis membrane," *Water Res.*, vol. 45, no. 20, pp. 6737–6744, 2011. doi:10.1016/j.watres.2011.10.037

concentration of draw solute, the membrane retained 99.5% of metal ions, including $Cr_2O_7^{2-}$, $HAsO_4^{2-}$, Pb^{2+}, Cd^{2+}, Cu^{2+}, and Hg^{2+}. When the concentration of the draw solute was increased by 0.5 M, the RO membrane absorbed 99.7% of the heavy metal ions, which indicates that increase in the concentration of feed solution and draw solute results in higher retention of pollutants. This recently developed RO process could be best technique for the industrial wastewater treatment when heavy metal ions are produced as the major contaminant [50].

8.5.2   REVERSE OSMOSIS

Reverse osmosis (RO) is a pressure-driven process that is used on large scale for removing diverse pollutants from wastewater. The major application of RO membranes is seen to produce potable water by desalination process. The filtration efficiency of reverse osmosis membrane with an average pore size of 0.01 μm is higher than the other filtration membranes like UF, MF and NF membranes. Various modifications in RO membranes have been done to meet the ever-increasing need for potable water.

A thin-film composite reverse osmosis membrane was tested for its efficiency against NF membrane to remove organic dye (methyl orange) and sodium sulfate salt. The membranes were placed in a cylindrical filtration unit made up of stainless steel, and the filtration performance of the membrane was tested at various feed concentrations at varying pressures. It was noted that the nanofiltration membrane absorbed 94–98.9% of methyl orange dye between 200 to 400 psi. However, the filtration performance of the nanofiltration membrane was slightly low as compared to the RO membrane because of larger pore size of the nanofiltration membrane. The RO membrane removed 99.9% of methyl orange dye under 400 psi. It also removed 96.03–97.97% of total dissolved solutes and 92.36–98.89% of salt from the given feed sample. The RO membrane can be used to generate potable water by eliminating unwanted contaminants from the wastewater on a large scale [51].

An ultra-low-pressure RO membrane was introduced to eliminate diverse organic pollutants from the water sample. The removal of organic pollutants such as phenolic compounds, acetic acids, sodium chloride salt, and urea was experimentally checked with the high flux polymeric RO membrane at various pH levels. The molecular weight cutoff for the membrane was approximately150 Dalton. It was observed that, when the molecular weight of RO membrane exceeded 150, the filtration performance of the membrane was greatly reduced. The membrane effectively rejected 98.5% NaCl and 99.7% acetic acid over a pH range of 3 to 9. It also removed <90% of phenolic compounds, including 2,4,5-trichlorophenol, 2,3- dichlorophenol, and 2,4-dichlorophenol, at a slightly alkaline pH. However, a low rejection of urea (~35%) was observed over a large range of pH levels. The experimental study suggests the use of an ultra-low-pressure RO membrane on an industrial scale for removing specific organic contaminants on an industrial scale [52].

A commercially available spiral-wound RO membrane was employed for purifying seawater containing a pharmaceutical product, namely, ciprofloxacin. All sets of the experiment were carried out at a steady pressure and a feed flow rate. The filtration experiment was carried out in two phases (varying temperature ([22–30°C] and constant temperature [25°C]). The varied ciprofloxacin concentration of 50,200 and 500 μg/l in the feed sample was tested in the experiment. The retention rate of ciprofloxacin was high over a large range of temperatures and feed concentrations. However, the best retention of ciprofloxacin, that is, 99.6%, was obtained at 500 μg/l ciprofloxacin at 30°C. The membrane also removed 98.5% of salt from the seawater. The simultaneous removal of salt and ciprofloxacin, along with high stability, recyclability, and the low cost of the membrane, proves its potential for industrial application [53].

A small-scale crossflow RO filtration process was carried out using a commercially available RO membrane to remove antibiotics like tetracycline (TC) and oxytetracycline (OTC) from the contaminated water. The various factors influencing the filtration performance of the membrane, including feed recovery, the concentrations of TC and OTC in the feed solution, and the salt concentration in the feed solution, was studied. A significant decrement in antibiotics retention was observed on increasing the feed recovery. Hence, in this experiment, the filtration was carried out at low feed recovery of approximately 15%. Similarly, the filtration performance of the membrane was decreased on the addition of NaCl and CaCl$_2$ to the feed solution. This is because the addition of salts in the feed solution causes pore swelling in the membrane, which decreases the amount of antibiotics filtered by the membrane. On the other hand, the rate of retention for antibiotics increased with an increase in the feed concentration. The higher efficiency of the membrane to eliminate both TC and OTC by approximately 90% defines its use on an industrial scale [54].

Commercially available RO membranes were used in a bench-scale filtration process to reject an herbicide called 2-methyl-4-chlorophenoxyaceticacid (MCPA) from acidic saline water. A cross-flow RO process was carried out at a constant velocity of 0.2 ms^{-2} with an effective membrane filtration area of 140 cm^2. When the saline wastewater containing MCPA was filtered through the membrane, approximately 99% of NaCl was retained by the membrane. A significant deduction total organic carbon equal to 93% was noted. The membrane successfully rejected 95% of MCPA from the acidic water sample. After the filtration process, when the filtrate was checked for its pH, a decrease in acid (H$^+$ ions) concentration was observed by about 25%. Furthermore, the rejection rate of MCPA by RO membranes was tested against nanofiltration membranes, and it was concluded that MCPA was best filtered by RO membranes. This mechanically stable membrane exhibits its potential application in water decontamination [55].

A complexation-RO process was carried out using a sulfonated polysulfone RO membrane to eliminate heavy metal ions from wastewater at low pressure, that is, less than 690 kPa. Ethylene diamine tetra acetic acid (EDTA) was used as a complexing agent. The wastewater containing CuCl$_2$ and ZnCl$_2$ was mixed with a complexing agent, and pH was altered from 3 to 5. When the wastewater was passed through the RO membrane in absence of EDTA, 93–96% of Cu^{2+} and Zn^{2+} were observed. In the presence of EDTA, the RO membrane rejected 99% of Cu^{2+} and Zn^{2+} from industrial wastewater at a very low pressure of 450 kPa. The complexation-RO process acts as the best candidate for water decontamination by removing unwanted metal ions from polluted water [56].

8.6 MEMBRANE BIOREACTOR

A bench-scale membrane bioreactor (MBR) plant was employed for the reduction of COD from textile-contaminated water. An MBR containing activated sludge was coupled with an UF unit that resided a tubular PVDF UF membrane in it. The UF membrane used had a molecular weight cutoff of 15 kDa and an active filtration area of 0.28 m^2. MBR rejected 60–75% of COD from the activated sludge. When the sludge with greater concentration of COD was passed through the filtration unit, it retained <90% of COD. The COD removal by the MBR was increased with an addition of inorganic salt nutrients to the sludge. Color rejection by an MBR was examined by spectral adsorption coefficient at various wavelengths (436,525 and 620 nm), and it was observed that color removal was maximum at 620 nm (57–98%). The membrane showed an antifouling ability over a long filtration period without any significant changes in its filtration performance. However, the low COD rejection rate of the process, compared to many other commercially used filtration MBRs, suggests further research for the improvement of this MBR method for water purification [57].

A highly dense polyethylene membrane in an MBR was impregnated with silica nanoparticles to eliminate the fouling that takes place due to the formation of cake on the membrane. Different HDPE membranes were developed by incorporation of different concentrations of silica nanoparticles (0.25/0.5/1 wt.%) in the membrane using thermally induced phase separation (TIPS). The physiological and chemical properties of the membrane were examined using FE-SEM and energy-dispersive X-ray analysis. The hydrophilicity of the membrane was significantly increased due to the impregnation of silica particles, which, in turn, increased the adsorption of COD by membrane. The membrane, with effective filtration area of 14 cm^2, reduced COD by 95%. It also decreased the fouling caused by cake formation. The antifouling capability of the membrane was determined using Hermia's model [58].

An MBR with a PES UF membrane was employed for the hospital water decontamination. The dynamic hospital wastewater contained diverse micropollutants and pharmaceutical compounds. A hospital effluent was passed through membrane from MBR using a peristaltic pump. The MBR successfully eliminated 94% of dissolved organic carbon and 92% of COD from hospital wastewater. It also removed some antibiotics (ciprofloxacin 96%), anti-inflammatory (mefenamic acid, paracetamol [acetaminophen], morphine, and metamizole are removed by 92%), anti-infectives, anesthetics (thiopental 91% and lidocaine 56%,), antiepileptics (levetiracetam 95%), and other pharmaceutical compounds (cilastatin 90% and ranitidine 71%) on large scale. The application of an MBR for hospital wastewater treatment proved useful for the primary treatment of various hospital wastewater. However, further processing of water is necessary before letting it into the other public water sources [59].

An ozone–oyster shell fix-bed bioreactor (OFBR) combined with an RO membrane was applied for the treatment of municipal contaminated water. The OFBR was connected through MBR, which included a polyvinylidene fluoride RO membrane. The ozone gas was realized into OFBR through the membranous structure. The municipal wastewater was passed through the membrane using a peristaltic pump. Ozonation of microbes such as Deinococcus-Thermus, Firmicutes, Actinobacteria, Planctomycetes, aerobic ammonia-oxidizing bacteria, nitrite-oxidizing bacteria, and others present in municipal wastewater during the filtration process led to a higher adsorption of ammonia by the membrane. The use of an OFBR MBR for purifying municipal wastewater successfully eliminated 99% of ammonium and 43% of total phosphorous. It also removed 73% of COD in wastewater. The membrane can be used on a large scale for the purification of municipal wastewater due to its remarkable ability to eliminate ammonia as well as the low cost of the process [60].

An OMBR was assimilated with an MF membrane to attain high biomass concentration which also increased the retention of ammonia nitrogen from a given water sample. Physiological and chemical characteristics of the membrane were determined by SEM and EXD, respectively. An FO polyamide membrane and a PVDF microfiltration membrane with effective filtration areas of 0.056 m^2 and 0.12 m^2, respectively, were placed in a bioreactor and subsequently subjected to activated synthetic sludge. The filtration process was carried out at neutral pH and room temperature. The OMBR rejected 93–100% of total organic carbon from the sludge. A high concentration of biomass lead to a greater reaction of ammonia-oxidizing bacteria with ammonia nitrogen ($NH4^+$-N), which caused a greater rejection of $NH4^+$-N (97%) by OMBR. The membrane also showed excellent antifouling ability and stability during the filtration period [61].

Bench-scale water purification was carried out using an OMBR to eliminate 27 organic pollutants, including most of the pharmaceutical products, hormones, pesticides, and so on. The activated sludge containing different organic pollutants was mixed with oxygen, which was obtained using an air pump. The peristaltic pump was used to pump activated sludge through the membrane. Out of the 27 pollutants present in the activated sludge, OMBR retained <80% of 25 high-molecular-weight contaminants. However, the retention rate of low-molecular-weight pollutants was highly dependent on biodegradation. The filtration performance of the membrane was slightly decreased due to the development of salinity in the OMBR [62].

A study was conducted to check the efficiency of a submerged MBR against a conventional activated sludge reactor for removal of BPA from synthetic wastewater. In the experiment, a polymeric membrane with an effective filtration area of 0.2 m^2 was placed in MBR. The activated sludge was filtered through both bioreactors, and the filtration efficiency of the MBR and conventional sludge reactor was noted. High-performance liquid chromatography was used to analyze the concentration of BPA in the filtrate. It was observed that the filtration efficiency of the MBR was much greater than conventional sludge reactor. This was due to the presence of metabolite, namely, 4-hydroxy-acetophenone in activated sludge of MBR, which causes degradation of BPA. The submerged MBR removed 99.3% of BPA, whereas the conventional sludge reactor removed 93.7% from polluted water [63].

TABLE 8.1

Different Types of Membrane and Membrane-Based Processes Used for Removing Different Pollutants from Wastewater

No.	Type of pollutant	Type of MF membrane	Applicability	Efficiency	Ref.
A.		Organic Pollutants			
1.	Turbidity and COD	Ceramic MF membrane tailored with magnesite	Lab scale	Removed ~99.9% of turbidity and 69.7% of COD	[13].
2.	Algal removal	Polyvinylidenefloride (PVDF) and cellulose-ester MF membranes	Lab scale	Very high (PVDF)	[17].
3.	Humic substances	Polyethersulfone electro-microfiltration membrane	Lab scale	Very high absorbance of total organic carbon, trihalomethane formation potential (THMFP) and UV absorbance	[18].
4.	Perfluorinated compounds (PFCs) including perfluorooctanoic acid (PFOA) and perfluorooctane sulfonate (PFOS)	Ceramic Microfiltration membrane	Lab scale	Rejected ~70% of perfluorooctanoic acid (PFOA) and perfluorooctane sulfonate (PFOS)	[19].
5.	Degeneration of p-chloroaniline (PCA)	Ceramic microfiltration membrane	Lab scale	Rejected ~87.1% of PCA and ~45.2% of total organic carbon	[22].
6.	Organic pollutant (Bisphenol A and COD)	Ceramic NF membrane	Lab scale	absorbed ~100% of BPA	[23].
7.	Phenol	Polyamide NF membranes (NF-97, NF-99 and DSS-HR98PP)	Lab scale	Rejection rate of the phenol (~80%)	[24].
8.		Micellar-enhanced ultrafiltration membrane (polyethersulfone)	Lab scale	High retention of phenol (95.8%)	[38].
9.	Organic pollutants like Carboxylic acids including acetic acid and butyric acid	Five different commercially available NF (NF270, HL, DL, DK and LF10) membrane	Lab scale	Comparative study (LF10 membrane absorbed – acetic acid (72.2%) and butyric acid (~70%)) and (NF270 membrane absorbed 52.6% of acetic acid and 69.7% of butyric acid)	[25].
10.	Organic pollutants	Ultra-low pressure reverse osmosis membrane	Lab scale	Rejected 98.5% Nacl and 99.7% acetic acid. It also removed <90% of phenolic compounds including as 2,4,5-trichlorophenol, 2,3- dichlorophenol, 2,4-dichlorophenol and urea (~35%)	[52].

No.	Target	Membrane/Technology	Scale	Result	Ref.
11.		Ozone oyster shells fix-bed bioreactors (OFBR) constitute of reverse osmosis membrane	Lab scale	Eliminated 99% of ammonium and 43% of total phosphorous. And 73% of chemical oxygen demand	[60].
12.		Hybrid membrane bioreactor (HMBR)	Lab scale	Rejected 90.2% and 92.75% of COD and color	[64].
13.	4-chlorophenol and other colloids	Commercial polymeric Ultrafiltration membrane	Lab scale	Rjected 90% of 4-chlorophenol	[36].
14.	Humic acid and certain dyes	Polyvinylidene fluoride (PVDF) Ultrafiltration membrane was impregnated with vermiculite nanoparticles	Lab scale	Rejected humic acid (94.5%) followed by Al_2O_3 (91.7)> SiO_2 (89)>CuO (88.3%)	[37].
15.	Chemical oxygen demand (COD) from textile-contaminated water	Membrane bioreactor (MBR): tubular PVDF UF membrane	Lab scale	Retained <90% of COD and color removal was maximum at 620 nm (57–98%)	[57].
16.	Chemical oxygen demand (COD)	HDPE membrane bioreactor by incorporation of silica nanoparticles	Lab scale	Reduced COD by 95%	[58].
17.	Bisphenol A	Submerged membrane bioreactor (MBR)	Lab scale	Removed 99.3% of bisphenol A	[63].
B.	**Dyes**				
18.	Organic dyes	MF membrane impregneated with zeolite and PEG	Lab scale	~95.55% of crystal violet and ~90.23% of methylene blue	[7].
19.		MF membrane fabricated with polyethylene glycol and tannic acid	Lab scale	Retain ~98.9% of Rhodamine	[11].
20.		Ultrafiltration	Lab scale	Removed 93–94% phenol	[35].
21.	Organic dyes and metal ions	PVDF membrane impregnated with cellulose nanofibers and meldrum's acid.	Lab scale	~99% Crystal violet dye and nanoparticles like iron (III) oxide.	[8].
22.	Anionic dyes	Polypropylene microfiltration membrane	Industrial scale	~100% Direct Red 2 and salt	[9].
23.	Carcinogenic dye (Direct Blue)	Polyurethane and cellulose acetate MF membrane	Lab scale	100% removal of direct blue	[10].
24.	Desalination and dye removal	Ultrathin nanofiltration membrane impregnated with reduced graphene oxide	Lab scale	removing 99.8% of methylene blue and 99.9% of direct red and 20–60 % of salts	[34].
25.	Dyes (Indigo dye/Black sulfur dye/)	UF membrane (Polysulfone/4,4-sulfoxylphenol/polystyrene/N,N-dimethylformamide (DMF))	Lab scale	Removed indigo and sulphur dye by 87.24 and 64.04% respectively	[40].

(Continued)

TABLE 8.1 (Continued)
Different Types of Membrane and Membrane-Based Processes Used for Removing Different Pollutants from Wastewater

No.	Pollutant/Application	Membrane	Scale	Result	Ref.
26.	Organic dye (Methyl orange) and sodium sulfate salt	Thin-film composite reverse osmosis membrane	Lab scale	Removed 99.9% of methyl orange dye and 96.03–97.97% of total dissolved solutes and 92.36–98.89% of salt	[51].
C.			**Salt removal**		
27.	Removal of salt from contaminated oily water	Poly-amide nanofiltration membrane impregnated with silica nanoparticles	Lab scale	Retain ~50% of the salts from the oily wastewater	[28].
28.	Removal of salt and pharmaceutical products	Polyethersulfone nanofiltration membranes (Commercial NF1 and NF2 membranes)		NF1 membrane rejected 5–15% whereas NF2 rejected 25–35% of sodium chloride and removed 60% of naproxen and diclofenac	[30].
29.	Removal of salt	Thin film composite Nanofiltration membrane		Retained calcium chloride (95.1%) salt	[29].
30.		Polyethersulfone (PES) membranes impregnated with silica or polypiperazine		Eliminated 97.3% Na_2SO_4, 91.1% $MgSO4$, 50.7%$MgCl2$ and 50.7% NaCl	[32].
31.		Thin composite polyamide membranous PSF FO membrane.		Rejected 97% of sodium chloride	[45].
32.		A robust nanocomposite polysulfone carbon nanotubes coated with thin polyamide membrane		Removed ~97% of NaCl	[46].
D.			**Herbicides**		
33.	Herbicides atrazine (ATZ)	MF membrane tailored with chitosan (CHI) and polystyrene sulphonate (PSS)	Lab scale	Absorbance of ATZ on the membrane i.e.92.23%	[12].
34.	Phenyl-urea herbicides including linuron, chlortoluron, diuron and isoproturon	Ultrafiltration membranes viz, thin film composite polyamide membrane (GK) and polyethersulfone membranes		maximum retention of linuron (~90%) (Retention rate: linuron>diuron>chlortoluron>isoproturon)	[43].
35.	Herbicide (2-methyl-4-chlorophenoxyaceticacid (MCPA)) from acidic saline water	Commercially available reverse osmosis membranes		99% of NaCl and total organic carbon 93% and rejected 95% of MCPA	[55].
E.			**Pharmaceutical/Antibiotics**		
36.	Antibiotics and hormones	Hydrophilic Nanofiltration membrane	Lab scale	Rejected 80% of tetracycline and 50% of doxcycline	[26].
37.	Pharmaceutical products such as acetaminophen, caffeine, diazepam, diclofenac, ibuprofen, naproxen, sulfamethoxazole, triclosan and trimethoprim	Commercial nanofiltration membranes (TFC-SR2, NF270, and MPS-34)		Rejected ~90% of the pharmaceutical compounds	[33].

#	Contaminant	Membrane/Technology	Scale	Remarks	Reference
38.	Pharmaceutical contaminants	Thin film composite polyamide (TFC)-polyamide membrane and cellulose triacetate FO membrane		Retention rate: diclofenac (99%) > carbamazepine (95%) > ibuprofen (93%) ≈ naproxen (93%)	[47].
39.	Pharmaceutical product viz. ciprofloxacin	Commercially available spiral-wound RO membrane		Retention of ciprofloxacin i.e.99.6% and removed 98.5% of salt	[53].
40.	Removal of antibiotics like tetracycline (TC) and oxytetracycline (OTC)	Cross-flow reverse osmosis membrane		Eliminated both tetracycline (TC) and oxytetracycline (OTC) by ~90%	[54].
41.	Micropollutants and pharmaceutical compounds	Membrane bioreactor with Polyethersulfone ultrafiltration membrane		Eliminated 94% of dissolved organic carbon and 92% of COD. Ciprofloxacin 96% and mefenamic acid, paracetamol (acetaminophen), morphine, and metamizole are removed by 92%, antiinfectives, anesthesics (thiopental 91% and lidocaine 56%), antiepileptics (levetiracetam 95%) and pharmaceutical compounds (cilastatin 90% and ranitidine 71%)	[59].
F.	**Bacteria/Viruses**				
42.	Industrial pharmaceutical waste.	Osmotic membrane bioreactor (OMBR)		Retained <80% of pharmaceutical products	[62].
43.	Pharmaceutical compounds namely, cyclophosphamide (CYC) and ciprofloxacin (CIP)	Nanofiltration membrane bioreactor		100% rejection of ciprofloxacin	[65].
44.	*E. coli*, bacterial viruses like MS2, and heavy metal ions including hexavalent chromium ions and lead ions	PAN/PET-drafted cellulose nanofibers MF membrane.	Lab scale	Absorbed ~100% of bacteria, virus as well as Cr (VI) and Pb (II)	[14].
45.	bacteriophages like MS2 and φX174	Cellulose acetate UF membrane	Lab scale	Very high	[44].
46.	Bacterial removal (*E.coli* and *S.aureus*)and dye elimination	A commercial FO membrane (Polyoxadiazole-co-hydrazide (PODH) and polytriazole-co-oxadiazole-co-hydrazide)	Lab scale	High dye retention and low anti-bacterial activity	[48].
G.	**Oil**				
47.	Emulsified oil	PVDF MF membrane	Lab scale	Very high	[15].
48.	Oil	Carbonized, coal-based microfiltration membrane		Absorbed ~97% of the oil	[16].
H.	**Heavy metal ions**				
49.	Chromium hydroxide	Electro-microfiltration acrylic polymer membrane	Lab scale	Very high	[20].

(Continued)

TABLE 8.1 (Continued)

Different Types of Membrane and Membrane-Based Processes Used for Removing Different Pollutants from Wastewater

No.	Pollutant	Membrane/Process	Scale	Result	Ref.
50.	Phosphate ions	Cellulose acetate MF membrane	Lab scale	Ejected ~100% of the phosphate ions	[21].
51.	Arsenic removal from drinking water	Thin film composite nanofiltration membrane (NF-300)	Lab scale	Rejected 99.80% of arsenate ions.	[27].
52.	Chloride and sulfate ions removal	Commercially available nanofiltration membrane	Lab scale	Removed 99.82 % and 77.50 % of the sulphate and chloride ions	[31].
53.	Heavy metals ions like Cd^{2+} and Zn^{2+}	A Commercial Ultrafiltration membrane	Lab scale	95% for Cd^{2+} and 99% for Zn^{2+}	[39].
54.	Arsenic (As(III) and As(V))	Hydrophilic polyacrylonitrile Ultrafiltration membrane	Lab scale	Arsenic rejection rate was over 79%.	[41].
55.	Heavy metal ions	UF membrane (polyethersulfone)	Lab scale	Rejected 97.6% of Cu (II), 99.5% of Cr (III), and 99.1% of Ni (II) ions	[42].
56.	Removal of certain hydrophilic-neutral, hydrophobic-neutral, and ionic micropollutants	A commercial Flat-sheet FO membranes	Lab scale	Rejected 96% of the micropollutants	[49].
57.	Heavy metal salts ($Na_2Cr_2O_7$, Na_2HAsO_4, $Pb(NO_3)_2$, $CdCl_2$, $CuSO_4$, $Hg(NO_3)_2$)	A commercial polyamide FO membrane coated with macrovoid-free Matrimid substrate	Lab scale	Retained 99.5% of metal ions including $Cr_2O_7^{2-}$, $HAsO_4^{2-}$, Pb^{2+}, Cd^{2+}, Cu^{2+}, and Hg^{2+}	[50].
58.	Heavy metal ions	Complexation-reverse osmosis membrane	Lab scale	Rejected 99% of Cu^{2+} and Zn^{2+}	[56].
59.	Ammonia nitrogen	Osmotic membrane bioreactor (OMBR) was assimilated with microfiltration membrane	Lab scale	Rejected 93–100% of total organic carbon and $NH4^+$-N (97%)	[61].
60.	Nitrate ions	Anaerobic packed-bed membrane bioreactor	Lab scale	Retained ~88.8% of nitrate ions	[66].

A hybrid MBR (HMBR) was employed for the purification of tannery-polluted water. The tannery-polluted water was, first, electro coagulated in an electric unit using direct current. Then the activated sludge was aerated with oxygen through air diffuser. Subsequently, a dead-end filtration process of the activated sludge was carried out using polyvinylidene fluoride MF membrane with active filtration area of 0.0143 m². After filtration process, the membrane was examined for its fouling and sludge deposition by SEM and energy-dispersive X-ray analysis, respectively. Furthermore, the efficiency of HMBR for removing COD and color from tannery water was compared with the filtration efficiency of an MBR. It was concluded that the HMBR was more efficient than the MBR as it rejected 90.2% and 92.75% of COD and color, respectively. On the other hand, the MBR removed only 72.69% of COD and 75.82% of color from tannery-contaminated water. The HMBR also exhibited 11% lower membrane fouling than MBR. Hence, the use of an HMBR is highly recommended for tannery wastewater treatment [64].

A crossflow nanofiltration process was carried out using a nanofiltration MBR for retaining two pharmaceutical compounds specifically, namely, cyclophosphamide (CYC) and ciprofloxacin (CIP). Synthetic wastewater containing COD, nitrogen, and phosphorous were kept in proportion of 100:10:1 along with pharmaceutical compounds. A volumetric pump was used to pass the effluent from the aeration tank to the filtration unit. The filtration performance was monitored for about 4 months. Complete (100%) rejection of ciprofloxacin was observed after the 35th day, whereas the rejection of cyclophosphamide decreased after the 20th day, with a significant increase in membrane fouling. However, the fouling of the membrane was controlled using physical membrane scraping. Further research is necessary to overcome the barriers like fouling and stability of the nanofiltration MBRs for its use on a large scale [65].

An anaerobic packed-bed MBR was used for eliminating nitrate ions from groundwater. A polypropylene MF membrane with pore size of 0.1 μm was placed in the MBR. Initially, the ground wastewater was filtered through a nylon membrane and then stored in an MBR by adding some salts, including KNO_3, KH_2PO_4, K_2HPO_4, $MgSO_4$, and $FeSO_4$. After the formation of a biofilm, methanol is added to the bioreactor to denitrify nitrate ions and reduce aerobic bacteria. After the filtration process, it was observed that the MBR successfully retained approximately 88.8% of the nitrate ions from the activated sludge at a hydraulic retention time of 5.3 h. The process of water filtration using a packed-bed MBR exhibits its capability for large-scale application by its excellent stability and high ion retention rate [66].

8.7   CONCLUSION

Membrane and membrane-based techniques are the most sustainable and promising methods for wastewater treatment. Today, a large range of membrane processes are applied to eliminate and absorb pollutants from contaminated water that are fatal to the ecosystem. Many UF, nanofiltration, MF, osmosis membranes, and MBRs have demonstrated excellent potential for removing selective waste from the water with greater stability and lower energy consumption rate. Many of the membranes alone or in association with other membranes or processes successfully produced potable water under low pressure, which also reduced the overall cost of the process. Using membrane technologies for wastewater treatment is an economical, environmentally friendly, and advantageous approach in wastewater treatment.

BIBLIOGRAPHY

1. X. Shi, K.Y. Leong and H.Y. Ng, "Anaerobic treatment of pharmaceutical wastewater: A critical review," *Bioresour. Technol.*, vol. 245, no. June, pp. 1238–1244, 2017.
2. S.F. Anis, R. Hashaikeh and N. Hilal, "Microfiltration membrane processes: A review of research trends over the past decade," *J. Water Process. Eng.*, vol. 32, no. December, pp. 100941, 2019.

3. Y. Gu, Y. Li, X. Li, P. Luo, H. Wang, Z.P. Robinson, X. Wang, J. Wu and F. Li, "The feasibility and challenges of energy self-sufficient wastewater treatment plants," *Appl. Energy.*, vol. 204, no. October, pp. 1463–1475, 2017.

4. T. Peters, "Membrane technology for water treatment," *Chem. Eng. Technol.*, vol. 33, no. 8, pp. 1233–1240, 2010.

5. B. Van der Bruggen, C. Vandecasteele, T. Van Gestel, W. Doyen and R. Leysen, "A review of pressure-driven membrane processes in wastewater treatment and drinking water production," *Environ. Prog.*, vol. 22, no. 1, pp. 46–56, 2003.

6. X. Fan, Y. Tao, L. Wang, X. Zhang, Y. Lei, Z. Wang and H. Noguchi, "Performance of an integrated process combining ozonation with ceramic membrane ultra-filtration for advanced treatment of drinking water," *Desalination*, vol. 335, no. 1, pp. 47–54, 2014.

7. S. Foorginezhad and M.M. Zerafat, "Microfiltration of cationic dyes using nano-clay membranes," *Ceram. Int.*, vol. 43, no. 17, pp. 15146–15159, 2017.

8. D.A. Gopakumar, D. Pasquini, M.A. Henrique, L.C. de Morais, Y. Grohens and S. Thomas, "Meldrum's acid modified cellulose nanofiber-based polyvinylidene fluoride microfiltration membrane for dye water treatment and nanoparticle removal,"*ACS Sustain. Chem. Eng.*, vol. 5, no. 2, pp. 2026–2033, 2017.

9. J.J. Porter and A.C. Gomes, "The rejection of anionic dyes and salt from water solutions using a polypropylene microfilter," *Desalination*, vol. 128, no. 1, pp. 81–90, 2000.

10. D.E. Zavastin, S. Gherman and I. Cretescu, "Removal of direct blue dye from aqueous solution using new polyurethane: Cellulose acetate blend micro-filtration membrane," *Rev. Chim.*, vol. 63, no. 10, pp. 1075–1078, 2012.

11. P. Shi, X. Hu, Y. Wang, M. Duan, S. Fang and W. Chen, "A PEG-tannic acid decorated microfiltration membrane for the fast removal of Rhodamine B from water," *Sep. Purif. Technol.*, vol. 207, no. December, pp. 443–450, 2018.

12. P.N. Chandra and K. Usha, "Removal of atrazine herbicide from water by polyelectrolyte multilayer membranes," *Mater. Today: Proc.*, vol. 41, no. 3, pp. 622–627, 2020.

13. A. Manni, B. Achiou, A. Karim, A. Harrati, C. Sadik, M. Ouammou, S.A. Younssi and A. El Bouari, "New low-cost ceramic microfiltration membrane made from natural magnesite for industrial wastewater treatment," *J. Environ. Chem. Eng.*, vol. 8, no. 4, pp. 103906, 2020.

14. R. Wang, S. Guan, A. Sato, X. Wang, Z. Wang, R. Yang, B.S. Hsiao and B. Chu, "Nanofibrous microfiltration membranes capable of removing bacteria, viruses and heavy metal ions," *J. Membr. Sci.*, vol. 446, no. November, pp. 376–382, 2013.

15. S.S. Madaeni and M.K. Yeganeh, "Microfiltration of emulsified oil wastewater," *J. Porous Mater.*, vol. 10, no. 2, pp. 131–138, 2003.

16. C. Song, T. Wang, Y. Pan and J. Qiu, "Preparation of coal-based microfiltration carbon membrane and application in oily wastewater treatment," *Sep. Purif. Technol.*, vol. 51, no. 1, pp. 80–84, 2006.

17. M.T. Hung and J.C. Liu, "Microfiltration for separation of green algae from water," *Colloids Surf B Biointerfaces*, vol. 51, no. 2, pp. 157–164, 2006.

18. Y.H. Weng, K.C. Li, L.H. Chaung-Hsieh and C.P. Huang, "Removal of humic substances (HS) from water by electro-microfiltration (EMF)," *Water Res.*, vol. 40, no. 9, pp. 1783–1794, 2006.

19. Y.T. Tsai, A. Yu-Chen Lin, Y.H. Weng and K.C. Li, "Treatment of perfluorinated chemicals by electro-microfiltration," *Environ. Sci. Technol.*, vol. 44, no. 20, pp. 7914–7920, 2010.

20. C. Visvanathan, R.B. Aim and S. Vigneswaran, "Application of cross-flow electro-microfiltration in chromium wastewater treatment," *Desalination*, vol. 71, no. 3, pp. 265–276, 1989.

21. G. Akay, B. Keskinler, A. Cakici and U. Danis, "Phosphate removal from water by red mud using cross-flow microfiltration," *Water Res.*, vol. 32, no. 3, pp. 717–726, 1998.

22. S. Xu, J. Zheng, Z. Wu, M. Liu and Z. Wang, "Degradation of p-chloroaniline using an electrochemical ceramic microfiltration membrane with built-in electrodes," *Electrochim. Acta*, vol. 292, no. December, pp. 655–666, 2018.

23. M. Zielińska, K. Bułkowska, A. Cydzik-Kwiatkowska, K. Bernat and I. Wojnowska-Baryła, "Removal of bisphenol A (BPA) from biologically treated wastewater by microfiltration and nanofiltration," *Int. J. Environ. Sci. Technol.*, vol. 13, no. 9, pp. 2239–2248, 2016.

24. A. Bódalo, E. Gómez, A.M. Hidalgo, M. Gómez, M.D. Murcia and I. López, "Nanofiltration membranes to reduce phenol concentration in wastewater," *Desalination*, vol. 245, no. 1–3, pp. 680–686, 2009.

25. M.P. Zacharof, S.J. Mandale, P.M. Williams and R.W. Lovitt, "Nanofiltration of treated digested agricultural wastewater for recovery of carboxylic acids," *J. Clean. Prod.*, vol. 112, no. 5, pp. 4749–4761, 2016.

26. I. Koyuncu, O.A. Arikan, M.R. Wiesner and C. Rice, "Removal of hormones and antibiotics by nanofiltration membranes," *J. Membr. Sci.*, vol. 309, no. 1–2, pp. 94–101, 2008.

27. R.S. Harisha, K.M. Hosamani, R.S. Keri, S.K. Nataraj and T.M. Aminabhavi, "Arsenic removal from drinking water using thin film composite nanofiltration membrane," *Desalination*, vol. 252, no. 1–3, pp. 75–80, 2010.

28. L.M. Jin, S.L. Yu, W.X. Shi, X.S. Yi, N. Sun, Y.L. Ge, and C. Ma, "Synthesis of a novel composite nanofiltration membrane incorporated SiO_2 nanoparticles for oily wastewater desalination," *Polymer*, vol. 53, no. 23, pp. 5295–5303, 2012.

29. H. Wang, Q. Zhang and S. Zhang, "Positively charged nanofiltration membrane formed by interfacial polymerization of 3, 3′, 5, 5′-biphenyl tetraacyl chloride and piperazine on a poly (acrylonitrile) (PAN) support," *J. Membr. Sci.*, vol. 378, no. 1–2, pp. 243–249, 2011.

30. M. Röhricht, J. Krisam, U. Weise, U.R. Kraus and R.A. Düring, "Elimination of carbamazepine, diclofenac and naproxen from treated wastewater by nanofiltration," *CLEAN – Soil, Air, Water*, vol. 37, no. 8, pp. 638–641, 2009.

31. A. Sinha, P. Biswas, S. Sarkar, U. Bora and M.K. Purkait, "Separation of chloride and sulphate ions from nanofiltration rejected wastewater of steel industry," *J. Water Process. Eng.*, vol. 33, no. February, pp. 101–108, 2020.

32. D. Hu, Z.L. Xu and C. Chen, "Polypiperazine-amide nanofiltration membrane containing silica nanoparticles prepared by interfacial polymerization," *Desalination*, vol. 301, no. September, pp. 75–81, 2012.

33. J. Garcia-Ivars, L. Martella, M. Massella, C. Carbonell-Alcaina, M.I. Alcaina-Miranda and M.I. Iborra-Clar, "Nanofiltration as tertiary treatment method for removing trace pharmaceutically active compounds in wastewater from wastewater treatment plants," *Water Res.*, vol. 125, no. August, pp. 360–373, 2017.

34. Y. Han, Z. Xu and C. Gao, "Ultrathin graphenenanofiltration membrane for water purification," Adv. Funct. Mater., vol. 23, no. 29, pp. 3693–3700, 2013.

35. M. Bielska and J. Szymanowski, "Removal of methylene blue from waste water using micellar enhanced ultrafiltration," *Water Res.*, vol. 40, no. 5, pp. 1027–1033, 2006.

36. G. Barjoveanu and C. Teodosiu, "Priority organic pollutants removal by ultrafiltration for wastewater recycling," *EEMJ*, vol. 8, no. 2, pp. 277–287, 2009.

37. H. Isawi, "Evaluating the performance of different nano-enhanced ultrafiltration membranes for the removal of organic pollutants from wastewater," *J. Water Process. Eng.*, vol. 31, no. October, p. 100833, 2019.

38. W. Zhang, G. Huang, J. Wei, H. Li, R. Zheng and Y. Zhou, "Removal of phenol from synthetic waste water using Gemini micellar-enhanced ultrafiltration (GMEUF)," *J. Hazard. Mater.*, vol. 235, pp. 128–137, 2012.

39. K. Trivunac and S. Stevanovic, "Removal of heavy metal ions from water by complexation-assisted ultrafiltration," *Chemosphere*, vol. 64, no. 3, pp. 486–491, 2006.

40. M. Berradi and A. El Harfi, "Purification of the textile finishing effluents by the ultrafiltration technique," *Int. J. Adv. Chem.*, vol. 2, no. 2, pp. 62–65, 2014.

41. L.H.C. Hsieh, Y.H. Weng, C.P. Huang and K.C. Li, "Removal of arsenic from groundwater by electro-ultrafiltration," *Desalination*, vol. 234, no. 1–3, pp. 402–408, 2008.

42. M.A. Barakat and E. Schmidt, "Polymer-enhanced ultrafiltration process for heavy metals removal from industrial wastewater," *Desalination*, vol. 256, no. 1–3, pp. 90–93, 2010.

43. J.L. Acero, F.J. Benitez, F.J. Real and C. García, "Removal of phenyl-urea herbicides in natural waters by UF membranes: Permeate flux, analysis of resistances and rejection coefficients," *Sep. Purif. Technol.*, vol. 65, no. 3, pp. 322–330, 2009.

44. A.M. ElHadidy, S. Peldszus and M.I. Van Dyke, "An evaluation of virus removal mechanisms by ultrafiltration membranes using MS2 and φX174 bacteriophage," *Sep. Purif. Technol.*, vol. 120, pp. 215–223, 2013.

45. N.Y. Yip, A. Tiraferri, W.A. Phillip, J.D. Schiffman and M. Elimelech, "High performance thin-film composite forward osmosis membrane," *Environ. Sci. Technol.*, vol. 44, no. 10, pp. 3812–3818, 2010.

46. Y. Wang, R. Ou, Q. Ge, H. Wang and T. Xu, "Preparation of polyethersulfone/carbon nanotube substrate for high-performance forward osmosis membrane," *Desalination*, vol. 330, pp. 70–78, 2013.

47. X. Jin, J. Shan, C. Wang, J. Wei and C.Y. Tang, "Rejection of pharmaceuticals by forward osmosis membranes," *J. Hazard. Mater.*, vol. 227, no. August, pp. 55–61, 2012.

48. M. Li, X. Wang, C.J. Porter, W. Cheng, X. Zhang, L. Wang and M. Elimelech, "Concentration and recovery of dyes from textile wastewater using a self-standing, support-free forward osmosis membrane," *Environ. Sci. Technol.*, vol. 53, no. 6, pp. 3078–3086, 2019.

49. R.V. Linares, V. Yangali-Quintanilla, Z. Li and G. Amy, "Rejection of micropollutants by clean and fouled forward osmosis membrane," *Water Res.*, vol. 45, no. 20, pp. 6737–6744, 2011.

50. Y. Cui, Q. Ge, X.Y. Liu and T.S. Chung, "Novel forward osmosis process to effectively remove heavy metal ions," *J. Membr. Sci.*, vol. 467, no. October, pp. 188–194, 2014.

51. S.K. Nataraj, K.M. Hosamani and T.M. Aminabhavi, "Nanofiltration and reverse osmosis thin film composite membrane module for the removal of dye and salts from the simulated mixtures," *Desalination*, vol. 249, no. 1, pp. 12–17, 2009.

52. H. Ozaki and H. Li, "Rejection of organic compounds by ultra-low pressure reverse osmosis membrane," *Water Res.*, vol. 36, no. 1, pp. 123–130, 2002.

53. J.J.S. Alonso, N. El Kori, N. Melián-Martel and B. Del Río-Gamero, "Removal of ciprofloxacin from seawater by reverse osmosis," *J. Environ. Manage.*, vol. 217, no. July, pp. 337–345, 2018.

54. W.Y. Li, Q. Wang, Q.F. Xiao and B.Z. Dong, "Removal of tetracycline and oxytetracycline in water by a reverse osmosis membrane," *In 2009 3rd Int. Conf. Bioinform. Biomed. Eng.*, pp. 1–4, 2009.

55. J. Zhang, G. Weston, X. Yang, S. Gray and M. Duke, "Removal of herbicide 2-methyl-4-chlorophenoxyacetic acid (MCPA) from saline industrial wastewater by reverse osmosis and nanofiltration," *Desalination*, vol. 496, no. December, pp. 114691, 2020.

56. Z. Ujang and G.K. Anderson, "Application of low-pressure reverse osmosis membrane for Zn^{2+} and Cu^{2+} removal from wastewater," *Water Sci. Technol.*, vol. 34, no. 9, pp. 247–253, 1996.

57. M. Brik, P. Schoeberl, B. Chamam, R. Braun and W. Fuchs, "Advanced treatment of textile wastewater towards reuse using a membrane bioreactor," *Process Biochem.*, vol. 41, no. 8, pp. 1751–1757, 2006.

58. M. Amini, H. Etemadi, A. Akbarzadeh and R. Yegani, "Preparation and performance evaluation of high-density polyethylene/silica nanocomposite membranes in membrane bioreactor system," *Biochem. Eng. J.*, vol. 127, no. November, pp. 196–205, 2017.

59. L. Kovalova, H. Siegrist, H. Singer, A. Wittmer and C.S. McArdell, "Hospital wastewater treatment by membrane bioreactor: performance and efficiency for organic micropollutant elimination," *Environ. Sci. Technol.*, vol. 46, no. 3, pp. 1536–1545, 2012.

60. J. Chen, S. Liu, J. Yan, J. Wen, Y. Hu and W. Zhang, "Intensive removal efficiency and mechanisms of carbon and ammonium in municipal wastewater treatment plant tail water by ozone oyster shells fix-bed bioreactor-membrane bioreactor combined system," *Ecol. Eng.*, vol. 101, pp. 75–83, 2017.

61. W. Zhu, X. Wang, Q. She, X. Li and Y. Ren, "Osmotic membrane bioreactors assisted with microfiltration membrane for salinity control (MF-OMBR) operating at high sludge concentrations: Performance and implication," *Chem. Eng. J.*, vol. 337, pp. 576–583, 2018.

62. A. Alturki, J. McDonald, S.J. Khan, F.I. Hai, W.E. Price and L.D. Nghiem, "Performance of a novel osmotic membrane bioreactor (OMBR) system: Flux stability and removal of trace organics," *Bioresour. Technol.*, vol. 113, pp. 201–206, 2012.

63. J. Chen, X. Huang and D. Lee, "Bisphenol A removal by a membrane bioreactor," *Process Biochem*, vol. 43, no. 4, pp. 451–456, 2008.

64. V. Suganthi, M. Mahalakshmi and B. Balasubramanian, "Development of hybrid membrane bioreactor for tannery effluent treatment," *Desalination*, vol. 309, pp. 231–236, 2013.

65. F. Zaviska, P. Drogui, A. Grasmick, A. Azais and M. Héran, "Nanofiltration membrane bioreactor for removing pharmaceutical compounds," *J. Membr. Sci.*, vol. 429, pp. 121–129, 2013.

66. E. Wąsik, J. Bohdziewicz and M. Błaszczyk, "Removal of nitrate ions from natural water using a membrane bioreactor," *Sep. Purif. Technol.*, vol. 22, pp. 383–392, 2001.

9 Technological Advancement of Membrane Treatment for Tannery Effluents Management

Sumalatha Boddu and Anoar Ali Khan

CONTENTS

DOI: 10.1201/9781003165019-9

9.1   INTRODUCTION

Water consumption has been increasing day by day during the last few decades due to rapid industrialization, urbanization and population explosion. With the limited availability of fresh water, it is mandatory to reuse wastewater as much as possible after treating it with water treatment techniques. Various industries discharge industrial effluents into the environment that pose adverse effects on biodiversity and aquatic ecosystem (which causes environmental pollution). The main manufacturing industries are leather industries which are responsible for water pollution. Among all the industrial waste, tannery effluents are ranked as the worst pollutants that produce phototoxic effects and a high accumulation of heavy metals.

The leather industry is now recognized as a major industry of significant economic importance on an international scale. Apart from the production of leather goods, leather has also been widely used for the production of various cosmetics, chemical and fertilizer industries. About 65% of the world's production of leather is estimated to go into leather footwear. Tanning is the process of transforming animal skins (a natural, renewable resource) into leather. Tanning is claimed to be the second-oldest profession in the world. Although the tannery industry has been recognized globally, this industry has received criticism on environmental grounds as this industry has been viewed as a major source of water pollution.

In tanneries, major expansions have taken place due to rapid industrialization and the globalization of world economies, and with increasing awareness of environmental conservation and protection, they are thus obligated to treat effluents to a level that causes less impact on the environment. Governments have been focusing on implementing strict regulations for the effective treatment of industrial effluents (including tannery effluents). Effluent treatment has become an important social issue as a result of the toxic and potential health risks from effluents. Collectively, Tanneries could form the basis for a state-of-the art technology for treating effluents from the tanning industry.

The tanning process is used to produce several leather goods from raw hides like bags, sandals and belts, among others, to fulfill consumers' daily needs. The sequential steps involved in the leather-processing industry are shown in Figure 9.1. Each of these processes results in a large amount of effluent, which contains appreciable organic materials like bones, flesh, fat and so on; inorganic chemicals like $CaOH$, $NaCl_2$, Na_2S, Na_2SO_4 and others; and high biochemical oxygen

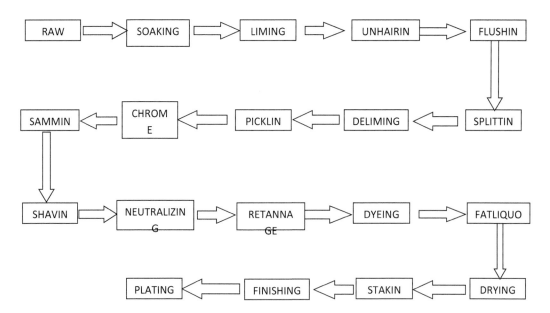

FIGURE 9.1   Steps involved in the processing of rawhide to leather.

demand (BOD) and chemical oxygen demand (COD) [1]. In processing of 1 kg raw hide to finished leather, approximately 40–45 liters of water are used. About 90% of water is discharged into the environment in the leather-making process [2]. However, the tanning process is specific to each end product and produces a significant number of various kinds of waste [3].

Tannery effluents carry heavy pollution loads due to a massive presence of highly colored compounds, sodium chloride and sulphate, various organic and inorganic substances, toxic metallic compounds, different types of tanning materials, which are biologically oxidizable, and large quantities of putrefying suspended matter. Tannery wastes are uniquely identified as activity-generated pollution of mixed character in the sense that both organic and inorganic constituents occur at concentrations higher than other wastes. During the processing of leather, the raw material is treated with various acids, alkalis, oils, fats, salts and tanning agents, among others, and as a result, toxic effluent is released as waste after processing. In the tannery industry, the processing treatments of hides and skins produce the biggest part of the effluent-loaded wastewater that originates from all the operations. It is either continuous from some operations or intermittent from few operations. Spent liquors, which are small in volume but highly polluted, from the soaking, liming, bating, pickling, tanning and finishing operation are discharged intermittently. Cr (III) salts have been used in tanning of leather. They are introduced with a variety of chemicals at each level of processing. At certain conditions, Cr (III) that can be oxidized as carcinogenic Cr (VI), which is the main cause for potential health risks in humans. The colored wastewater released from the tannery industry after processing is toxic and responsible for mutagenic impacts in living organisms. The chemical waste released from processing industries has huge COD and BOD, suspended and dissolved solids, chromium, surfactants and more, as shown in Table 9.1. Various effluent treatment technologies

TABLE 9.1
Tannery Wastewater Physical-Chemical Characteristics per the World Health Organization

S.No	Parameters	Permissible limits
1	pH	5.5–9
2	BOD	100
3	COD	250
4	Total suspended solids (mg/L)	600
5	Total dissolved solids (mg/L)	2100
6	Total hardness (mg/L)	600
7	Total alkalinity (mg/L)	600
8	Electrical conductivity (μS/cm)	1200
9	Chromium (mg/L)	2
10	Lead (mg/L)	0.1
11	Copper (mg/L)	0.1
12	Chlorine (mg/L)	1000
13	Iron (mg/L)	10
14	Chloride (mg/L)	1000
15	Calcium (mg/L)	200
16	Sulfate (mg/L)	650
17	Nitrates (mg/L)	80
18	Potassium (mg/L)	150
19	Zinc (mg/L)	1
20	Nickel (mg/L)	3
21	Cadmium (mg/L)	2
22	Oil & grease (mg/L)	10

have been proposed globally for treating tannery wastewater. The treatment of tannery wastewater is carried out by physical or chemical or biological or combination of these methods.

Tannery wastewaters produced from processing are mainly characterized by their measurements of COD, BOD, total suspended solids (TSS), total dissolved solids (TDS), chromium salts and sulfides, among other elements. Tannery wastewater is dark brown in color and rich in toxic chemical constituents.

Various chemical, physicochemical and biological technologies were investigated for treating highly loaded, toxic tannery effluent. Most advanced technologies were tried for the maximum removal efficiency of pollutants. The higher amounts of chemicals utilized, toxic sludge produced and treatment areas are problems for conventional methods. For treating wastewater, there are many conventional water treatment techniques like filtration, coagulation and flocculation, precipitation, ion exchange, adsorption and membrane separation. Except membrane separation, other techniques have several disadvantages like biomass generation, excessive chemical utilization, inefficient removal and the more.

Among the several conventional techniques, the membrane separation process has several advantages, such as effective separation, cost-effectiveness and less space being required [4]. In this regard, the treatment of tannery wastewater using membrane separation have been an active field research in past few years. The only disadvantage of the membrane separation process is fouling, which reduces the efficiency of separation. Integrated membrane technology was developed to overcome membrane fouling and achieve good separation. The main aim of the current chapter is to focus on the treatment of effluent from tannery industries using membrane separation processes.

9.2 WASTEWATER TREATMENT TECHNIQUES

9.2.1 Coagulation and Flocculation

The well-known conventional techniques for wastewater treatment are coagulation and flocculation, which are used for removing turbid and suspended particles. Aluminum and ferric sulfates are the most frequently used coagulants in wastewater treatment. After adding the coagulant to wastewater, the impurities are entrapped and floc formation takes place. These flocs are removed by sedimentation and filtration.

Flocculation is the process where impurities are agglomerated together to form floc. These flocs are removed by flotation. Polyacrylamide and polyferricsulfate, among others, are commonly used flocculant materials in wastewater treatment [5]. The main disadvantages of coagulation and flocculation are sludge formation and high operational costs.

9.2.2 Precipitation

Precipitation is a simple, well-known and effective technique for the separation of the dissolved particles (heavy metals) by using reagents into insoluble precipitates that are further removed by filtration, floatation and sedimentation. The main limitations of this method are large quantities of silt formation, chemical consumption and its inefficiency for heavy metal ions removal present at low concentrations [6].

9.2.3 Ion Exchange

The ion exchange method is the most frequently used technique for heavy metal removal from industrial wastewater. This method is usually utilized for demineralization or water softening. Ion-exchange resins are used to remove positively and negatively charged ions from wastewater using ion exchange process. Cation exchanger or cation exchange resins have ionizable groups such as $-OH$, $-COOH$ and $-SO_3H$. An anion exchanger or anion exchange resins have ionizable groups such as $-NHCH_3$, $-OH$ and $-NH_2$. This process has some limitations like large volume requirements, a high initial investment cost, being inefficient for certain target pollutants and being a pH-sensitive process.

Advanced oxidation processes (AOPs) are used for removing organic pollutants present in wastewater by oxidation through reactions with hydroxyl radicals. AOPs comprise several oxidation processes, such as Fenton, ozone, ultraviolet (UV) radiation and hydrogen peroxide. The efficiency of AOP processes depends on the concentration of the pollutant, operating temperature and the oxidant. High investment costs and unknown intermediates formation are problems with this method.

9.2.4 ADSORPTION

Adsorption is the most economical and efficient method for removing organic and inorganic pollutants from industrial wastewater. It is a surface phenomenon in which a substance (adsorbate) accumulates on the surface of the solid (adsorbent). The most commonly used adsorbent is activated carbon, which is used for removing organic pollutants industrial effluents owing to its large surface area and high-volume micro- and mesopores. This process has some limitations, that is, removing adsorbent that needs regeneration and incineration. Regeneration is a costly process in which sometimes a loss of material takes place [7].

9.2.5 ACTIVATED SLUDGE PROCESS

The **activated sludge process** is a more efficient method for treating industrial wastewater in which several microorganisms adsorb toxic organic pollutants having strong decomposition characteristics. Sludge formation and the requirement of needing a large space are major limitations in this method [8–9].

9.2.6 AERATED LAGOONS

Aerated lagoons are the most commonly used biological method for removing BOD and suspended solids. Wastewater from primary treatment is sent through large aeration tanks for 2–6 days. Large space requirements, the contamination of microorganisms and cost are the shortcomings of this method.

9.2.7 TRICKLING FILTER

Trickling filters are a cost-effective method for separating organic pollutants. It is an aerobic process in which organic matter is oxidized into methane, CO_2 and water by the action of microorganisms. Clogging and odor are the main drawbacks of this method.

Traditional methods are utilized to reduce the different organic compounds and metal ion concentrations to the required regulatory standards. Physical treatment techniques, such as adsorption and filtration, are not efficient for attaining discharge limits. Floatation and coagulation create a large quantity of sludge whereas chemical oxidation needs storage and the transportation of reactants and has a low-capacity rate. A high investment cost is required for AOPs. To overcome the drawbacks associated with conventional treatment methods, researchers are trying to develop novel technologies. In this regard, membrane technology has gained importance and is more effective in cases of lower levels of pollutant concentration.

9.3 MEMBRANE SEPARATION PROCESSES

In the past few years, membrane separation processes have been used to recover chromium from leather tanning industry effluents. Large-scale membrane separation processes are feasible and cost-effective. The applications of membrane separation are gradually increasing due to the decrease in cost of membranes. The main technological developments of membrane separation processes over conventional techniques are the ease of separation progress (better separation performance) and fewer energy requirements. The best available technology for treating different industrial separation processes is membrane technology.

To attain desired separation through membrane-based separation techniques, selecting a suitable process with the appropriate driving force, size, shape and membrane is required. Membrane separation processes are classified into pressure-, concentration-, electrical- and thermal-driven processes, which are shown in Table 9.2.

TABLE 9.2

Classification of Membrane Separation Processes

Membrane separation process	Type of membrane	Pore size	Driving force	Mechanism of separation (principle of separation)	Application
Pressure-driven membrane separation process					
Microfiltration	Porous	0.05–10 µm	Pressure difference (0.1–2 bar)	Sieving	Food, pharmaceutical industries, water treatment
Ultrafiltration	Porous	1–100 nm	Pressure difference (1–10 bar)	Sieving	Textile, food, pharmaceutical industries, dairy, water treatment
Nanofiltration		10–1 nm	Pressure difference (10–25 bar)	Solution-diffusion	Brackish water desalination, wastewater treatment
Reverse osmosis		<2 nm	Pressure difference (15–80 bar)	Solution-diffusion	Brackish and seawater desalination, concentration of juice and milk
Concentration-driven separation process					
Pervaporation	Nonporous		Vapor pressure difference (0.001–1 bar) Concentration difference	Solution-diffusion	Hydrogen, helium recovery
Gas separation	Porous/ nonporous	<1 µm	Partial pressure difference Concentration difference	solution/diffusion (nonporous membranes) Knudsen flow (porous membranes)	Removal of organic components from water
Dialysis			Concentration	Solution-diffusion	Hemodialysis, paper and pulp industry
Electrical-driven membrane separation process					
Electrodialysis	nonporous		Electrical potential difference	Donnan exclusion mechanism	Seawater desalination, separation of amino aids
Temperature-driven membrane separation process					
Membrane distillation		0.2–1 µm	vapor pressure difference	vapor–liquid equilibrium	Seawater desalination, semiconductor industry

9.3.1 Microfiltration

Microfiltration (MF) is used for removing microorganisms and colloidal/suspended pollutants from industrial effluents. The effluents are passed through a porous membrane (0.1–10 μm) with 0.1–2 bar pressure. In this process, separation is based on a sieving mechanism, with particles bigger than the pore size being retained on the membrane and smaller particles passing through the membrane. Initially, MF was used for removing microorganisms in drinking water; since then, it has been used in the food, pharmaceutical, petroleum and biotechnology industries [10].

9.3.2 Ultrafiltration

Generally, MF and ultrafiltration (UF) are used as pretreatments for reverse osmosis (RO) and nanofiltration (NF) techniques. UF membranes have smaller pore sizes (1–100 nm) compared to MF. Because of the smaller pore sizes, high pressure is essential for attaining maximum permeability. So it requires more pressure (1–10 bar) for getting the desired output. UF is used for removing macromolecules, with larger ones (>300 kilo Dalton) being retained by the membrane and smaller ones permeating freely through membrane [11]. UF is applied for in the water purification, food, dairy and textile industries.

9.3.3 NF

NF is used for removing low-molecular-weight organic composites, colloids and divalent salts with molecular weights of 100–350 Dalton. NF membranes have pore sizes of 1–10 nm, which is smaller than UF membranes. NF requires low pressure of 10–25 bar for attaining a higher flux. NF is used for desalinating brackish water and wastewater treatment [12].

9.3.4 RO

RO is used for excluding all dissolved solids and suspended solids (smaller pollutants). RO is a pressure-driven process with an applied range of 20–80 bar, which exceeds the osmotic pressure and eliminates smallest particles (<350 Da) with high separation efficiency. RO membranes are dense membranes with pore sizes <2 nm. RO is mainly used for seawater desalination and is applied to treating wastewater from the tanning, leather, textile, food and petroleum industries [13].

9.3.5 Pervaporation

Pervaporation is used for the removal of trace elements of volatile components present in liquid mixtures by vapor pressures through a porous/nonporous membrane [14–15]. It is applied in the separation of hydrocarbons (petrochemical industries) and volatile organic compounds. In this technique, the concentration difference is the driving force. In this method, separation is achieved based on a solution–diffusion mechanism that results in vapor as the permeate, which may be removed by either applying low pressure or flowing an inert medium.

9.3.6 Electrodialysis

Electrodialysis is used to remove selective ionic components from aqueous solutions by applying electric potential through ion-exchange membranes. Ion-exchange membranes (IEMs) are made from polymeric materials with fixed ionic charge groups in the polymeric matrix, and these are dense in nature. IEMs are classified into two types: cation exchange membranes (CEMs) and anion exchange membranes (AEMs). CEMs contain negatively charged groups in their polymer matrix while AEMs contain positively charged groups that can selectively pass oppositely charged ions based on the Donnan exclusion principle. It is mostly used for desalinating seawater and removing organic acids from the food and pharmaceutical industries.

9.3.7 MEMBRANE DISTILLATION

From many years, membrane distillation (MD) has been a promising method for desalinating seawater and treating wastewater. Almost all macromolecules, colloids, volatile and nonvolatile substances and salts are removed by hydrophobic membranes as compared to hydrophilic membranes. In MD, only vapor molecules are transported through a hydrophobic membrane from a feed mixture as permeate due to the vapor pressure difference, which is the driving force. This process has some limitations, such as a higher energy consumption, high sensitivity to temperature polarization, membrane wetting and poor separation of organic solvents with less surface tension [16–17].

9.3.8 MEMBRANE BIOREACTOR

Membrane bioreactors (MBRs) are activated sludge treatment processes used for effluent treatment wastes released from the industries after processing. In the past, MBRs have emerged as efficient techniques for industrial wastewater treatment, in which a permeable selective membrane, for example, MF or UF, is integrated with a biological process – specifically a suspended growth bioreactor. The main disadvantages of MBRs are their cost and energy requirements. However, concentration polarization and membrane fouling, which reduce the performance and lifetime of membrane, are also major challenges in this technique. Fouling requires frequent membrane cleaning. After cleaning, MBR requires fresh water and chemicals.

Almost all the commercial membrane processes available today use the membranes as filters because they reject solid materials, and result in clarified or disinfected effluent. It is now being widely used for the treatment of municipal and industrial wastewater. Therefore, MBR technology is regarded as a key element of advanced wastewater treatment. It is a focus when moving toward sustainable water management across the municipal and industry sectors.

9.4 INTEGRATED/HYBRID MEMBRANE SEPARATION PROCESSES

Membrane integrated/hybrid techniques (HMPs) are in advance for treating complex wastewater. Such integrated techniques may have numerous permutations and combinations of techniques, such as physicochemical, chemical, biological and membrane separation techniques. In the modern era, the membrane market is continually growing with admirable prospects in the future with typical use in industrial research and development; meanwhile, membrane technology offers ability, versatility and compactness to be merged with other separation techniques to give integrated/hybrid techniques.

Limitations in operations, stream concentration and fouling, which reduces the life span and performance of membranes, are the main shortcomings of membrane separation processes. In order to overcome the obstacles in membrane separation techniques, integrated technologies are proposed to achieve maximum productivity of targeted separation processes. It is difficult to separate highly polluted wastewater using membrane separation processes. Furthermore, reusing water by recycling it is not sufficient after the physical chemical treatment of waste streams. In order to overcome the limitations of standalone techniques and performance enhancement, an integrated technology was proposed. The main objective of the HMP technique is to minimize the operating cost of the process. Many researchers recommend hybrid treatment processes for water reuse. Today HMP systems have the tangible possibility of decreasing energy requirements, reducing harmful effects on environment and reusing and recycling by-products, leading to the concept of zero liquid discharge.

Tannery wastewater contains more concentrations of ammonium, chromium, sulfate, sulfide, chloride and high weight of organic matter; due to this load, it is difficult to treat tannery effluent

[18]. However, the high load of organics has not been removed by primary and secondary treatments, which inhibit microorganisms based on the size of the molecules, the nature of the functional groups and the solubility in water [19]. In order to reuse tannery wastewater, the RO technique is essential for removing high concentrations of chloride. With the application of a simple RO membrane, the permeate produced from the system is reusable in the tannery production unit, which greatly reduces the requirement of groundwater consumption. Hence, tailored biological treatment–RO hybrid treatment techniques provide satisfactory results compared to single biological treatment processes [20].

Suganthi et al. [21] investigated a hybrid membrane process that coupled electrocoagulation, the activated sludge process and MF for the exclusion of color and COD from tannery effluents. The resulting HMBR provided high-quality treated water. Hence, the HMBR gave a good result (increasing permeate flux, fouling and treatment efficiency) compared with MBR. The color and maximum COD removed by the HMBR and the MBR are 90% and 93% and 73% and 76%, respectively.

Bhattacharya et al. [22] proved that two-stage membrane processes MF followed by RO provide good results compared to conventional processes. The resulting water was suitable for reuse in the process of tanning, minimizing the freshwater requirement during the tanning process. The cost of the proposed hybrid technology by membrane processes was less than the conventional processes. After tanning with treated water, the finished product was analyzed and the results compared with leather produced by fresh water. Hence, the two-stage process was successfully decreasing the organic and inorganic loading such as BOD, COD, heavy metals etc.

Fababuj-Roger et al. [23] verified the integration of physical chemical treatment with MF, UF and RO processes for water reuse in tannery industries. Initially, the water contains COD ranges from 3000 g/ml to 4000 g/ml and conductivities of nearly 20 ms/cm. For the exclusion of soluble COD, coupling of physical-chemical treatment with UF was not efficient when used for removing concentrations greater than 2000 mg/l. For attaining reusable water, the embedding of filtration, UF and RO was used in this study. The RO permeate flux got to 40 l/(m2 h) at 30 bar. During the RO process, membrane fouling was not observed.

Dasgupta et al. [24] studied the efficacy of a coagulation and NF integrated technique for removing chromium from tannery wastewater. Furthermore, an individual optimized coagulation pretreatment process designed for treating raw wastewater was conducted by using response surface methodology. Furthermore, a comparative study was carried out between the hybrid coagulation-NF and a coagulation technique in terms of qualities of permeate in order to estimate the feasibility as showed by the integrated process in producing treated effluent. Hence, the treated water from the hybrid process was suitable for reuse. Additionally, the membrane fouling tendency in coagulation method and hybrid coagulation-NF process were compared. Finally, this study concluded that a reduction in membrane fouling was attained by the tailored coagulation-NF technique.

An MBR is an integrated process that is the combination of a membrane separation process and a biological treatment. As discussed by Faisal et al. [25], an MBR method is used when organic loads and suspended solids are present. The MBR technique produces a high-quality effluent that is suitable for reuse.

9.5   MEMBRANE FOULING

During filtration, membrane pores or surfaces are covered with compounds; this phenomenon is called fouling. Simply, it is the blockage of the membrane's pores. Fouling is achieved by the combination of the adsorption of particulates and sieving onto the membrane's surface. Membrane fouling leads to a rise in complexity of filtration and hydraulic resistance and decreases the rate of permeate production. As the pores of the membrane get blocked and the filtration efficiency decreases, the energy requirement increases. Membrane fouling is the most challenging issue in membrane separation processes. Due to the continuous usage of membrane, the pores become blocked, which further leads to low filtration, high time requirements, damage to the membrane and more, so to overcome all these problems and increase the life expectancy of the membrane, the membrane needs to be cleaned.

9.5.1   Membrane Cleaning

There are several membrane cleaning techniques like backward flush, forward flush, chemical cleaning and air/water flush.

9.5.1.1   Forward Flush

In a forward flush, the membrane is flushed with feed water at a high velocity, which creates turbulence and helps in removing particles. In this process, the particles that are absorbed to the membrane alone are released, and particles absorbed to membrane pores are not discharged.

9.5.1.2   Backward Flush

It is a reverse filtration process. The pressure on the permeate side of the membrane is higher than the pressure within the membranes, causing the pores to be cleaned. A backward flush is executed under a pressure of about 2.5 times greater than the production pressure.

9.5.1.3   Air or Air/Water Flush

This is a forward flush during which air is injected into the supplier pipe. Because air is used (while the water speed remains the same), a much more turbulent cleaning system is created. Here, the flushing occurs inside the membrane using an air/water mixture. In this process, the air/water mixture leads to the formation of bubbles, which create further turbulence. Due to this turbulence, the fouling of the membrane is removed.

When forward and backward flushes do not sufficiently restore the membrane, then a chemical cleaning process is used. In a chemical cleaning process, the membranes are soaked with a solution of chlorine bleach, hydrochloric acid or hydrogen peroxide. After soaking the membrane for a certain period, the membrane is then rinsed by forward or backward flushing.

9.6   MEMBRANE MODULES

To achieve the required separation, industrial membrane plants require hundreds to thousands of square meters of membrane. There are many ways for economic membrane packages to provide huge surface area for effective and efficient separation. Usually, the designs for membrane modules are interrelated the efficiency of prevention of membrane fouling. Commonly used membrane modules are plate and frame, spiral wound, tubular and hollow fiber. The typical characteristics of membrane modules are listed in Table 9.3.

TABLE 9.3
Typical Characteristics of Membrane Modules

Membrane module	Packing density (m^2/m^3)	Resistance to fouling	Ease of cleaning	Relative cost	Application
Plate and frame	30–500	Good	Good	High	Microfiltration, ultrafiltration, dialysis, reverse osmosis, pervaporation
Spiral wound	200–800	Moderate	Fair	Low	Microfiltration, ultrafiltration, dialysis, reverse osmosis, gas permeation
Tubular	30–200	Very good	Excellent	High	Ultrafiltration, reverse osmosis
Hallow fiber	500–2000	Poor	Poor	Low	Ultrafiltration, dialysis, reverse osmosis, gas permeation

9.6.1 PLATE-AND-FRAME MODULES

One of the initial types of membrane systems is the plate-and-frame module, which is substituted by spiral-wound modules and hollow-fiber modules because they are relatively cheaper than plate-and-frame modules. At present, plate-and-frame modules are used minimally in RO and UF processes with highly fouling conditions.

9.6.2 TUBULAR MODULES

Tubular modules are used especially when a high resistance to membrane fouling is necessary, which are usually bounded to UF applications. These membranes contain small tubes with diameters of 0.5–1 cm embedded inside a single large tube. The large number of tubes is held in series inside a tubular membrane system.

9.6.3 SPIRAL-WOUND MODULES

Commercial-scale modules contain a few membrane envelopes each having area of 10–20 ft^2, enclosed around the axial collection pipe. The typical commercial spiral wound is 0.66 ft in diameter and is 3.33 ft long. The pressure drop is reduced by multi-envelope designs in which permeate travels through a central pipe.

9.6.4 HOLLOW-FIBER MODULES

Usually, hollow-fiber modules are 10–20 cm in diameter and have a height ranging 3–5 ft. They are mostly operated with the feed stream on the exterior of the fiber. Water traverses into the lumen of the fiber inside the membrane. A large number of fibers are composed together and "potted" in an epoxy resin at two ends and placed into an outer shell.

9.7 FUTURE PERSPECTIVES

At present, the application of hybrid/integrated processes can change the process economics and performance even though the feasibility and accessibility of designs are deficient. So much attention is required when applying HMP processes to real tannery effluents. HMP/integrated techniques have several advantages; still, there are some demerits, such as high energy and pressure requirements. To overcome the energy requirement for HMP processes, future investigations should be done on waste heat recovery and utilizing it a proper direction that results in the technique being economical. Another future perspective is to design HMP systems with the aim of decreasing environmental harmful effects; only then will sustainable development be possible.

BIBLIOGRAPHY

1. Das, C., Das Gupta, S., & De, S. (2007). Selection of membrane separation processes for treatment of tannery effluent. Journal of Environmental Protection, 1, 75–82.
2. Gadlula, S., Ndlovu, L. N., Ndebele, N. R., & Ncube, L. K. (2019). Membrane technology in tannery wastewater management: A review. Zimbabwe Journal of Science & Technology, 14, 57–72.
3. Zouboulis, A. I., Peleka, E. N., & Ntolia, A. (2019). Treatment of tannery wastewater with vibratory shear-enhanced processing membrane filtration. Separations, 6(2), 20. doi:10.3390/separations6020020.
4. Sarkar, S., & Chakraborty, S. (2020). Nanocomposite polymeric membrane a new trend of water and wastewater treatment: A short review. Groundwater for Sustainable Development, 100533.
5. Fu, F., & Wang, Q. (2011). Removal of heavy metal ions from wastewaters: A review. Journal of Environmental Management, 92, 407–418.

6. Zinicovscaia, I. (2016). Conventional methods of wastewater treatment. In Cyanobacteria for bioremediation of wastewaters (pp. 17–25). Cham: Springer.

7. Sivagami, K., Sakthivel, K. P., & Nambi, I. M. (2018). Advanced oxidation processes for the treatment of tannery wastewater. Journal of Environmental Chemical Engineering, 6(3), 3656–3663.

8. Haydar, S., Aziz, J. A., & Ahmad, M. S. (2007). Biological treatment of tannery wastewater using activated sludge process. Pakistan Journal of Engineering and Applied Sciences, 1, 61–66.

9. Elmagd, A. M. A., & Mahmoud, M. S. (2014). Tannery wastewater treatment using activated sludge process system (Lab Scale Modeling). International Journal of Engineering and Technical Research (IJETR), 2, 21–28. ISSN: 2321–0869.

10. Van Der Bruggen, B., Vandecasteele, C., Van Gestel, T., Doyen, W., & Leysen, R. (2003). A review of pressure-driven membrane processes in wastewater treatment and drinking water production. Environ. Prog., 22, 46–56. doi:10.1002/ep.670220116.

11. Christensen, E. R., & Plaumann, K. W. (1981). Waste reuse: Ultrafiltration of industrial and municipal wastewaters. Journal Water Pollution Control Federation, 53, 1206–1212.

12. Choi, J.-H., Fukushi, K., & Yamamoto, K. (2008). A study on the removal of organic acids from wastewaters using nanofiltration membranes. Separation and Purification Technology, 59, 17–25.

13. Bódalo-Santoyo, J. L., Gómez-Carrasco, E., Gómez-Gómez, F., Máximo-Martín, & Hidalgo-Montesinos, A. M. (2003). Application of reverse osmosis to reduce pollutants present in industrial wastewater. Desalination, 155, 101–108.

14. Crespo, J. G., & Brazinha, C. (2015). Fundamentals of pervaporation. In Pervaporation, vapour permeation and membrane distillation (pp. 3–17). Oxford: Woodhead Publishing.

15. Jyoti, G., Keshav, A., & Anandkumar, J. (2015). Review on pervaporation: Theory, membrane performance, and application to intensification of esterification reaction. Journal of Engineering, Article ID 927068. doi.org/10.1155/2015/927068.

16. Alkhudhiri, A., Darwish, N., & Hilal, N. (2012). Membrane distillation: A comprehensive review. Desalination, 287, 2–18.

17. Lee, J. G., Alsaadi, A. S., Karam, A. M., Francis, L., Soukane, S., & Ghaffour, N. (2017). Total water production capacity inversion phenomenon in multi-stage direct contact membrane distillation: A theoretical study. Journal of Membrane Science, 544, 126–134.

18. De Gisi, S., Galasso, M., & De Feo, G. (2009). Treatment of tannery wastewater through the combination of a conventional activated sludge process and reverse osmosis with a plane membrane. Desalination, 249(1), 337–342.

19. Kennedy, L. J., & Sekaran, G. (2004). Integrated biological and catalytic oxidation of organics/inorganics in tannery wastewater by rice husk based mesoporous activated carbon. Bacillus sp. Carbon, 42(12–13), 2399–2407.

20. Fababuj-Roger, M., Mendoza-Roca, J. A., Galiana-Aleixandre, M. V., Bes-Pia, A., Cuartas-Uribe, B., & Iborra-Clar, A. (2007). Reuse of tannery wastewaters by combination of ultrafiltration and reverse osmosis after a conventional physical-chemical treatment. Desalination, 204(1–3), 219–226.

21. Suganthi, V., Mahalakshmi, M., & Balasubramanian, B. (2013). Development of hybrid membrane bioreactor for tannery effluent treatment. Desalination, 309, 231–236.

22. Bhattacharya, P., Roy, A., Sarkar, S., Ghosh, S., Majumdar, S., Chakraborty, S., . . . Bandyopadhyay, S. (2013). Combination technology of ceramic microfiltration and reverse osmosis for tannery wastewater recovery. Water Resources and Industry, 3, 48–62.

23. Fababuj-Roger, M., Mendoza-Roca, J. A., Galiana-Aleixandre, M. V., Bes-Pia´, A., Cuartas-Uribe, B., & Iborra-Clar, A. (2007). Reuse of tannery wastewaters by combination of ultrafiltration and reverse osmosis after a conventional physical-chemical treatment. Desalination, 204, 219–226.

24. Dasgupta, J., Mondal, D., Chakraborty, S., Sikder, J., Curcio, S., & Arafat, H. A. (2015). Nanofiltration based water reclamation from tannery effluent following coagulation pretreatment. Ecotoxicology and Environmental Safety, 121, 22–30.

25. Hai, F. I., Yamamoto, K., Nakajima, F., & Fukushi, K. (2010). Recalcitrant industrial wastewater treatment by membrane bioreactor (MBR). In S. Gorley (Eds.), Handbook of membrane research: Properties, performance and applications (pp. 67–104). New York: Nova Science Publishers.

10 The Application of Pressure-Driven Membrane Technology for the Treatment of Pulp and Paper Industrial Bleaching Wastewater

Anusha Chandra and Anoar Ali Khan

CONTENTS

10.1 INTRODUCTION

Wastewater discharged from the pulp and paper industry is one of the major sources of industrial water pollution. The paper and pulp industry is one of the largest consumers of water that consumes large quantities of fresh water, that is, approximately 270–455 m^3 per ton of paper produced [1–3]. This industry requires a huge amount of water for processing operations, internal cleaning and cooling purposes. Fresh water is mainly required in various unit operations and processes like raw material processing (washing and dissolution), pulping (pulp digesting and washing), pulp bleaching (washing between bleaching stages, extraction) and paper making [2,4]. A detailed flow sheet of various unit processes carried out in pulp and paper industry is given in Figure 10.1.

Almost 70% of the water is utilized as process water [2]. In the same way, the paper industry releases large volumes of liquid (aqueous) effluent, 220–380 m^3 per ton of paper [1,2,5,6]. The amount and nature of wastewater generated can vary depending on types of raw material (wood, agricultural biomass), processes like pulping (chemical – kraft, sulfite; mechanical), bleaching and extraction, papermaking, effluent internal recirculation, quantity of water used and finished paper product grade [1,2]. In particular, in pulping and papermaking, the use of different varieties of

DOI: 10.1201/9781003165019-10

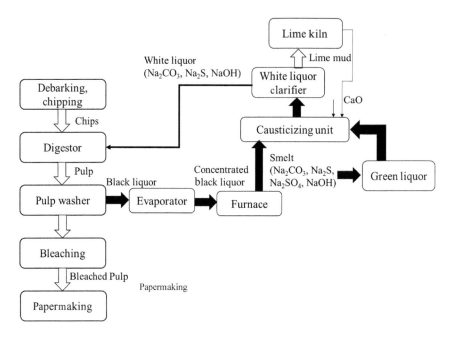

FIGURE 10.1   Detailed flowsheet of various unit processes in the pulp and paper industry.

chemicals alters the effluent's characteristics. Chemical pulping and bleaching produce larger quantities of toxic effluents that impinge the environment.

However, over the past few years, water consumption in paper industries has been reduced drastically by the internal recirculation of treated water. While in the case of a closed water circuit in the paper industry, recycling minimizes the consumption of water, where the presence of contaminants or pollutants in process water makes the water unsuitable for reuse. The poor quality of water affects the quality of the final product, scale deposition and corrosion of equipment [2,4,7,8]. Therefore, using recycled water by treating effluents could be a choice for reducing water consumption. For sustainable water management, wastewater treatment and the reuse of water would substantially decrease freshwater usage and the pollution load to the environment. The selection of effective and economical treatment methods for purifying the effluent before recycling should be required.

The effluents emitted from the paper industry are complex in nature containing organic compounds, mainly the degradation products of lignocellulosic materials of plants such as lignin and its extractives, cellulose, hemicellulose, carbohydrates and solid materials [2,6]. Globally, pulp production is increasing and accounted for 178.8 million metric tons in 2015. The main raw materials for the pulp and paper industry are wood and agricultural residues, which mainly contains lignin (wood – 25–30%, agricultural residue – 15–20%). The lignin content in the raw material causes pulp darkening during the cooking of pulp, and this has to be removed by bleaching. The main chemicals for bleaching in India are chlorine and chlorine-related compounds. During the bleaching process, elemental chlorine breaks down the lignin and results in chlorinated organic compounds such as 2,3,7,8-tetrachlorodibenzodioxin, which is carcinogenic and bio-accumulated [9]. The kraft or sulfite process is used for pulp production and mainly contains lignin [10]. The presence of color-imparting phenolic compounds, that is, lignin compounds and their derivatives, are responsible for the dark color of wastewater [11]. During the bleaching process, lignin and its derivatives are removed from pulp, imparting a dark color to the wastewater. In addition, chlorinated organics and chlorine-related toxic compounds are released into paper mill wastewater. These colored, toxic compounds degrade the quality of water, which affects the aquatic life in the water bodies receiving paper mill effluents. The color present in the water bodies lowers the passage of sunlight and thereby decreases the photosynthesis of flora and fauna and makes them aesthetically unpleasant. Moreover, the organic matter and color present in effluents reduce the dissolved oxygen content in water bodies, leading to the death of aquatic life [2].

Most often, in the bleaching process, chlorine and its related bleaching agents are used because they are effective and cheap. During the process of bleaching the pulp with hypochlorite, chlorine or chlorine dioxide, chloroform would be formed and is discharged as bleaching effluent from kraft pulp mill. Not only chloroform but also 17 related volatile organochlorines are formed during kraft pulp bleaching process (or ECF bleaching). The effluent containing chloroform and related organochlorine compounds causes adverse effects on human beings and the environment. Chloroform is carcinogenic, meaning it can damage human organs (kidney, liver, heart, etc.) and cause cancer [12]. During chlorination (bleaching process), lignin, carbohydrates and chlorine-related compounds are degraded and dissolved into spent bleaching effluent [13]. The bleaching effluents containing chlorinated phenols and lignins impose an acute toxicity to aquatic life and humans, such as growth retardation, infertility, endocrine disruption, improper liver functioning and more [2].

Paper mill effluents include high amounts of color; total suspended solids (TSS); turbidity; biochemical oxygen demand (BOD); organochlorinated compounds, namely, absorbable organic halides (AOX); and chemical oxygen demand (COD). As per the environment standards and guidelines, effluent containing toxic and recalcitrant compounds must be treated before being discharged into the environment. Conventionally, much of the wastewater generated in the pulp and paper industry is incinerated in a boiler to recover chemicals and produce energy. Every year, nearly 49.5 million metric tons of lignin present in pulp mill wastewater is burned [10]. This generates solid waste and gaseous effluents, that is, particulate matter. From the boiler/furnace, the smelt obtained contains valuable chemicals (Na_2CO_3, Na_2S, Na_2SO_4 and $NaOH$) used for cooking pulp in digestors. During the conversion of green liquor to white liquor in a causticizing unit, a large amount of lime has to be added, and in return, a huge quantity of lime mud (solid waste) is generated. This process is substantially an energy-intensive process. The other available physical, chemical and biological treatment methods used for removing color and toxic AOX compounds are coagulation-flocculation; precipitation; adsorption; ozonation; electrochemical methods, that is, electrocoagulation; wet air oxidation; Fenton's process; and more [14–16]. However, over several years, researchers have studied the effectiveness of the conventional incineration of paper mill wastewater (black liquor and green liquor), physical and chemical treatment methods; these are not implemented at an industrial scale as the methods are energy-intensive and very expensive for treating the unit volume of the effluent. Physical and chemical treatment methods are effective and able to reduce the color, TSS, turbidity, higher-molecular-weight lignins and COD whereas 80% of the biodegradable fraction of BOD and low-molecular-weight lignins could be reduced by biological methods. However, biological methods are not effective for reducing COD, color and chlorinated phenols. The traditional biological methods such as activated sludge, aerated lagoons and anaerobic degradation techniques reduce 40% of AOX and other stringent chlorinated compounds at high cost and with an excess generation of sludge [2,11]. The use of a bioremediation process with white-rot fungi for industrial treatment of large amounts of bleach plant wastewater needs suitable cultivation procedures, aeration and co-substrate, which would be costly [11]. Apart from these regular treatment methods, membrane processes are gaining attention for the cost-effective filtration of industrial effluents and for minimizing freshwater consumption. Pressure-driven membrane processes like microfiltration (MF), ultrafiltration (UF), nanofiltration (NF) and reverse osmosis (RO) are widely used to reduce COD and BOD and remove the toxic compounds (chlorinated compounds and lignin derivatives) from the paper industry effluent [1]. This chapter mainly focus on the application of pressure-driven membrane separation processes and discusses membrane fouling and ceramic membrane utilization for treating highly colored, alkaline and toxic bleaching plant effluents.

10.2 OVERVIEW OF EFFLUENTS IN THE PAPER INDUSTRY

10.2.1 SOURCES OF WASTEWATER EMITTING FROM DIFFERENT PROCESSES

The composition of wastewater generated from paper mills varies, which is mainly dependent on the raw material selected, the type of pulping process, bleaching and the quality of finished paper [2,17,18]. The freshwater consumption and effluents (pollutants) released from various processes in the paper industry are shown in Figure 10.2.

10.2.1.1 Raw Materials Processing (Debarking and Chipping of Wood)

The initial step is raw material washing and processing. In this step, the wood logs are soaked in water to clean the dirt, and then they are debarked and chopped into small wood chips. The wastewater generated from this step contains dirt, suspended solids, bark, small solids (wood), BOD and COD.

10.2.1.2 Pulping (Cooking/Digesting of Wood)

Subsequently, the chips are sent to digestor where wood chips are cooked (digesting) in water and chemicals (Na_2CO_3, Na_2S and $NaOH$) at 150–200°C. Based up on the raw material, different pulping techniques are used: (1) mechanical pulping or (2) chemical pulping: the kraft and sulfite process. After digestion, the fiber (cellulose) from the pulp is separated, and the spent liquid, known as black liquor, contains highly pollutants such as alkali, Na_2S, lignin and its derivatives, fatty acids, BOD and COD. Due to the presence of lignin and hemicellulose, pulp is in dark color, and in order to produce white pulp, bleaching is used. The bleaching process reduces the lignin and improves the pulp's color. The lignin components containing aromatic rings and unsaturated structures present in wood are responsible for color in paper industry wastewater. Usually, the natural lignin present in wood is light in color, and during pulping process, the dark color develops [2], which is released into

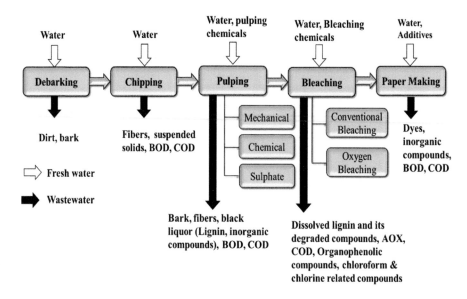

FIGURE 10.2 Pictorial representation of freshwater consumption and effluents (pollutants) released from various processes in the paper industry.

Source: Modified from [1,2].

wastewater. The colored compounds present in the pulping and bleaching processes are recalcitrant, which are resistant to biodegradation and treatment methods [19].

10.2.1.3 Bleaching Process in Paper Industry

Pulp bleaching is a chemical process in which digested pulp color is lightened by removal of the color-constituted compound lignin. Bleaching technology began in 1774 when Karl Wilhelm Scheele bleached natural fibers using chlorine, and later calcium hypochlorite (bleaching powder) was used in paper industry for bleaching. Later, chlorine dioxide and oxygen were used as bleaching agents to overcome the shortcomings of chlorine and calcium hypochlorite, such as slow mixing and a loss of pulp strength during the bleaching process. In paper mills, a multistage sequential bleaching followed by alkaline extraction was used to improve the pulp brightness by removing color imparting compound lignin [13,17]. In the modern world, the kraft process is mostly used for manufacturing pulp, and the obtained kraft pulp has an approximate brightness of 10–30, which is not suitable for producing high-grade white paper. Pulp brightness is related to the color of organic compound lignin (where chromophoric groups present in lignin are responsible for brown or dark brown color of pulp), and pulp bleaching is used to increase the brightness of pulp maximum up to 90% GE [17]. Moreover, paper made with bleached pulp is more stable and durable due to the removal of lignin. Otherwise, if lignin is not removed and is present in pulp, the paper produced with this will undergo color change with age and embrittlement on exposure to sunlight.

Pulp bleaching takes in two stages: (1) chlorination process, where delignification takes place by the addition of bleaching agents (chlorine/chlorine dioxide), and (2) alkali extraction to remove the dissolved lignin and soluble colored components. Removing lignin compounds by using bleaching agents after the pulping process is known as delignification. Delignification stages include oxidation by chlorine (C), alkali extraction of dissolved lignin (E), brightening of the pulp with sodium hypochlorite (H) and chlorine dioxide (D). The schematic of detailed bleaching stages are given in Figure 10.3. In a pulp and paper mill, bleaching process takes in 5–7 multistages that are carried out in bleaching towers. The common sequence followed in bleaching kraft process pulp is given as CEDED, where C is chlorination, D is chlorine dioxide and E is alkali. After every stage of bleaching (or in stages of pulp bleaching), washing the pulp takes place in drum washers. The number of bleaching stages required will depend on the pulp and the type of pulping method used. Shorter bleaching sequences are required for sulfite pulps compared to kraft [19].

The discharge of bleaching effluents, especially those originating from the sludge of bleached pulp mill, contains chlorinated organics such as chloroform and dioxins, which cause concern to the

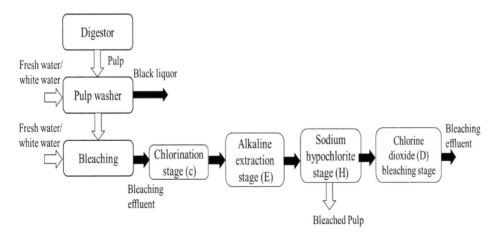

FIGURE 10.3 Illustration of sequential bleaching stages followed in paper industry.

environment. Wastewater released from paper mill bleach plants contains high amounts of AOX, one of the major pollutants that cause/pose environmental problems. Chlorinated compounds produced during bleaching process are soluble in alkali, and these are extracted in the extraction phase [11]. Low-molecular-weight chlorolignins are toxic and mutagenic in nature and are highly resistant to biodegradation. These compounds accumulate in fatty tissues of aquatic organisms like fish when these compounds are discharged into water streams [11,13]. In addition to chlorinated organics and phenolics, high concentrations of chlorides are responsible for corrosiveness. The recalcitrance nature of chlorolignin compounds make the effluent treatment difficulty. Under aerobic conditions, low-molecular-weight chlorinated organics may be methylated and cause harm to fish.

During the kraft pulp bleaching process, high- and low-molecular-weight compounds (almost 250 small chlorinated compounds) are formed and released into wastewater [2]. High-molecular-weight (>30,000 Da) lignin compounds from the alkaline extraction stage in bleaching units imparts the color load to the wastewater. Generally, wastewater released from the bleach unit contains 80% of color, 60% of COD and 30% of BOD of the total pollution load originated from the paper mill [11]. Low-molecular-weight compounds (chlorinated phenols, chloroaliphatics, chloroform, chloroacetic acid) having molecular mass <1000 g/mol are highly toxic to aquatic animals [2,11]. These compounds bioaccumulate in body fat of aquatic animals. Bleached kraft's process units discharge large amounts of colored (brown) effluents with a high quantity of lignin and its degradation products, COD and so on, which are resistant to biological treatment methods [17].

10.3 TREATMENT METHODS

Effluents from kraft and sulfite bleach plants contain BOD, COD, suspended solids, AOX, color, phosphorous, nitrogen, bleaching chemicals (inorganic salts of sodium and magnesium) and organic materials (lignin compounds). With the aim of reducing discharged pollutant load on the environment as per the regulations and consumption of fresh water, effluents need to be treated before being discharged or reused in industry. The proper treatment methods have to be chosen so that the treated effluent water can be reused. If the poorly treated effluent is reused or recycled in the industry it would lead to problems like degradation of paper quality during pulp and papermaking; deposition of scales in digestors, boilers, tanks and the like; the corrosion of pipes and equipment; and the consumption of larger quantities of cleaning chemicals [10]. Therefore, the increase in environmental restrictions, a shortage of freshwater resources and an increase in effluent quality that is to be discharged lead the search for suitable treatment method for removing lignin, color and AOX from kraft bleaching plant effluents [20].

Primary treatment techniques like filtration, sedimentation, coagulation, flocculation and floatation are used to remove the settleable solids, suspended solids and 30% of BOD. Depending upon the nature of effluent, secondary and tertiary treatment methods would be used. Biological methods like activated sludge, activated lagoon, trickling filters and others and electrooxidation, ion exchange and membrane filtration are used for treating AOX, COD and BOD from paper mill effluent [15,16]. The disposal of sludge, the use of harmful solvents and the regeneration of adsorbents are the main drawbacks associated with coagulation, flocculation and adsorption. In a similar manner, microbial contamination, sludge disposal, cost, large installation space, the presence of pollutant residuals in treated water, recalcitrance and the degrading nature of AOX in biological methods limit their usage. Membrane processes are promising compared to conventional wastewater treatment techniques and these processes are widely used for treating pulp and paper industrial effluents [21]. The possibility of continuous separation, no requirement of chemical additives during separation, easy upscaling and flexibility in integration with other processes seems to be advantageous over other treatment methods. Membrane is a selective barrier between two phases from which a particular species or more from the feed is transported through it by the application of driving force [22] (shown in Figure 10.4). In the membrane process, flux is proportional to the driving force or gradient (concentration, pressure, temperature and potential).

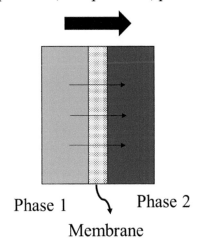

Driving force (Concentration, pressure, temperature, potential)

Phase 1 Phase 2

Membrane

FIGURE 10.4 Schematic representation of transport through a membrane.

This chapter mainly emphasizes pressure-driven membrane separation processes for treating the paper mill effluent. Here, different types of techniques, their separation mechanisms and the membrane properties are elaborately discussed. Furthermore, detailed descriptions of suitable pressure-driven technique/hybrid (combination) methods, different cases in which they have been successfully applied and the challenges in applying these methods in paper industries are provided.

10.3.1 Pressure-Driven Membrane Process

Pressure-driven membrane separation techniques are well established, economical and competent techniques that can replace the convectional separation methods for treating industrial wastewater. Usually, these processes are used to concentrate or purify the solution. These techniques differ mainly in the nature and characteristics of membranes through which the species are to be separated and operating conditions [22]. Based on the particle size of solute to be separated, the membrane pore size and the structure, pressure-driven membrane separation techniques are classified into MF, UF, NF and RO. In these processes, applied pressure is the driving force for separation of different-sized particles through membranes. The extent of separation efficiency depends on the membrane pore size, morphology, the size of species and the applied pressure. For these processes, in the case of porous membranes, pore size determines the separation flux and rejection, while for nonporous or dense membranes, the membrane's intrinsic properties govern the flux. A decrease in membrane pore size from MF to RO results in the separation of smaller-sized solute particles passing through the membrane. As the membrane pore size is smaller, an increase in resistance to the mass transfer demands a higher applied pressure for separation. The membrane structure and thickness are also responsible for mass transport and flux. Different types of membrane structure (porous/dense, symmetric/asymmetric), pore size, material, applied pressure and separation mechanisms of various membrane-based filtration techniques are tabulated in Table 10.1 [21,22].

10.3.1.1 MF

MF is almost similar to normal filtration, but here, filtration occurs through a membrane. Generally in this method, a porous membrane is used. The pore size of an MF membrane is large, 10 μm–0.1 μm. The mechanism of separation is by the difference in the particle size by a sieving mechanism;

TABLE 10.1

Characteristics of Pressure-Driven Separation Techniques

Technique	Type of membrane	Membrane materials	Membrane pore size	Filtration mechanism	Applied transmembrane pressure (TMP)	Applications
Microfiltration (MF)	Microporous symmetric/ asymmetric	Polymeric (cellulose acetate, poly vinylidene fluoride, polysulfone), ceramic (Al_2O_3, ZrO_2, TiO_2)	10 µm – 0.1 µm	Sieving (based on particle size)	0.1–2 bar	Food, Beverages, pharmaceuticals, wastewater treatment for the separation of suspended particles, bacteria
Ultrafiltration (UF)	Microporous asymmetric	Polymeric (polyacrylonitrile, polysulfone), ceramic (Al_2O_3, ZrO_2)	0.1 µm – 0.01 µm	Sieving (based on particle size)	0.1–5 bar	Separation of bacteria, yeast, macromolecules from food, textile, phramaceuticals, dairy, water treatment
Nanofiltration (NF)	Composite asymmetric	Polyamide	10 nm – 1 nm	Sieving, Donnan exclusion	8–30 bar	Removal of Inorganic salts (Multivalent ions) from seawater, industrial wastewater
Reverse Osmosis (RO)	Dense Thin film composite asymmetric	Cellulose triacetate, polyamide & poly (ether urea)	< 1 nm	Solution-diffusion (based on differences in diffusivity & solubility)	20–80 bar	Industrial wastewater treatment, seawater desalination and concentration of fruit juices, milk etc.

that is, larger particles are retained, and smaller particles passed through these membranes [22–24]. If the particle size is relatively larger than the membrane pore size, high separation selectivity can be achieved. A smaller range of applied pressure (0.1–2 bar) is sufficient for the separation of particles due to the large pore size. Polymeric (cellulose acetate, poly vinylidene fluoride, polysulfone), ceramic (inorganic materials, i.e., Al_2O_3, ZrO_2, TiO_2) and hybrid composite materials are widely used for MF membrane preparation. Membranes prepared with inorganic membranes have good thermal and chemical resistance with a narrow pore size distribution. Apart from membrane material, the thickness and type of membrane may also be responsible for transport resistance through a porous symmetric MF membrane. Most of the used MF membranes are asymmetric with 1-µm top-layer thickness.

10.3.1.2 UF

UF is used to separate low-molecular-weight compounds, suspended solids, bacteria and viruses. UF membranes containing smaller pores, 0.1 µm–0.01 µm, relative to MF, and because of this, to achieve higher permeability, a high pressure in the range of 0.1–5 bar is required [23]. Usually UF membranes are asymmetric in nature with a dense, porous top layer with a thickness of ≤1 µm supported by a porous sublayer. This top layer determines the transport resistance. Most of the UF membranes are prepared from polymers such as poly ether ether ketone, cellulose acetate, polyimide, polysulfone and polyacrylonitrile and ceramics. UF is also used as a pretreatment in some wastewater treatments, desalination, textile and pharmaceutical industries, among others.

10.3.1.3 NF

NF is used to separate low-molecular-weight organic particles, suspended solids, inorganic and multivalent salts. The basic difference between NF and UF is the size of the membrane pore and the applied pressure. Consequently, there would be a difference in size of solute particles that are retained and separated by these two processes. NF membranes have smaller membrane pore sizes (10 nm–1 nm) relative to UF. Even a higher applied pressure (8–30 bar) is required for higher flux compared to UF and MF. Based on the size difference by sieving mechanism (in the case of a noncharged membrane) and through Donnan exclusion (charge based separation in the case of a charged membrane), the species passing through NF membranes are transported and separated [22,25]. Usually, NF membranes are composite asymmetric membranes with top-layer thickness of ≤1 μm and a sublayer thickness of 150 μm. These are prepared with polyamide by an interfacial polymerization technique. This process is widely used for removing salts from sea/brackish water, organic matter, heavy metals and dyes from the paper, textile and leather industries.

10.3.1.4 RO

RO is used for industrial wastewater treatment, seawater desalination, concentration of fruit juices in the food industry, milk in the food industry, wastewater in the galvanic industry and more. RO membranes are dense (nonporous) asymmetric or composite membranes with a membrane pore size of <1 nm that are prepared from cellulose triacetate, poly (ether urea) and polyamide by the phase-inversion technique. RO separates all the colloids, suspended solids and mono- and multivalent salts and allows water (solvent) through the membrane when high pressures of 20–80 bar are applied. In the RO process, pressure greater than osmotic pressure ($\Delta P > \Delta \pi$) must be applied, and the amount of pressure required will depend on solute concentration. In nonporous, dense membranes, due to a difference in the diffusivity or solubility, separation takes place, which is known as a solution–diffusion mechanism. Here, mainly the membrane's intrinsic properties regulate the selectivity and flux [22,26]. High operating cost (high energy consumption), concentration polarization and membrane fouling are the major challenges for RO technique. In order to avoid fouling pretreatments like MF and UF, and periodic cleaning is required [26].

A schematic diagram representing various pressure-driven membrane processes and particles separated is shown in Figure 10.5.

Pressure-driven membrane processes are used to treat pulp and paper effluents produced from bleaching plants [27–29], the deinking process, white water (process water) [7], the separation or recovery of organic compounds and the bleaching chemicals from pulping process spent liquor [10,30]. Bleaching effluents contain 70–95% of chlorolignins with molecular weight >1000 Da, which for accounts 50–80% total solids content and 85–90% of color. While the low-molecular-weight organochlorine compounds are responsible for toxicity and BOD. Organochlorinated compounds, especially low-molecular-weight compounds, are the major toxic pollutants that are challenging the environment. These compounds are preferably removed by using biological treatment methods, but higher-molecular-weight chlorolignins present in effluents could affect biological methods. Membrane separation processes (e.g., UF) are used as pretreatments to remove high-molecular-weight compounds before biological methods are used so that the COD, color and AOX levels are reduced [31]. The application of MF subsequent to UF and RO for purification of wastewater results in poor performance in terms of COD due to the presence of bacterial waste and organic matter (fats, proteins etc.,) during the biological treatment, which is performed before membrane filtration. Therefore, Pizzichini et al. [4] performed membrane filtration before the biological treatment. They have compared the performances of ceramic MF membrane and polymeric spiral-wound MF, UF and RO modules. Higher performance was observed with the ceramic MF module with a cutoff of 0.14 μm, whereas low performance and high fouling were noted with polymeric MF and UF. When tubular ceramic MF was used followed by RO, 80% of the wastewater was recovered and reused as pure water.

Earlier researchers focused much on the application of membrane processes for treating different bleaching stages effluents since the acidic filtrate from chlorination (C) or chlorine dioxide stage (D)

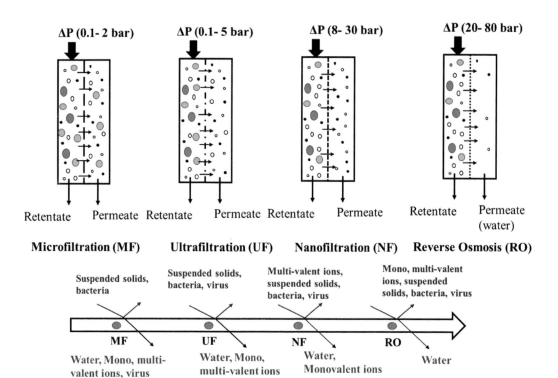

FIGURE 10.5 Schematic representation of various pressure-driven techniques and transport through membrane.

and alkaline filtrate from extraction stage (E) are the most polluting effluents released from bleach plants. These effluents contain chlorinated compounds, color, chlorides, phenols, lignin, resin acids and terpenes [27]. Mainly the largest pollution load is from the first extraction stage, and the number of chlorinated organics depends on the lignin content in pulp and the quantity of bleaching agent (chlorine) used. Shukla et al. [27] studied the treatment of effluents from chlorination and extraction bleaching stages with a series of thin-film polyamide/polysulfone spiral-wound UF, NF and RO membranes. At a higher transmembrane pressure, initially, a high permeate flux was observed, and thereafter, flux rapidly declined, which may be possibly due to concentration polarization and the fouling phenomena. In the case of low-molecular-weight pollutants from the chlorinated stage of bleach effluents, higher removal performances were noted with NF and RO compared to UF. Fifty to 89% of COD removal and 80–91% of AOX were achieved with the thin-film polyamide spiral-wound RO membrane with an optimum operating pressure of 13.7 bar. Rosa and Pinho [32] evaluated the performance of UF and NF for reducing color and organochlorinated compounds from two different first alkaline extraction effluents. Using UF, 72% of total organochlorinated compounds (TOX) and 92% color were removed, whereas with NF, 90% of TOX removal and total color were achieved.

Shukla et al. [33] investigated the effectiveness of membrane filtration techniques (UF, NF, RO) for treating hardwood pulp bleaching (CEHH) effluent of an Indian paper mill. UF removed 80% COD, 52% AOX and 93% color; 93% COD, 75 % AOX and 95% color were removed with NF, while 99% of COD, 95% AOX and almost 100% color were removed with RO at higher inlet pressures of 13.7–17.2 bars. Higher fouling was observed in the case of RO. Onate et al. [28] investigated the sequential use of UF–NF–RO to separate alkaline extraction bleaching effluents generated from kraft cellulose production. Elemental chlorine-free (ECF) bleaching was done to produce high-brightness cellulose in a pine wood mill, and the effluents contained chemicals, inorganic and organic load. A sequential UF–NF–RO system removed 99% color, COD, 98% chloride, total phenols and AOX and 97% conductivity. Most of the organic fractions present in alkaline effluents were removed by UF membranes of 10 kDa molecular-weight cutoff (MWCO). With UF membrane having MWCO > 10 kDa operated at 6 bar, 25°C retained 78% total phenol and COD, 98% color and 82% adsorbable organic halogens (AOX); 10/1 or 5/1 kDa UF was used as a pretreatment to NF/RO. Chloride ions present in UF permeate were removed by RO, and the produced high-quality water (permeate) could be reused.

In order to reduce the discharge of effluents from bleach plants, paper mills are adopting modern bleaching sequences without using chlorine. Falth et al. [29] studied the seven different kraft paper mill alkaline bleach filtrates (total chlorine free [TCF], ECF and bleach plants) with UF. They observed that UF flux depends on bleach filtrate concentration, especially when the reduction of lower-molecular-weight compounds from effluent resulted in higher fluxes. The influence of lower-molecular-weight compounds on flux is higher than higher-molecular-weight compounds. The organic substances retention was efficient and higher for effluents collected from the first alkaline stage of traditional ECF mills compared to effluents from modern ECF and TCF mills, which consist of larger quantities of low-molecular-weight compounds.

The main drawbacks associated with these membrane techniques are low fluxes and membrane fouling. Membrane flux and fouling are mainly dependent on effluent characteristics which need to be treated, membrane properties (e.g., material of membrane, pore size, surface [hydrophilic/hydrophobic, charge, roughness], thickness, etc.), membrane module and operating conditions. The membrane performance varies from case to case, and no clear-cut conclusion can be drawn without testing for each case. In particular, in bleaching units, the alkali extraction stage effluent treatment by membrane filtration is directly affected by the effluent characteristics (pH, temperature, amount of organic load) and membrane selection [20]. The removal efficiency of membrane separation techniques depends on the membrane's properties, especially pore size for pressure-driven membrane processes, the concentration and nature of feed that needs to be treated and the operating pressure [27,31]. Yao et al. [31] analyzed the performance of different MWCO UF membranes in series (MWCO of 1, 30,000, 15,000 and 5,000 Da, respectively; UF membranes in series) for kraft pulp bleach effluents' pretreatment. Tighter UF and NF membranes with MWCO < 1500 Da resulted in 90% TOC and 99% AOD reduction. Quezada et al. [20] studied the performance; removal of color, COD and permeate flux; and cost of three different membrane processes – (1) tight UF, (2) open UF + NF and (3) NF – for treating kraft paper mill EPO (oxygen and peroxide-reinforced extraction) bleaching plant filtrate. They observed that usage of tight UF was the best option with 79% COD removal and 86% color reduction.

Membrane properties are found to be the determining factors. Usually, higher fluxes are possible with a more hydrophilic membrane, and if the membrane surface is more hydrophobic, more hydrophobic materials (fatty acids, resins) are adsorbed, causing irreversible membrane fouling. Smooth hydrophilic membranes are less prone to fouling compared to rough hydrophobic membranes. In addition to this, the membrane module (tubular, spiral wound and flat sheet) is crucial for fluxes and fouling. In the case of spiral-wound membrane modules, feed containing solids could clog membranes, and to avoid this, pretreatments are necessary. It is difficult to clean the clogged SW module, and moreover, very low fluxes are acquired due to the availability of very small flow channels in spiral-wound modules [8].

In addition, it is noteworthy that the temperatures of effluent coming out from mechanical pulping is at 80°C, and these high temperatures limit the use of polymeric membranes for filtration on a long-run basis. Therefore, in this case, it is advised to use ceramic membranes, which can tolerate the high temperatures of mechanical pulping [8].

Ceramic membrane separation seems to be advantageous over polymeric membrane filtration in terms of thermal, mechanical and chemical stability; long-term durability for continuous operations; easy back flushing; cleaning; and toleration of bacterial fouling. Most of the membranes are composite in nature with a porous asymmetric structure consisting of thin top layer, an intermediate layer and porous support layer [34]. Flat-sheet, spiral and tubular ceramic membrane modules are available. Ebrahimi et al. [1] investigated the use of ceramic tubular UF membranes for treating sulfite mill pulp bleaching effluent. The COD of treated effluent was reduced effectively with ceramic UF, but the fouling issues raised during a single UF treatment could be resolved with MF followed by UF. Nataraj et al. [35] studied a pilot-scale hybrid membrane separation with UF-electrodialysis (ED) for the treatment of paper mill effluent. In their process, a tubular ceramic MF module was used as pretreatment to ED at 60°C for 120 min. The application of MF in batch mode resulted in clear permeate that was free from suspended particles with a stable flux of 121 L/m²h at 4 bar pressure. They achieved 546 mg/L of TDS, 0.61 mS/cm conductivity and <20 mg/L COD, and more than 80% of wastewater was treated and reused. In addition to this, energy from biomass (wood residuals, bark and black liquor residue) was also recovered with this hybrid process.

Other prominent applications of membrane filtration in the paper industry are for recovery of valuable chemicals and compounds. Lignin is an important raw material for production of biofuels, synthetic tannins, vanillic acid and carbon nanotubes. Black liquor from the kraft process and spent sulfite liquor are the main sources of lignin in the paper industry. Spent sulfite liquor (effluent) produced during the sulfite pulping process in the paper industry is conventionally concentrated by evaporators, and later, it is burned to recover chemicals used during pulping. Spent liquor mainly contains monosaccharides, polysaccharides, hemicellulose, lignosulfates, lignin and pulping chemicals, among others. Various methods are used not only for removing harmful carcinogenic compounds present in effluents that need to be discharged but also for recovering the compounds (lignin, hemicellulose) and chemicals (NaOH, Na_2SO_4, Na_2CO_3) from effluents so that they can be reused. Lignin compounds, which are widely present in paper mill effluents, can be used for the development of plasticizers, biofuels and adhesives. In the same way, hemicelluloses are used for the production of hydrogels, biofuels, surfactants and barrier films. Therefore, the recovery of lignin, hemicellulose from paper effluent could increase the commercialization of hemicellulose-based products and reduce the effluent load that needs to be treated [30]. Over the last decades, membrane filtration techniques have been used for the concentration and recovery of lignin. Most often, organic polymer membranes such as cellulose acetate, polyether sulfone, polysulfone, cellulose and others are used with different MWCO and membrane geometries (flat sheet, spiral and tubular). Ceramic membranes could withstand the high temperatures and pH of pulp wastewater compared to polymer membranes [10,34].

Rudainy et al. [30] studied the removal of hydrophobic compound lignin from spent sulfite liquor by polymeric ion exchange resins before UF. They observed reduced UF membrane fouling, a 38% increase in UF flux and an increase in separation from 17% to 59%. This resulted in a high-purity hemicellulose-rich retentate. Niortilla et al. [8] studied membrane filtration in an integrated mechanical paper mill total effluent generated from different processes like grinding room circulation water, clarified white water from paper machine and so on. They evaluated fluxes and the removal of suspended solids, multivalent salts, polysaccharides and lignin compounds by UF and NF. Thirty percent of the organic load and microorganisms were removed by UF while NF removed most of the organic load and multivalent ions like SO_4 ²⁻, Al ³⁺, Ca ²⁺, Mg ²⁺ and others. The permeate (purified water) or product water obtained from UF/NF could be internally recycled back to paper machines for washing so as to replace fresh water. NF permeate could be recirculated even as shower water in the paper machine that is free of multivalent ions may reduce the corrosion significantly since the fresh water used in the paper industry may contain some of the multivalent ions (calcium, magnesium etc.) that lead to slime deposition and corrosion.

Jonsson et al. [36] investigated the effect of tubular and polymeric UF membrane cutoff (4–100 kDa) on lignin and hemicellulose retention from hardwood and softwood cooking liquor. In addition to this, the lignin was removed from hardwood black liquor using a ceramic UF membrane of 15-kDa cutoff, and permeate was concentrated with a polymeric NF membrane of 1-kDa cutoff. Humpert et al. [10] investigated the recovery and concentration of lignin from spent sulfite liquor using ceramic hollow-fiber membranes. They have tested the performance of three ceramic membranes of 3, 8, 30 nm. The highest retention of lignosulfonate (69%) was achieved by 3-nm ceramic hollow-fiber membrane due to higher transmembrane pressure (TMP), higher wall shear stress and permeate flux because of the geometry (smaller inner diameter). They also concluded that ceramic hollow-fiber membranes performed better than tubular membranes for spent sulfite liquor concentration. Costa et al. [37] evaluated the fractionation of hardwood Kraft liquor by sequential UF with three tubular membranes of 5-, 15- and 50-kDa cutoffs. Lower TDS and higher inorganic content were achieved with decrease in membrane cutoff. Low-purity lignin was separated with a 50-kDa membrane.

In a similar way, pressure-driven membrane separation processes are used for treating white water and effluent from papermaking. Oliveira et al. [7] evaluated the UF of white water (process water from different process phases) removed the suspended solids. The UF-treated white water could be reused in papermaking process like water for cleaning the devices, dilution water, sealing water and so on. They have combined UF with chemical precipitation such that the calcium ions present in the white water were removed, and this treated water could possibly be reused in pulp washers of a bleach plant. Otherwise, the calcium ion increased the hardness content of water and resulted in scale formation. The study concluded that reused low-hardness water in a bleach plant was feasible and that the mechanical and optical properties of bleached pulp with this reused water did not vary relative to fresh water. Conductivity, TDS and pH filtrate after the bleaching process with reused white water are likely to be increased compared with fresh water.

10.4   CHALLENGES AND FUTURE PERSPECTIVES

Fouling, concentration polarization are the main aspects affecting the membrane performance, lifetime and flux. Fouling is a phenomenon in which the unwanted substances deposit on the surface of the membrane or within the membrane pore, increasing the membrane's resistance during separation. This substantially decreases the flux and requires more pressure for separation, thereby increasing the overall cost and reducing the membrane's life span. In order to reduce the fouling and restore the permeate flux, membrane cleaning is required. Cleaning with alkalis, acids and cleaning agents could remove the substances deposited on surface but not the clogged substances in pores. Deionized water, alkali, Decon 75, Ultrasil 10 and Ultratide solutions are used as cleaning solutions to remove the color and absorbed pollutants from membranes [29,31]. Moreover, membranes are cleaned with different cleaning solutions at higher temperatures (55 or 60°C). Periodic membrane cleaning might not solve this issue completely.

Back flushing is a process that is widely used in industries to clean fouled membranes. The parameters governing back flushing are back pressure, flux and time. Another solution for preventing fouling in pressure-driven membrane separation processes is the application of a ceramic membrane, especially for treating bleaching effluents that contain large amounts of organic matter [1].

Membrane filtration performance could be enhanced by applying different pretreatment methods like physical/chemical methods (coagulation, flocculation), biological methods (activated sludge, aerobic digestion, etc.) and membrane processes (looser membrane process for tighter membranes, that is, UF for NF). Biological digestion can be used as pretreatment method for membrane filtration processes to remove the low-molecular-weight organic compounds present in paper mill effluent. Thus, NF flux can be improved [8].

Another strategy for reducing the pollutant load from the paper industry or increasing the removal efficiency of pollutants is using of new process technologies in pulping or bleaching stages. Today, some paper industries have started implementing biopulping and biobleaching processes that are more energy efficient and environmentally friendly. In these methods, the raw materials and pulps undergo enzymatic or fungal treatments instead of chemical treatments. Moreover, bleaching processes are becoming TCF and ECF in which delignification is carried out using ozone, oxygen and hydrogen peroxide, making bleaching processes environmentally friendly [6]. These strategies would help improve pulp brightness and final paper quality and reduce toxic AOX compounds so that the pollution load decreases. White-rot fungi is the commonly used enzyme for biobleaching, but these biological methods are slow, and maintaining process conditions is difficult. Even though the previously mentioned strategies (biopulping/biobleaching, ECF/TCF) seem economic and environmental friendly, where the release of toxic chlorinated compounds is reduced relative to conventional bleaching process, the production of low-quality paper is the main drawback.

10.5 CONCLUSION

Membrane processes are techniques that are economically feasible for treating pulp and paper industry effluents. In paper industries, colloidal, suspended solids, polysaccharides, high-molecular-weight lignin-related compounds and multivalent salts will be mostly removed by using pressure-driven-based membrane separation techniques like MF, UF and NF. Predominantly, bleaching plants require large quantities of fresh water and generate huge amounts of effluents with high inorganic and high-molecular-weight organic loads, which pose serious environmental problems. The extraction of lignin with alkali (NaOH) in bleaching pulp resulted in colored effluent having a pH of 7–10 and a conductivity of 4–7 mS/cm. The presence of high-molecular-weight lignins causes the dark color of effluent, and low-molecular-weight chlorinated lignins impart toxicity. Apparently, bleaching effluents from the kraft process corresponds to high levels of pollution from the paper industry. Bleaching effluents attribute 80–90% of color and 60–70% of COD load of wastewater and release of chlorinated dioxins in terms of AOX, causing environmental problems. Precipitation, adsorption, ion exchange and membrane filtrations have been used for several years to concentrate the colored effluents. Biological treatment methods are mostly not suitable due to recalcitrant chlorinated phenols which are not degraded by microorganisms whereas, slow degradation is observed for high molecular weight lignin compounds. Therefore, removal of chlorinated organic and colour of bleaching effluent is difficult with biological methods relative to membrane filtration methods.

In order to lower the AOX pollutants during bleaching process, chlorine can be replaced by oxygen, ozone and others. Using oxygen before pulp bleaching and in the alkali extraction stage removes lignin, lowers AOX generation and improves pulp brightness. From most of the available literature, it is evident that UF and NF are suitable to treat pulp and paper effluents compared to RO. In addition to this, ceramic membranes are superior to polymeric membranes for pulp and paper effluent treatment because of their high stability (chemical, thermal, mechanical), operational durability in harsh environments and easy cleaning. Even though the former case seems beneficial, high capital costs, large installation space requirements and low packing densities are the limiting parameters and key challenges for wider application in the paper industry. The development of cheaper ceramic membranes and modules could overcome the present challenges to implementation in industries.

BIBLIOGRAPHY

1. M. Ebrahimi, N. Busse, S. Kerker, O. Schmitz, M. Hilpert, P. Czermak, Treatment of the bleaching effluent from sulfite pulp production by ceramic membrane filtration, Membranes (Basel). 6 (2015) 1–15. doi:10.3390/membranes6010007.
2. M.A. Hubbe, J.R. Metts, D. Hermosilla, M.A. Blanco, L. Yerushalmi, F. Haghighat, P. Lindholm-Lehto, Z. Khodaparast, M. Kamali, A. Elliott, Wastewater treatment and reclamation: A review of pulp and paper industry practices and opportunities, BioResources. 11 (2016) 7953–8091. doi:10.15376/biores.11.3.hubbe.
3. S. Ciputra, A. Antony, R. Phillips, D. Richardson, G. Leslie, Comparison of treatment options for removal of recalcitrant dissolved organic matter from paper mill effluent, Chemosphere. 81 (2010) 86–91. doi:10.1016/j.chemosphere.2010.06.060.
4. M. Pizzichini, C. Russo, C.D. Di Meo, Purification of pulp and paper wastewater, with membrane technology, for water reuse in a closed loop, Desalination. 178 (2005) 351–359. doi:10.1016/j.desal.2004.11.045.

5. C.H. Ko, C. Fan, Enhanced chemical oxygen demand removal and flux reduction in pulp and paper wastewater treatment using laccase-polymerized membrane filtration, J. Hazard. Mater. 181 (2010) 763–770. doi:10.1016/j.jhazmat.2010.05.079.

6. S.K. Garg, M. Tripathi, Strategies for decolorization and detoxification of pulp and paper mill effluent, Rev. Environ. Contam. Toxicol. 212 (2011) 113–136. doi:10.1007/978-1-4419-8453-1_4.

7. C.R. Oliveira, C.M. Silva, A.F. Milanez, Application of ultrafiltration in the pulp and paper industry: Metals removal and whitewater reuse, Water Sci. Technol. 55 (2007) 117–123. doi:10.2166/wst.2007.219.

8. J. Nuortila-Jokinen, M. Mänttäri, T. Huuhilo, M. Kallioinen, M. Nyström, Water circuit closure with membrane technology in the pulp and paper industry, Water Sci. Technol. 50 (2004) 217–227. doi:10.2166/wst.2004.0199.

9. I.P. mills association Agro, Report on oxygen bleaching study for agro based paper mills, 2002.

10. D. Humpert, M. Ebrahimi, A. Stroh, P. Czermak, Recovery of lignosulfonates from spent sulfite liquor using ceramic hollow-fiber membranes, Membranes (Basel). 9 (2019) 1–16. doi:10.3390/membranes9040045.

11. L. Christov, B. Van Driessel, Waste water bioremediation in the pulp and paper industry, Indian J. Biotechnol. 2 (2003) 444–450.

12. K. Nakamata, Y. Motoe, H. Ohi, Evaluation of chloroform formed in process of kraft pulp bleaching mill using chlorine dioxide, J. Wood Sci. 50 (2004) 242–247. doi:10.1007/s10086-003-0553-7.

13. R. Nagarathnamma, P. Bajpai, Decolorization and detoxification of extraction-stage effluent from chlorine bleaching of kraft pulp by Rhizopus oryzae, Appl. Environ. Microbiol. 65 (1999) 1078–1082. doi:10.1128/aem.65.3.1078-1082.1999.

14. K. Mehmood, S.K.U. Rehman, J. Wang, F. Farooq, Q. Mahmood, A.M. Jadoon, M.F. Javed, I. Ahmad, Treatment of pulp and paper industrial effluent using physicochemical process for recycling, Water (Switzerland). 11 (2019) 1–15. doi:10.3390/w11112393.

15. R. Sridhar, V. Sivakumar, V.P. Immanuel, J.P. Maran, Treatment of pulp and paper industry bleaching effluent by electrocoagulant process, J. Hazard. Mater. 186 (2011) 1495–1502.

16. A. Saadia, A. Ashfaq, Environmental management in pulp and paper industry, J. Ind. Pollut. Control. 26 (2010) 71–77.

17. Pulp Bleaching Technology, in: Pulp Bleach. Technol., n.d.: pp. 41–49.

18. Caetano Ana, M.N. De Pinho, E. Drioli, H. Muntau, Membrane Technology: Applications to Industrial Wastewater Treatment, 1st ed., Springer International Publishing, Dordrecht, 1995.

19. A. Latha, M.C. Arivukarasi, C.M. Keerthana, R. Subashri, V. Vishnu Priya, Paper and pulp industry manufacturing and treatment processes a review, Int. J. Eng. Res. 6 (2018). doi:10.17577/ijertcon011.

20. R. Quezada, M.C. Silva, A.A.P. Rezende, L. Nilsson, M. Manfredi, Membrane treatment of the bleaching plant (EPO) filtrate of a kraft pulp mill, Water Sci. Technol. 70 (2014) 843–851. doi:10.2166/wst.2014.304.

21. A. Zirehpour, A. Rahimpour, Membranes for wastewater treatment, in: Nanostructured Polym. Membr., John Wiley & Sons, Inc., Hoboken, NJ and Scrivener Publishing LLC, Beverly, MA, 2016: pp. 159–207. doi:10.1002/9781118831823.ch4.

22. M. Mulder, Basic Principles of Membrane Technology, 2nd ed., Kluwer Academic Publishers, Dordrecht, The Netherlands, 1996.

23. B. Van Der Bruggen, C. Vandecasteele, T. Van Gestel, W. Doyen, R. Leysen, A review of pressure-driven membrane processes in wastewater treatment and drinking water production, Environ. Prog. 22 (2003) 46–56. doi:10.1002/ep.670220116.

24. J. Dasgupta, J. Sikder, S. Chakraborty, S. Curcio, E. Drioli, Remediation of textile effluents by membrane based treatment techniques: A state of the art review, J. Environ. Manage. 147 (2015) 55–72. doi:10.1016/j.jenvman.2014.08.008.

25. M.N. Chollom, S. Rathilal, V.L. Pillay, D. Alfa, The applicability of nanofiltration for the treatment and reuse of textile reactive dye effluent, Water SA. 41 (2015) 398–405. doi:10.4314/wsa.v41i3.12.

26. S.P. Dharupaneedi, S.K. Nataraj, M. Nadagouda, K.R. Reddy, S.S. Shukla, T.M. Aminabhavi, Membrane-based separation of potential emerging pollutants, Sep. Purif. Technol. 210 (2019) 850–866. doi:10.1016/j.seppur.2018.09.003.

27. S.K. Shukla, V. Kumar, T. Kim, M.C. Bansal, Membrane filtration of chlorination and extraction stage bleach plant effluent in Indian paper Industry, Clean Technol. Environ. Policy. 15 (2013) 235–243. doi:10.1007/s10098-012-0501-6.

28. E. Onate, E. Rodríguez, R. Bórquez, C. Zaror, Membrane treatment of alkaline bleaching effluents from elementary chlorine free kraft softwood cellulose production, Environ. Technol. (United Kingdom). 36 (2015) 890–900. doi:10.1080/09593330.2014.966765.

29. F. Fälth, A.S. Jönsson, R. Wimmerstedt, Ultrafiltration of effluents from chlorine-free, kraft pulp bleach plants, Desalination. 133 (2001) 155–165. doi:10.1016/S0011-9164(01)00094-7.

30. B. Al-rudainy, M. Galbe, O. Wallberg, Hemicellulose recovery from spent-sulfite-liquor: Lignin removal by adsorption to resins for improvement of the ultrafiltration process, Molecules. 25 (2020) 3435.

31. W.X. Yao, K.J. Kennedy, C.M.I. Tam, J.D. Hazlett, Pre-treatment of kraft pulp bleach plant effluent by selected ultrafiltration membranes, Can. J. Chem. Eng. 72 (1994) 991–999.

32. M.J. Rosa, M.N. de Pinho, The role of ultrafiltration and nanofiltration on the minimisation of the environmental impact of bleached pulp effluents, J. Memb. Sci. 102 (1995) 155–161. doi:10.1016/0376-7388(94)00283-5.

33. S.K. Shukla, V. Kumar, M.C. Bansal, Treatment of combined bleaching effluent by membrane filtration technology for system closure in paper industry, Desalin. Water Treat. 14 (2010) 464–470. doi:10.5004/dwt.2010.1879.

34. S.M. Samaei, S. Gato-Trinidad, A. Altaee, The application of pressure-driven ceramic membrane technology for the treatment of industrial wastewaters: A review, Sep. Purif. Technol. 200 (2018) 198–220. doi:10.1016/j.seppur.2018.02.041.

35. S.K. Nataraj, S. Sridhar, I.N. Shaikha, D.S. Reddy, T.M. Aminabhavi, Membrane-based microfiltration/electrodialysis hybrid process for the treatment of paper industry wastewater, Sep. Purif. Technol. 57 (2007) 185–192. doi:10.1016/j.seppur.2007.03.014.

36. A.-S. Jonsson, A.-K. Nordin, O. Wallberg, Concentration and purification of lignin in hardwood kraft pulping liquor by ultrafiltration and nanofiltration, Chem. Eng. Res. Des. 86 (2008) 1271–1280.

37. C.A.E. Costa, P.C.R. Pinto, A. Rodrigues, Lignin fractionation from E. globulus kraft liquor by ultrafiltration in a three stage membrane sequence, Sep. Purif. Technol. 192 (2018) 140–151.

11 Future Scope of Membrane Technology in Pineapple Juice Processing
A Review

Jagamohan Meher, Sivakumar Durairaj, Sandeep Dharmadhikari, and Rajanandini Meher

CONTENTS

11.1   INTRODUCTION

Pineapple (*Ananas comosus*) is a nutritious tropical fruit with a lot of juice, a strong tropical flavor, and several health benefits. Pineapple has become increasingly popular in recent years across the world because of its adaption to a wide range of soil and climatic conditions. It is the world's third most important tropical fruit, next to bananas and citrus. In comparison to other pineapple-derived products, pineapple juice, powder, and functional beverages are in high demand in the food industry (Ali et al., 2020). Juices have also become more popular as a result of their high vitamin, mineral, antioxidant, and dietary fiber content, which improve digestion (Chaudhary et al., 2019). However, for a long time, cost reduction, quality improvement, nutritional value, customer acceptance, and cost minimization have all been major concerns in pineapple juice production

DOI: 10.1201/9781003165019-11

(Sant'Anna et al., 2012). Membrane technology is becoming one of the most widely used separation techniques for processing and marketing juices that enhance the nutritional and sensory qualities of fresh vegetables and fruits (Conidi et al., 2018). High performance, simple machinery, comfortable operations, and minimal operational usage are only a few of the benefits. Membrane technology has several benefits over traditional separation methods of juice separation (Figure 11.1), including low operating temperatures, unique separation mechanisms, no chemical additives, processed fluids not being subjected to any temperature stress, rapid scale-up, versatile construction lightweight, and low energy consumption.

Microfiltration (MF), nanofiltration (NF), ultrafiltration (UF), osmotic distillation (OD), reverse osmosis (RO), forward osmosis (FO), and membrane distillation (MD) have all been

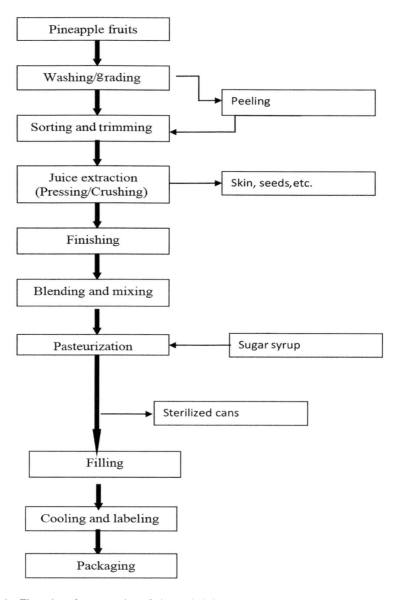

FIGURE 11.1   Flow sheet for processing of pineapple juice.

Source: Adapted from FAO (1995).

successfully used as alternatives to conventional fruit juice processing in the phases of clarification, stabilization, depectinization, and concentration (Basile et al., 2016). Thermal evaporation causes heat-sensitive chemicals to degrade, lowering the efficiency of the final product significantly. Membrane-based operations are a potential option. Fining agents including bentonite, diatomaceous earth, silica sol, and gelatin are used to filter a wide range of fluids, but when disposed of, they constitute an environmental risk. Integrating or replacing a variety of existing techniques with novel membrane-based technologies minimizes direct and indirect energy usage while improving the finished product's organoleptic qualities. The integration of these membrane technologies can significantly minimize the price of various processes by minimizing waste creation and energy use (Sotoft et al., 2012).

This study aims to produce a special overview of recent advancements in pineapple juice–processing membrane operations for aroma compound clarity, concentration, and recovery. The first half of the chapter covers the driving forces and how they relate to membrane separation operations; the second half examines and explains specific applications as well as notable technological developments and enhancements over previous methods.

11.2   JUICE'S COMPOSITION AND HEALTH BENEFITS

Understanding pineapple's chemical composition and nutritional values should be a vital indicator for monitoring the fruit's quality and evaluating whether it needs to be processed. Pineapple content is determined by several factors, including the ripening process and cultivar. Pineapple juice is a light yellow liquid made up of protein, carbohydrates, vitamins, and minerals. It comprises 89.5% water and 9.5% dry matter. The nutritional content present in juices produced from fully matured fresh fruits is listed in Table 11.1.

11.3   MEMBRANE SEPARATION PROCESSES

The following section explains the numerous membrane techniques utilized in the food-processing industry, as well as how they are used in the production of pineapple juice. Membrane separation techniques have grown in popularity in manufacturing applications since the late 1960s, and they are now good alternatives to more conventional processes such as distillation, extraction, and evaporation. Using driving factors such as pressure, chemical potential, electrical potential, and temperature difference, several membrane processes could be separated to complete the separation are shown in Table 11.2

TABLE 11.1

Pineapple Juice Composition at Full Ripe Stage (per 100 g; adapted from (Ali et al., 2020)

Pineapple juice constituents					
Proximate composition		**Minerals**		**Vitamins**	
Carbohydrate	12.1	Magnesium	13.6	Ascorbic acid	14.0
Protein	0.4	Calcium	8.1	Folate	23.0
Fat	0.1	Potassium	134.0	Niacin	0.3
Fibre	0.2	Manganese	1.2	Thiamin	0.1
Total sugars	12.1	Sodium	5.2	Riboflavin	0.02
Ash	0.4	Phosphorus	9.8		

TABLE 11.2

Membrane Separation Processes and Their Driving Forces

Driving force Membrane process	Pressure difference	Chemical potential difference	Electrical potential difference	Temperature difference
	• UF • MF • RO • NF	• LM • GS • Dialysis • PV • Vapor permeation	• Membrane electrolysis • Electrodialysis • Membrane electrophoresis	• MD

TABLE 11.3

Pressure-Driven Membrane Operations (adapted from Ilame and Singh, 2015)

Pressure- driven membrane	Pore size (µ)	Operating pressure (bar)	Basis of rejection	Solutes to be separated	Purpose
MF	10^2–10^4	0.5–2	Absolute's size of particles (0.02–10 µm)	Suspended matters, oil droplets, microorganisms	Clarification and turbidity removal
UF	1–10^2	1–10	MWCO (10^3–10^5 Da)	Viruses, salts, sugars, polyphenols, colloids, and enzymes	
NF	1–10	20–40	MWCO (200–1000 Da)	Sugars, low-molecular-weight polyphenols, dyes	Decolorization and purity increases
RO	10^{-1}–1	30–60	MWCO	Salts, electrolytes	Concentration and desalination

11.3.1 PRESSURE-DRIVEN MEMBRANE OPERATIONS

Pressure-driven membrane filtration is a popular method for clarifying and concentrating in the juice business due to its low energy and temperature usage and good selectivity. Pressure-driven membrane technologies that improve selectivity include UF, MF, NF, and RO (Table 11.3). Because of the pressure differential across the semipermeable membrane, substances in the feed solution are selectively separated.

Hence, the feed solution is separated into two sections: a filtrate or permeate, which contains all particles that have traveled through the membrane, and retentate, which contains all substances that have been refused by the membrane. The degree of rejection is determined by membrane characteristics such as charge, pore size, and surface features. The electrostatic repulsion between ions or charged molecules and the membrane surface is influenced by the membrane charge, which is critical for membrane performance.

11.3.2 CHEMICAL POTENTIAL-DRIVEN MEMBRANE OPERATIONS

A concentration gradient across the membrane is described as a difference in a molecule's concentration on both sides of the membrane. Due to their net thermal mobility, molecules will shift from a high-concentration to a low-concentration site. The net force that moves molecules along a concentration gradient is known as the stop chemical driving force. This force is directly proportional to the concentration gradient. The force is proportionate to the gradient, in other words. Each molecule has its concentration gradient or chemical driving force if a cell membrane contains more than one type of molecule. Pervaporation (PV), dialysis, gas separation (GS), liquid membranes (LM), and vapor permeation are chemical potential-driven membrane processes that can improve selectivity.

11.3.3 ELECTRICAL POTENTIAL-DRIVEN MEMBRANE OPERATIONS

These membranes are made to allow specific ions to pass through while prohibiting water molecules from doing so. An ion is classed as a "cation" if it contains one or more positive charges and as an "anion" if it has one or more negative charges. Membranes for electrical potential-driven technologies are made with ion exchange resins. Cation exchange resin is cast onto fabric or ground up within a plastic matrix to generate a cation exchange membrane in electrical potential is driven water treatment technologies. A cation exchange membrane allows only cations to pass through. Negatively charged anions are repelled by the resin's negative charge, therefore they cannot flow through a cation exchange membrane.

11.3.4 TEMPERATURE-DRIVEN MEMBRANE OPERATIONS

MD is a phase-shift-driven thermal separation method. A hydrophobic barrier blocks the liquid phase, while the vapor phase (such as water vapor) can pass through the holes. The process is driven by a partial vapor pressure differential, which is usually caused by a temperature difference.

11.4 FRUIT JUICE CLARIFICATION

Pineapple juice contains pectin in its natural state. As a result, before concentration, clarification may be required. However, a clearing process is essential to avoid the formation of a foggy look during storage. In addition, the clarifying stage reduces the bitterness of the juice due to the high tannin concentration in the juice (Tao and Yun, 2017). These polyphenols contribute to the formation of haze by mechanisms such as prior polymerization or condensation, resulting in polymeric complexes that settle in the bottom of storage containers. Dietitians recommend that these components be preserved throughout the production of fruit juice since they have a protective effect on human health (Poh and Abdul Majid, 2011). Enzyme treatment, flocculation (bentonite, gelatin, diatomaceous, silica sol), cooling, filtration, and decantation are all traditional fruit juice clearing processes that have drawbacks in terms of treatment and disposal, prolonged operating timeframes, limited returns, and increased cost.

11.4.1 BY USING MF AND UF MEMBRANES

Pressure-driven membrane technologies like MF and UF have proved to be effective at clarifying fruit juices and are commercially viable. These procedures are a potential alternative to traditional pineapple juice clarification and stabilization methods since they save time and energy while boosting juice yield and eliminating the use of filter aids and fining agents (bentonite, silica sol, gelatin; Severcan, 2018). It is possible to reduce the amount of enzyme required for macromolecule hydrolysis, and enzymes can be reused and recycled. MF and UF are particularly good at retaining juice flavor, nutritional value, and freshness while delivering natural fresh-tasting, additive-free meals and high quality because the extraction process does not require chemical agents or heat. The juice is divided into a fibrous concentrated pulp (retentate) and a cleared fraction (permeate) that is devoid of spoilage microorganisms and stable during these operations. Solutes of low molecular weight (sucrose, acids, salts, fragrances, taste components) pass through these membranes, but large-molecular-weight molecules (pectin or proteins) are retained.

Youravong et al. (2010) assessed the effect of hydrophobicity and membrane pore size on the quality of clarified pineapple wine and fouling characteristics using stirred cell dead-end MF. The test membranes were mixed cellulose acetate (pore sizes 0.45 and 0.22 Î1/4m), modified polyvinylidene fluoride (0.22 Î1/4m), and polyethersulfone (PES) (0.22 Î1/4m). All types of membranes were found to clarify the pineapple wine. Membrane pore size and hydrophobicity both played a role in the reversible and irreversible fouling of membranes. For pineapple wine clarification, 0.45 Î1/4m MCE showed to be the optimum choice in terms of permeate flow and fouling.

Jiraratananon et al. (1997) investigated the formation of self-forming dynamic membranes when pineapple juice (12 °Brix) was pumped over the porous ceramic membrane module at 25°C for 1 hour with crossflow velocities (CFV) (1.30–3.95 m s^{-1}) and applied pressures (100, 200, and 300 kPa). After 30 minutes of circulation at 2.0 m s^{-1} and 300 kPa, the dynamic membrane was well formed, having a consistent flow of 6.0103 m^3/m^2 h with 84–87% and 6% of macromolecules and sugars rejection produced in the filtering mode. However, in the case of UF by alumina membrane with a molecular weight cutoff (MWCO) of 50,000, having a consistent flow of 15.8 × 10^{-3} m^3/m^2 h, and macromolecules and sugars rejection were 91%, and 10.5%, respectively. Hence among these, the UF membrane was identified as the most promising method. When subjected to a change infiltration condition, the self-forming dynamic membrane's stability was satisfactory. The permeation flux rose with CFV when the applied pressure was reduced and decreased when it was reduced.

De Carvalho et al. (2008) examine the loss of sugars in pineapple juice following pretreatment with commercial pectinase alone and in combination with a cellulase, as well as clarification by crossflow MF and UF, using two different module geometries to determine which membrane process preserves these nutrients the best. Polysulfone (PS) membranes with pore sizes of 0.1, 0.45, and cutoffs of 50–100 KDa, as well as PES and polyvinylidene fluoride (PVDF) membranes with pore diameters of 0.3 lm and cutoffs of 30–80 KDa, were studied at 25°C and various TMP At the 5% level, High-pressure liquid chromatography measurement of the sugar content of the clarified pineapple juices revealed significant variances. These studies demonstrated that membrane pore diameters or cutoffs, along with module geometry, influenced the cleared juice sugar concentration. The sugar content was found to be lower when the pineapple juice was clarified using a 30–80 KDa tubular membrane at 1.5 bar. Although juices cleared with PS membranes (50 KDa – 7.5 bar) have the best total sugar recoveries, due to their tubular construction and module geometry, the use of 0.3 lm PES is more appealing and acceptable.

Carneiro et al. (2002) studied the cold sterilization and clarity of pineapple juice via MF in 10 trials under the same working conditions of 25°C and 100 kPa. A tubular PES membrane with a pore size (0.3 m) and an effective filtering area (0.05 m^2) was used in the pilot system. After 15 minutes of processing, the permeate flow had barely changed. It was brought down to about 100 L/hm^2. Because of the great reduction in haze and viscosity, as well as the absence of significant changes in the juice's pH, acidity, sugar, and soluble solid content, the clarity technique was rated very efficient. The permeate from the procedure was collected in sterile bottles and stored refrigerated for 28 days (8°C) inside a laminar flow station. At 7-day intervals, the samples were subjected to microbiological examinations.

Jaeger de Carvalho et al. (1998) performed a study using MF and UF systems with ceramic and PS membranes, three types of clarified liquids were obtained from concentrated pineapple juice reconstituted to 12 °Brix. The best volume recovery was achieved with 50000 Da PS membranes. With the 0.22-m ceramic membrane, component recovery was improved. The 50000 Da PS membrane was more effective at removing tannins and pectin. In terms of lowering turbidity, both membranes with a 50,000-Da cutoff performed similarly. Overall, the ceramic membrane with a thickness of 0.22 m performed best. The maximum flow rate of clarified juice was attained with the 0.22 m ceramic membrane (52.02 L m^{-2} h^{-1}).

Laorko et al. (2010) studied the membrane fouling, permeate flux, and quality of clarified pineapple juice as a function of membrane property. For UF and, MF membranes with pore diameters of 0.2 m and 0.1 and MWCOs of 100 and 30 kDa were utilized, respectively. Membrane filtering had little influence on the pH of clarified juice, but it did reduce sugar and acidity and entirely removed suspended particles and microorganisms. The permeate flux, irreversible fouling value, total phenolic content, vitamin C content, and antioxidant capacity were all maximum with the 0.2-m membrane. Based on these findings, the membrane with a hole size of 0.2 m was determined to be the best choice for pineapple juice clarity. The CFV of 3.4 ms^{-1} and the TMP of 0.7 bar was the best operating parameters for the clarification of pineapple juice by membrane filtration. During the MF of pineapple juice under optimal conditions, an average flow of about 37 lm^2 h^{-1} was achieved using batch concentration mode.

Barrosi et al. (2004) used a mixture of cellulase, hemicellulose, and pectinase at doses of 300, 100, 20 mg/L at 40°C to investigate the effects of enzymatic treatment in cherry juice and pineapple juice. They used a PS hollow-fiber membrane for filtration and a ceramic tubular membrane for UF. The permeate flow rate of the PS hollow fiber membrane is lower. Ultra-filtering depectinized juice treated with a 20-mg/L enzyme concentration is economically advantageous because the increases in permeate flow rate with the 100- and 300-mg/L enzyme concentrations were not significant.

Youravong et al. (2010) employed a tubular ceramic membrane and an MF technology to clarify pineapple wine. It has a membrane pore size, TMP, and a CFV of 0.2 lm, 2 bar, and 2.0 m/s, respectively. Gas sparging's effects on clarified wine quality, fouling, and permeate flow were studied. Permeate flow was found to be improved by up to 138% using a rather low gas sparging rate. Increasing the gas sparging rate did not enhance the permeate flux when compared to the permeate flux without gas sparging. Gas sparging was used to modify the density of the cake layer. The cake resistance rose as the gas sparging rate was raised. An increased gas sparging rate only improves reversible fouling according to studies. After MF, the turbidity of pineapple wine was reduced, resulting in a clear product with a bright yellow color. The turbidity of pineapple wine was reduced after MF, resulting in a clear product with vivid yellow color. Gas sparging reduced the alcohol (%) of the wine, resulting in a loss of alcohol content.

Yu (2005) investigated the effects of UF and MF membranes on the clarity of pineapple juice. The effects of operational variables (such as pressure, temperature, and time) on membrane separation efficiency, membrane washing, and juice permeate quality were studied in this study. A pressure of 0.06 MPa and a temperature of 45°C were found to be the best operating parameters for this study. The PVDF-MF membrane had a greater antipollution ability for pineapple juice after cleaning than the PS-UF membrane, with a water penetration flux recovery rate of 97.8%. The nutrition composition of the original pineapple juice was kept in permeation juice, while macromolecules, bacteria, and some pigments were considerably decreased by a membrane, resulting in a vast increase in pineapple juice sensory quality.

Pérez-Carvajal et al. (2005) investigate how to obtain clarified pineapple juice using crossflow MF. At 30°C, the experiments were conducted on a semi-industrial scale utilizing pilot equipment and a tubular alumina membrane with an average pore diameter of 0.2 gm. Enzymes were used to pretreat all of the samples. At a TMP of 200 kPa, acidity, pH, turbidity, soluble solids, ascorbic acid (%), and total carotenoids (%) were all assessed. Finally, the impact of clarity on the profile of volatile components was investigated using gas chromatography with mass spectrometry. When compared to other pineapple varieties, the Golden™ variety contains more sugar and vitamin C, as well as less acidity. Except for carotenoids, which were kept by the ceramic membrane, crossflow MF enabled the preservation of physicochemical properties in the clarified pineapple juice. The permeate fluxes (approximately 75 L h^{-1} m^{-2}) and process yield (85%) are compatible with a potential industrial application.

TABLE 11.4

Clarification of Pineapple Juice by UF and MF Membranes

Process	Membrane used	Operating parameters	Reference
MF & UF	PS plate and frame 50000 Da	TMP- 72.5 psi, Flow rate – 25.0 L m^{-2} h^{-1}, Temp – room temperature	Jaeger de Carvalho et al. (1998)
	PS plate and frame 50000 Da without cleaning	TMP – 72.5 psi, Flow rate – 22.7 L m^{-2} h^{-1}, Temp – room temperature	
	0.22 μm Ceramic tubular	TMP – 15.0 psi, Flow rate – 52.0 L m^{-2} h^{-1}, Temp – room temperature	
	Ceramic tubular 50000 Da	TMP – 56.0 psi, Flow rate – 46.8 L m^{-2} h^{-1}, Temp – room temperature	
UF	Ceramic tubular	TMP – 0.8 bar, Tangential flow rate – 570 L/h, Temp – 30°C	Barrosi et al. (2004)
	PS hollow fiber	TMP – 4 bar, Tangential flow rate of – 570 L/h, Temp – 30°C	
UF	Multichannel monolith alumina	TMP- 100–300 kPa CFV – 1.30–3.95 m s^{-1}, Temp – 25°C	Jiraratananon et al. (1997)
MF	0.3 μm Tubular PES	TMP – 1.5 and 3.0 bar, Temp – 25°C, Effective filtration area – 0.05 m^2	Carvalho and Silva (2010)
MF	0.2 μm Single-channeled tubular ceramic	TMP – 2 bar, CFV – 2.0 m/s, Temp – 25°C	Youravong et al. (2010)
MF & UF	0.3 μm Tubular PS	TMP – 1.5 bar, Temp – 25°C	Carvalho et al. (2008)
MF	0.3 μm Tubular PES	TMP – 2 bar, CFV –0.5 m/s, Temp- – 25°C	Carvalho et al. (2010)
MF	0.2 μm Hollow fiber PS	TMP – 1 bar, CFV – 1.2 m/s, Temp – 20°C	Laorko et al. (2010)
MF	0.2 μm Hollow fiber PS	TMP – 10–70, CFV – 1.5–3.4, Flow rate –25–70 L/m^{2}h	Laorko et al. (2011)
UF	0.01 μm Tubular, α-Al$_2$O$_3$ / TiO$_2$	TMP – 2.0–6.0 bar, CFV – 4.17 m/s, Temp – 30–50°C	De Barros et al. (2003)
	Hollow fiber PS, 100 kDa	TMP – 0.2–2.0 bar, CFV – 1.19 m/s, Temp 20–40°C	
MF	0.2 μm PS hollow fiber	TMP – 1.0 bar, CFV – 1.2 m/s, Temp – 20°C	Laorko et al. (2013)
MF	0.3 μm tubular PES	TMP – lOOkPa, CFV – 6 m/s, Temp – 25°C, Effective filtration area – 0.05 m^2	Carneiro et al. (2002)

11.5 FRUIT JUICE CONCENTRATION

Fruit juice concentrates are useful in the industry since they may be utilized in ice creams, fruit jelly, jellies, and fruit juice drinks. Fruit juice concentration has several advantages, including reduced weight and volume, as well as decreased packing, shipping, handling, and storage expenses. The lack of water movement makes the product even more homogeneous. Finally, the concentration stage assists in the product's preparation for final drying.

Water is removed at elevated temperatures; then volatile flavors are recovered and concentrated before being reintroduced into the concentrated product. In the industrial concentration of fruit juices, multistage vacuum evaporation processes are widely used. Traditional evaporation processes, on the other hand, have several disadvantages, including off-flavor development, high energy consumption, nutritional content loss, and color changes due to thermal impacts. Cryoconcentration, an alternative to thermal evaporation, extracts water as ice rather than vapor.

Freshly squeezed juices are concentrated by thermal evaporation up to 90°C in most marketed juices. Thermal treatments can have a major impact on the nutritional quality and flavor of fruit juices because heat-sensitive compounds confer these attributes. Fruit juices with distinct fresh fruit characteristics are in high demand. Researchers are looking for breakthrough technologies that could increase the quality of fruit juices as a result of rising demand. Because of its capacity to operate at moderate temperatures and pressures, membrane technology is a feasible alternative for processing fruit juices (Jiao et al., 2004). Juices can be concentrated at low temperatures using membrane processes including NF, RO, MD, and OD, saving energy while maintaining aroma, nutritional, and bioactive elements.

This constraint can now be overcome due to rapid technological advances in the development of novel membranes and improvements in process engineering as well as integrated membrane processes that could help in the production of concentrated pineapple juice (Bowden and Isaccs, 1989; Hongvaleerat et al., 2008; Jiao et al., 2004). This section gives a summary of recent significant breakthroughs in pineapple juice concentration membrane processes, such as NF, RO, FO, OD, and integrated membrane processes.

11.5.1 NF

LIU et al. (2009) investigate the effect of operational variables like pressure and time affect membrane separation efficiency, membrane washing, and the quality of processed juice. The ideal operating pressure for UF was 0.12 MPa, whereas the optimal operating pressure for RO and NF was 0.50 MPa, according to the data. The spiral-wound membrane had a better antipollution ability for pineapple juice than the hollow-fiber membrane, and after cleaning with 0.2% NaOH solution, the membrane flux could be recovered to 96%. Because the majority of two nutritious components in the original pineapple juice were identified as reserved in permeation juice, while macromolecules were eliminated by UF, the sensory quality of pineapple juice was increased. To concentrate pineapple juice, RO and NF could be utilized.

11.5.2 RO

RO concentrates juices at low temperatures, saving energy while keeping the aroma, bioactive, and nutritional content intact. The excellent selectivity and solute retention of RO membranes are well recognized. Fruit juices' viscosity and osmotic pressure grow rapidly as the sugar concentration rises, dramatically lowering the process's productivity. Furthermore, high working pressures may degrade the quality of the juice. Because its final concentration is limited (usually up to 30 °Brix), RO can be employed as a pre-concentration step before a final concentration using another method.

Bowden and Isaccs (1989) concentrated pineapple juice from 130- to 250-g/kg soluble solids in a pilot-scale tube and plate-and-frame RO devices. The operating temperature of the clarifier, the types of membranes used, the flow rates, the pressure, and the concentration level were all investigated. Permeate flux, which averaged 20 L/m²h, was impacted by all parameters except clarity. The concentrated juice had a flavor that was comparable to single strength, and the permeate lost very few soluble components. Pineapple juice may be concentrated up to 250-g/kg soluble solids with high-quality retention at 6000 kPa, 40°C, and a velocity of 3 m/s.

Salleh. et al. (2020) examine the sensory attributes of concentrated pineapple juice produced using RO. Fresh pineapple juices were concentrated at four distinct pressure (20–60 bar) and temperature (20–60°C) combinations, and their sensory attributes were evaluated in terms of color, aroma, sweetness, sharpness/sourness, overall acceptability, and buying intent. After that, the juice with the greatest overall approval score was compared to store-bought pineapple juice. The majority of the panelists agreed that pineapple juice prepared at 60 bar and 20°C was the best treatment of all. Furthermore, when treated pineapple juice was compared to commercial pineapple juice, the majority of panelists favored RO pineapple juice, especially in terms of buying intent.

Couto et al. (2011) used RO to determine the concentration of single-strength pineapple juice. The concentration was carried out in a 0.65 m² plate and frame module with 60-bar TMP at 20°C using polyamide composite membranes. The flow rate of the permeate was 17 L h m⁻². With a volumetric concentration factor of 2.9 °Brix, the total soluble solids (TSS) content in the juice increased from 11 to 31 °Brix.

11.5.3 FO

FO is a cutting-edge membrane technology used in the food sector to concentrate liquid foods while protecting heat-sensitive components. The primary advantages of FO over both thermal and traditional membrane processing are the low hydraulic pressure, low treatment temperature, reduced fouling tendency, and high solid content.

During pineapple juice FO, Babu et al. (2006) investigated how flow velocity, feed temperature, and osmotic agent concentration affect transmembrane flux. The flow rate of both the osmotic agent and the juice had a considerable impact on the transmembrane flux, with the effect being stronger at higher feed concentrations. The temperature of the feed was boosted, which increased the flow. The optimum alternative for boosting process performance while keeping the sensory features of the juice was found to be a combination of sodium chloride and sucrose as an osmotic agent.

Using three cycles of operation, Nayak et al. (2011) concentrated the pineapple juice up to 12-fold (from 4.4 to 54 °Brix) using the FO technique. In this process, because of reverse solute diffusion, transports of osmotic agents (salt or sugar) to the product take place. The use of sugar solution in FO for the concentration of fruit juices results in a reduced flow due to its high viscosity and lower osmotic pressure. Because of the minimal salt transfer to the product side, the use of NaCl solution as a draw solution imparts a salty flavor to the fruit juices. As a result, the concentration of pineapple juice was determined using a combined osmotic agent of salt and sugar. As the concentration of NaCl rose (from 0% to 16%, w/w), migration surged to 1.28%. The concentration of sucrose was increased (from 0% to 40% w/w), and the migration of NaCl was reduced (from 1.87 to 0.58 %). As the proportion of osmotic agent solution increased, the overall mass transfer coefficient (K) dropped.

Pineapple juice was concentrated to 60 °Brix while retaining a high level of ascorbic acid (Babu et al., 2006). NaCl solution is commonly often used as a draw solution because it is simple to maintain, inexpensive, and nontoxic. But, due to salt dispersion from the DS, it may result in salty juice. Babu et al. (2006) developed a sucrose–NaCl DS mixture for pineapple juice concentration to resolve this issue. When a 40% sucrose and 12% NaCl solution was administered, the original TSS of 12.4 °Brix was boosted to 60 °Brix. A mixed DS solution was used to prevent salt from diffusing into the juice, resulting in a less salty taste.

11.5.4 OD

OD, also known as membrane evaporation, isothermal MD, osmotic evaporation, or gas membrane extraction, is an athermal membrane process based on microporous hydrophobic membranes. Two water solutions (feed and osmotic solution) with varied solute concentrations are used to divide each side of the membrane in OMD. Water evaporates at a greater vapor pressure from the solution's surface (feed), diffuses through the membrane pores, and condenses on the solution's surface at a lower vapor pressure. The feed becomes more concentrated as a result, while the OD solution becomes more diluted (Hogan et al., 1998).

Chutintarasri (1991) examined the usage of OD and UF to concentrate pineapple juice and clarify pineapple mill juice. In the OD process, the effects of process parameters such as flow rates and temperatures were investigated, whereas in the UF process, the methods used in extracting mill juice, hydraulic press, and blender, as well as the aforementioned process parameters combined with enzyme pretreatment, were investigated. In the OD process, a feed flow rate of 4 l/min at 50°C produced a maximum processing capacity of 2.87 kg/m^2/hr. UF pilot-plant tubular HFM 180 membranes with a surface area of 0.4 m^2 and an MWCO of 18,000 Daltons were utilized to clear pineapple mill juice. The product is translucent and has a decreased viscosity (from 1.27 to 1.04 cps), turbidity (from 875 to 725 NTU), and viscosity (from 1.27 to 1.04 cps). The enzyme-treated juice produced 46.73 l/m^2/hr of filtrate at a 4-bar operating TMP and ambient temperature. The flow rate was increased via enzyme pretreatment and high TMP. The viscosity of the enzyme-treated mill juice pineapple syrup was lower than that of the control syrup.

According to Ravindra Babu and Sambasivarao (2015), in the OD concentration of pineapple juice, concentration polarization adds more to transmembrane flow than temperature polarization. Moreover, flux deterioration is mostly caused by dilution of the stripping solution at low TSS levels of the feed juice, and it is primarily caused by juice viscosity (viscous polarization), juice concentration, and temperature.

Hongvaleerat et al. (2008) exploited OD to concentrate both clarified and single-strength pineapple juice. The OD tests were conducted in a laboratory with two circuits: one for juice and the other for brine solution. As an extraction phase, a saturated calcium chloride solution with a concentration of 5.5–0.6 mol/l was utilized. A 0.2-m flat-sheet hydrophobic membrane with a PTFE layer and a porous polypropylene (PP) substrate served as the membrane. Increasing the temperature of the juice from 20 to 35°C nearly doubled the evaporation flux, while increasing the circulation velocity of the salt solution increased it by about 7%, according to the data. Temperature-related flux increases are caused by an increase in water partial pressure at the liquid–gas interface, which boosts the water transfer driving force. Evaporation fluxes were larger in the cleared juice (8.5 kg/m^2 h) than in the single-strength juice (6.1 kg/m^2 h), indicating that pulp has a substantial impact on OD performance characteristics. Furthermore, there were no significant changes in color or other important quality markers after the juice was analyzed. Both clarified and single-strength juices benefited from the TSS concentration factor, which boosted titrable acidity and phenolic content.

For osmotic evaporation testing, Shaw et al. (2002) concentrated pasteurized pineapple juice into a 51 °Brix concentrate, which was then reconstituted to single-strength juice. According to headspace gas chromatography, the concentrate preserved an average of 62 % of the volatile components found in the original juice. For numerous reasons, including the loss of appealing flavor top notes and the appearance of some processed flavor in the concentrate, a sensory panel picked the fresh juice above the reconstituted concentrate. According to an HSGC examination of four independent commercial juice samples with a wide range of quantitative values for volatile components, the original juice matched the weakest of these commercial juices. Other less-volatile components were discovered in concentrated juice extracts that were not initially recognized by HSGC of the juice. In the beginning, these components were present in the juice in small amounts. Despite the fact that this nonthermally produced concentrate contains more volatile components than concentrates created using typical thermal processing methods, adding aqueous fragrance may be necessary for better flavors.

Hongvaleerat et al. (2005) investigated how osmotic evaporation affects the concentration of pineapple juice in this investigation. The tests were conducted in a lab unit with two separate circuits: pineapple juice and brine. A calcium chloride solution is used as a brine. The concentration of clarified and single-strength pineapple juices was investigated. The evaporative flux ranged from 8.5 kg/hm^2 to 5.5 kg/hm^2 for single-strength juice concentration, enabling the juice to be concentrated to 55 °Brix. After osmotic evaporation, the clarified juice's concentration reached 53 °Brix. The evaporative flux in this scenario ranged from 6.6 to 9.9 kg/hm^2.

Babu et al. (2008) used an osmotic MD in a plate and frame membrane module to concentrate clarified pineapple juice. During the osmotic MD process, concentration and temperature polarization impacts are found to have a considerable impact on flow reduction. At various operating parameters, such as feed, osmotic agent flow rate (25–100 ml/min), and osmotic agent concentration (2–10 mol/kg), the impact of these polarization impacts on decreasing the driving force. Temperature polarization has a stronger influence on flux drop than concentration polarization. When both concentration and temperature polarization effects were considered, the observed fluxes were in strong agreement with theoretical fluxes. At room temperature, the pineapple juice was concentrated to a TSS of 62 °Brix.

11.5.5   INTEGRATED MEMBRANE OPERATIONS IN PINEAPPLE JUICES PRODUCTION

Membrane technology has been shown in several studies to be a viable replacement for traditional unit operations at several stages of fruit juice production (such as stabilization, fractionation, clarification, concentration, and aroma compound recovery; Conidi et al., 2018). Membrane technology is a promising option for modernizing the pineapple juice industrial transformation cycle. With low obstacle volume, greater automation possibilities, modularity, remote control, reduced energy consumption, and waste creation, this method strives to incorporate contemporary technology into production cycles. The following section examines and discusses many integrated membrane systems used in pineapple and other juice processing applications.

Naveen (2004) investigated integrated membrane technologies such as UF/RO followed by osmotic MD for large-scale pineapple juice processing. UF pre-clarification of pineapple juice was followed by RO concentrations of up to 25 °Brix at various stages of processing. The OMD approach was used to achieve the juice's final concentration (>60 °Brix). Quantitative descriptive analysis was used to analyze the sensory attributes of the resulting juice concentrate, which demonstrated that the quality of the juice was quite similar to that of the original pineapple juice. According to the research, integrated treatment systems such as UF, RO, and OMD have a lot of promise for improving overall product quality.

The quantity of pineapple juice used by Hongvaleerat et al. (2008) to explore the impact of pulp on OD performance. The fluxes recorded for unclarified and clarified juice at a concentration of 13 to 56 wt.% TDS were investigated to evaluate this. A tubular ceramic membrane with a nominal pore width of 0.1 m and a feed pressure of 2 bar was used to expose pineapple juice with a pulp concentration of 5.6 wt.% TDS to MF. A Pall-Gelman PTFE membrane with a nominal pore diameter of 0.2 m was used in a plate-and-frame module for coupled OD–MD method. The feed and strip temperatures were set to 35 and 20°C, respectively. During the concentration range investigated, the concentration of unclarified juice steadily fell from 8.6 to 3.7 kg m² h1. As TDS concentrations increased, this decline was related to a reduction of water resulting in a significant increase in viscosity.

In clarified juice, a similar flow pattern was identified. Due to pulp removal alone, the flow in the latter case was 17% higher than in the unclarified juice. Furthermore, unlike thermal evaporation, during concentration, whether with or without clarifying, there were no notable changes in the juice's major properties. pH, organic acid concentration, TPC, and hue were among them. Based on these preliminary findings, it was determined that producing pineapple juice concentrate by combining integrated MF–combined OD–DCMD processing is a viable option.

Several various methods for preparing and processing pineapple juice to keep the characteristics of the original fresh fruits have been proposed in recent years. Álvarez et al. (2000) developed a process architecture based on membrane and classic separation technologies for the clarification and concentration of pineapple juice. The preconcentration of clarified juice up to 25 °Brix using RO, the extraction and concentration of aroma constituents by PV, and a final concentration up to 72 °Brix by traditional evaporation are all explained using an enzymatic membrane reactor. When compared to the previous process, the integrated membrane approach resulted in a 14% lower overall capital cost and a 5% higher process yield. Because less energy was utilized to concentrate the juice, total production expenses decreased by 8%. Membrane replacement accounted for only 2% of total operating costs, with UF, PV, and RO membranes having life expectancies of 2, 2, and 3 years, respectively.

Aguiar et al. (2012) proposed a new way to make high-quality pineapple juice concentrate. The enzyme-treated juice was clarified with MF, preconcentrated with RO (29 °Brix), and concentrated with OD (53 °Brix) before being concentrated with OD. The concentration stage of the integrated membrane process resulted in an 18% drop in phenolic compounds, as well as a loss of more volatile molecules.

Sensory testing confirmed that the reconstituted concentrated juice smelled and tasted great and that customers were quite happy with it. Onsekizoglu et al. (2010) looked at the impact of a variety of integrated membrane technologies on the clarity and concentration of pineapple juice, as well as the product quality. A mixture of fining agents (bentonite and gelatin) and UF were used to clarify fresh pineapple juice with an initial TSS level of 12 °Brix. The clarified juice was then concentrated using MD/OD, MD, and OD membranes, as well as conventional thermal evaporation up to 65 °Brix.

11.6   RECOVERY OF AROMA COMPOUNDS

The recovery of aroma components from diverse fruit juices is a significant food processing operation, and membrane separation technologies have several new applications in the pipeline. In the pharmaceutical, food, and cosmetics industries, the need for flavor and fragrance components is on the rise (Tylewicz et al., 2017). Flavors and perfumes include alcohols, ketones, lactones, esters, short-chain n-alkanes and alkenes, thiols, aldehydes, and other organic acids. Terpenes are particularly important because they are the most prevalent chemical group in nature and are responsible for the vital smells found in plants (flowers) and some fruits. Pineapple juice's fragrance complex contains volatile compounds such as esters, aldehydes, and alcohols, as well as ethers, lactones, terpenes, and ketones in lower amounts. The overall aroma component concentration in pineapple juice is 100–1000 parts per million. Each pair of smell components determines the flavor of a normal juice. The esters, for example, give the juice a pleasant fruity flavor that corresponds to the sensation of ripeness. The aldehydes impart a fresh green flavor to the pineapple, which is characteristic of unripe pineapple. Both the fruity and immature flavors are influenced by the alcohols, which make up the largest group of components in terms of amount. Ethyl butyrate is the major aroma element in pineapple juice that gives it its characteristic flavor (Flath and Forrey, 1970; Lamer et al., 1994). In the preparation of pineapple juice, pasteurization and evaporation are utilized, which can result in the chemical and physical loss of heat-sensitive aroma components, resulting in a loss of scent component and a change in flavor. The aroma recovery process can be aided by PV, a membrane-based approach that does not require heat treatment.

11.6.1   PV

Aromas, essential oils, and fragrances are categorized and manufactured using aldehydes, esters, alcohols, ketones, terpenoids, carotenoid-based derivatives, and lactones (El Hadi et al., 2013). Because they are linked to sugars via glycosides (Sarry and Günata, 2004), glycosylated aromatic molecules or bonded volatile molecules have high stability in fruits and vegetables and hence do not produce any aroma. For their extraction from natural sources, chemical agents (e.g., acidity), biochemical approaches (e.g., enzymes), and physical extraction techniques (e.g., temperature) are required (Joana Gil-Chávez et al., 2013). Aromas are present in their volatile condition once released by glycoside hydrolysis due to their limited reactivity, volatility, and thermal stability, making recovery difficult. PV was found to be a potential method for recovering and selectively separating compounds in this study.

PV is a separation technique in which a liquid feed combination is partially vaporized and then passed through a nonporous permselective membrane. PV can handle separation difficulties that normal, equilibrium-dependent separation techniques cannot since it is based on a solution diffusion process (Karlsson and Tragardh, 1996). Despite its successes and potential, PV has struggled to acquire traction in the food industry. PV has also been used to recover aroma components in various fruit juices including grape (Rajagopalan and Cheryan, 1995), pineapple (Pereira et al., 2005), orange juice (Aroujalian and Raisi, 2007), strawberry (Isci et al., 2006).

The use of PV to restore the flavor of pineapple juice was examined by Pereira et al. (2005). They used clarified single-intensity pineapple juices for the experiment. Using a hybrid ethylene–propylene–diene monomer hollow fiber, the researchers were able to achieve extremely high enrichment of the most volatile components. Their research demonstrated that utilizing a highly selective polymer is useful when the number of organic solutes in the feed decreases.

Sampranpiboon et al. (2000) extracted aroma compounds from mixtures of ethyl hexanoate and ethyl butanoate, which are prevalent in pineapple and banana juice, utilizing PDMS and POMS membranes. According to their observations, the POMS membrane was found to become more perm selective to aroma molecules than that of the PDMS membrane. PV efficiency was impacted by hydrophobicity, with the more hydrophobic ETH having higher efficiency.

11.7   CONCLUSION

Because of their growing popularity as strategies for protecting the juice's overall quality, membrane-based procedures utilized in pineapple juice processing have been reviewed. In terms of assuring the juice's microbiological stability and preventing spoilage of the finished product, UF and MF have been demonstrated to be comparable to pasteurization. Nutritional content, aroma, and juice freshness are preserved in comparison to the use of fining chemicals, resulting in fresh-tasting, high-quality, additive-free-clarified, and natural items. Juice fractionation possibilities are boosted by tight NF and UF membranes, which allow bioactive components of interest to be recovered and purified for use in functional ingredient synthesis. RO and OD are acceptable alternatives to thermal evaporation for juice concentration. TSS can be produced by OD at low pressures and temperatures while minimizing thermal and mechanical stress on the processed juice. Overall, the results demonstrate that the technique keeps the unique properties of fresh juice, such as total antioxidant activity, color, organic acids, and phenolic components, quite well. PV is a fast-growing alternative technology for extracting aroma compounds from juice because of its efficiency and low cost. All these methods are effective in improving profitability and minimizing production losses in the juice processing of other underused fruits while also delivering environmental advantages. Finally, today's emerging concept of revamping the traditional flow diagram of pineapple juice manufacturing with significant advantages in terms of restoration of health-promoting compounds, quality, environmental impact, and energy consumption is the integration of different membrane operations among themselves and with other conventional technologies.

ACKNOWLEDGMENTS

The authors thank anonymous coworkers and reviewers for their insightful remarks and recommendations, which helped to improve this manuscript.

ABBREVIATIONS

CFV   cross-flow velocity
FO   forward osmosis
LM   liquid membranes

MD	membrane distillation
MWCO	molecular weight cutoff
NF	nanofiltration
OD	osmotic distillation
PES	polyethersulfone
PS	polysulfone
PV	pervaporation
PVDF	polyvinylidene fluoride
RO	reverse osmosis
TMP	transmembrane pressure
TSS	total soluble solids
UF	ultrafiltration

BIBLIOGRAPHY

Aguiar, I.B.;Miranda, N.G.M.;Gomes, F.S.;Santos, M.C.S.;Freitas, D.d.G.C.;Tonon, R.V.;Cabral, L.M.C., (2012). Physicochemical and sensory properties of apple juice concentrated by reverse osmosis and osmotic evaporation. Innovative Food Science & Emerging Technologies, 16: 137–142.

Ali, M.M.;Hashim, N.;Abd Aziz, S.;Lasekan, O., (2020). Pineapple (Ananas comosus): A comprehensive review of nutritional values, volatile compounds, health benefits, and potential food products. Food Res. Int.: 109675.

Álvarez, S.;Riera, F.A.;Álvarez, R.;Coca, J.;Cuperus, F.P.;Th Bouwer, S.;Boswinkel, G.;van Gemert, R.W.;Veldsink, J.W.;Giorno, L.;Donato, L.;Todisco, S.;Drioli, E.;Olsson, J.;Trägårdh, G.;Gaeta, S.N.;Panyor, L., (2000). A new integrated membrane process for producing clarified apple juice and apple juice aroma concentrate. J. Food Eng., 46(2): 109–125.

Aroujalian, A.;Raisi, A., (2007). Recovery of volatile aroma components from orange juice by pervaporation. J. Membr. Sci., 303(1–2): 154–161.

Babu, B.R.;Rastogi, N.;Raghavarao, K., (2006). Effect of process parameters on transmembrane flux during direct osmosis. J. Membr. Sci., 280(1–2): 185–194.

Babu, B.R.;Rastogi, N.;Raghavarao, K., (2008). Concentration and temperature polarization effects during osmotic membrane distillation. J. Membr. Sci., 322(1): 146–153.

Barrosi, S.T.;Mendes, E.S.;Peres, L., (2004). Influence of depectinization in the ultrafiltration of West Indian cherry (Malpighia glabra L.) and pineapple (Ananas comosus (L.) Meer) juices. Food Science and Technology, 24: 194–201.

Basile, A.;Iulianelli, A.;Liguori, S., (2016). Membrane reactor: An integrated "membrane+reaction" system. Integrated Membrane Systems and Processes; Basile, A., Charcosset, C., Eds: 231–253.

Bowden, R.;Isaccs, A., (1989). Concentration of pineapple juice by reverse osmosis.

Carneiro, L.;dos Santos Sa, I.;dos Santos Gomes, F.;Matta, V.M.;Cabral, L.M.C., (2002). Cold sterilization and clarification of pineapple juice by tangential microfiltration. Desalination, 148(1): 93–98.

Chaudhary, V.;Kumar, V.;Vaishali, S.;Sing, K.;Kumar, R.;Kumar, V., (2019). Pineapple (Ananas cosmosus) product processing a review. Journal of Pharmacognosy and Phytochemistry, 8(3): 4642–4652.

Chutintarasri, B., (1991). Concentrated pineapple juice production by osmotic distillation and clarification of mill juice production ultrafiltration process.

Conidi, C.;Drioli, E.;Cassano, A., (2018). Membrane-based agro-food production processes for polyphenol separation, purification and concentration. Current Opinion in Food Science, 23: 149–164.

Couto, D.S.;Cabral, L.M.C.;Matta, V.M.d.;Deliza, R.;Freitas, D.d.G.C., (2011). Concentration of pineapple juice by reverse osmosis: Physicochemical characteristics and consumer acceptance. Food Science and Technology, 31: 905–910.

De Carvalho, L.M.J.;De Castro, I.M.;Da Silva, C.A.B., (2008). A study of retention of sugars in the process of clarification of pineapple juice (Ananas comosus, L. Merril) by micro-and ultra-filtration. J. Food Eng., 87(4): 447–454.

El Hadi, M.A.M.;Zhang, F.-J.;Wu, F.-F.;Zhou, C.-H.;Tao, J., (2013). Advances in fruit aroma volatile research. Molecules, 18(7): 8200–8229.

Flath, R.A.;Forrey, R., (1970). Volatile components of smooth cayenne pineapple. Journal of Agricultural and Food Chemistry, 18(2): 306–309.

Hongvaleerat, C.;Cabral, L.M.C.;Dornier, M.;Reynes, M.;Ningsanond, S., (2005). Concentration of clarified and pulpy pineapple juice by osmotic evaporation. CEMAGREF.

Hongvaleerat, C.;Cabral, L.M.C.;Dornier, M.;Reynes, M.;Ningsanond, S., (2008). Concentration of pineapple juice by osmotic evaporation. J. Food Eng., 88(4): 548–552.

Ilame, S.A.;Singh, S.V., (2015). Application of membrane separation in fruit and vegetable juice processing: A review. Crit. Rev. Food Sci. Nutr., 55(7): 964–987.

Isci, A.;Sahin, S.;Sumnu, G., (2006). Recovery of strawberry aroma compounds by pervaporation. J. Food Eng., 75(1): 36–42.

Jaeger de Carvalho, L.M.;Bento da Silva, C.A.;Pierucci, A.P.T.R., (1998). Clarification of pineapple juice (Ananas comosus L. Merryl) by ultrafiltration and microfiltration: Physicochemical evaluation of clarified juices, soft drink formulation, and sensorial evaluation. Journal of Agricultural and Food Chemistry, 46(6): 2185–2189.

Jiao, B.;Cassano, A.;Drioli, E., (2004). Recent advances on membrane processes for the concentration of fruit juices: A review. J. Food Eng., 63(3): 303–324.

Jiraratananon, R.;Uttapap, D.;Tangamornsuksun, C., (1997). Self-forming dynamic membrane for ultrafiltration of pineapple juice. J. Membr. Sci., 129(1): 135–143.

Joana Gil-Chávez, G.;Villa, J.A.;Fernando Ayala-Zavala, J.;Basilio Heredia, J.;Sepulveda, D.;Yahia, E.M.;González-Aguilar, G.A., (2013). Technologies for extraction and production of bioactive compounds to be used as nutraceuticals and food ingredients: An overview. Comprehensive Reviews in Food Science and Food Safety, 12(1): 5–23.

Karlsson, H.O.;Tragardh, G., (1996). Applications of pervaporation in food processing. Trends Food Sci. Technol., 7(3): 78–83.

Lamer, T.;Rohart, M.;Voilley, A.;Baussart, H., (1994). Influence of sorption and diffusion of aroma compounds in silicone rubber on their extraction by pervaporation. J. Membr. Sci., 90(3): 251–263.

Laorko, A.;Li, Z.;Tongchitpakdee, S.;Chantachum, S.;Youravong, W., (2010). Effect of membrane property and operating conditions on phytochemical properties and permeate flux during clarification of pineapple juice. J. Food Eng., 100(3): 514–521.

Laorko, A.;Li, Z.;Tongchitpakdee, S.;Youravong, W., (2011). Effect of gas sparging on flux enhancement and phytochemical properties of clarified pineapple juice by microfiltration. Sep. Purif. Technol., 80(3): 445–451.

Liu, Z.-J.;Wang, X.-M.;Hu, X.-Y.;Huang, H.-H., (2009). Study on the application of membrane separation technology in processing of pineapple juice [J]. Science and Technology of Food Industry, 3.

Naveen, N., (2004). Integrated Biotechnological Approaches for the Purificatiion and Concentratiion of Liquid Foods Proteins and Food Colors. University of Mysore.

Nayak, C.A.;Valluri, S.S.;Rastogi, N.K., (2011). Effect of high or low molecular weight of components of feed on transmembrane flux during forward osmosis. J. Food Eng., 106(1): 48–52.

Onsekizoglu, P.;Bahceci, K.S.;Acar, M.J., (2010). Clarification and the concentration of apple juice using membrane processes: A comparative quality assessment. J. Membr. Sci., 352(1): 160–165.

Pereira, C.C.;Rufino, J.R.M.;Habert, A.C.;Nobrega, R.;Cabral, L.M.C.;Borges, C.P., (2005). Aroma compounds recovery of tropical fruit juice by pervaporation: Membrane material selection and process evaluation. J. Food Eng., 66(1): 77–87.

Pérez-Carvajal, A.M.;Badilla, M.;Moler, J.;Vaillant, F.;Perez, A.;Hernandez, L., (2005). Effect of cross-flow microfiltration on physico-chemical quality of clarified pineapple juice. CEMAGREF.

Poh, S.;Abdul Majid, F., (2011). Thermal stability of free bromelain and bromelain-polyphenol complex in pineapple juice. International Food Research Journal, 18(3).

Rajagopalan, N.;Cheryan, M., (1995). Pervaporation of grape juice aroma. J. Membr. Sci., 104(3): 243–250.

Ravindra Babu, B.;Sambasivarao, J., (2015). Osmotic membrane distillation in downstream processing. Current Biochemical Engineering, 2(1): 33–38.

Sampranpiboon, P.;Jiraratananon, R.;Uttapap, D.;Feng, X.;Huang, R., (2000). Pervaporation separation of ethyl butyrate and isopropanol with polyether block amide (PEBA) membranes. J. Membr. Sci., 173(1): 53–59.

Sant'Anna, V.;Marczak, L.D.F.;Tessaro, I.C., (2012). Membrane concentration of liquid foods by forward osmosis: Process and quality view. J. Food Eng., 111(3): 483–489.

Sarry, J.-E.;Günata, Z., (2004). Plant and microbial glycoside hydrolases: Volatile release from glycosidic aroma precursors. Food Chem., 87(4): 509–521.

Severcan, S.Ş., 2018. Fabrication of New Generation Membranes and Their Applications in Fruit Juice Industry. Abdullah Gül Üniversitesi.

Shaw, P.E.;Lebrun, M.;Ducamp, M.N.;Jordan, M.J.;Goodner, K.L., (2002). Pineapple juice concentrated by osmotic evaporation 1. J. Food Qual., 25(1): 39–49.

Sotoft, L.F.;Christensen, K.V.;Andrésen, R.;Norddahl, B., (2012). Full scale plant with membrane based concentration of blackcurrant juice on the basis of laboratory and pilot scale tests. Chemical Engineering and Processing: Process Intensification, 54: 12–21.

Tao, L.;Yun, Z., (2017). Research on clarification of pineapple juice with compound clarifier. Science and Technology of Food Industry, (09): 164.

Tylewicz, U.;Inchingolo, R.;Rodriguez-Estrada, M.T., (2017). Chapter 9: Food aroma compounds. Nutraceutical and Functional Food Components; Galanakis, C.M., Ed. Academic Press: 297–334.

Youravong, W.;Li, Z.;Laorko, A., (2010). Influence of gas sparging on clarification of pineapple wine by microfiltration. J. Food Eng., 96(3): 427–432.

Yu, C., (2005). Study on the application of membrane separation technology in clarification of pineapple juice [J]. Science and Technology of Food Industry, 9.

12 Membrane Processes

Vandana Gupta and J. Anandkumar

CONTENTS

12.1 INTRODUCTION

Water scarcity is one of the major problems associated with wastewater generation. Most of anthropogenic activities are water-dependent; therefore, wastewater generation is increasing day by day with the increase in the human and industrial populations. Wastewater generation is inevitable as it is a vital part in all sectors of life. Hence, wastewater reclamation is one of the major concerns today. Wastewater can be treated with some efficient treatment techniques, and treated water can supplement freshwater resources and potable water. One of the most efficient techniques for wastewater reclamation is the membrane process that offers many prospects in wastewater treatment, such as low capital cost, less energy requirement, easy handling, reduced size of equipment and so on [1].

In the current era, membrane separation processes are well developed and widely used processes with several applications, such as wastewater reclamation, drinking water purification, desalination, electrolysis, hemodialysis, gas separation and purification, electrochemical industry, food and beverages industries and more.

A membrane is basically a thin layer or semipermeable barrier that separates two phases in a selective manner by allowing one phase to percolate through it [2]. The existence of membranes was noticed in the early 18th century. Since then, continuous innovation of membranes has been taken place in order to make them more efficient, economical and more suitable for

DOI: 10.1201/9781003165019-12

a wide spectrum of applications. Membrane industries are currently employing an easily scalable production process and selection of appropriate materials. Membrane innovations involve the production or improvement of membranes that must meet the desired intrinsic properties for specific applications [3]. In the last couple of decades, membrane processes have grown significantly due to the benefits that offer in water reclamation and treatment. It offers many prospects with significant reduction in equipment size, power consumption and cost compared to conventional processes.

Presently, membrane processes have gained a wide range of applications in different fields due to several of their intrinsic properties. They have become an essential part of our daily life, from drinking water to purification of wastewater. Biotechnological and biomedical applications of membrane processes include extracting, recovering, concentrating, fractionating and purifying valuable components. It has an important role in dairy plants, as well as the food and beverages industries, as a product property booster. The use of membranes in flue cell operation gave an alternative for producing energy using wastewater with less use of chemicals. One of the most important applications of membranes is as an artificial kidney, which gives new life to a person dealing with kidney failure. Membrane technology is also gaining a place in drug delivery systems. One of the typical applications of the membrane process is associated with the treatment of industrial, municipal and agricultural wastewater, as well as polluted air. It can potentially remove the metal, ions, organic, hazardous and toxic containments from it. Several precious components can also be recovered from wastewater by using membrane processes. Specific applications of different membrane processes have been discussed further in this chapter.

12.2 MEMBRANE SEPARATION PROCESS VERSUS CONVENTIONAL PROCESS

There are several conventional techniques for wastewater treatment such as adsorption, electrocoagulation, wet air oxidation, biodegradation, catalytic ozonation and others [4–6]. Water treatment methodology is divided as primary, secondary and tertiary treatments. Primary water treatment methods involve filtration, ion exchange, flocculation, coagulation and adsorption that remove the suspended and floatable materials form wastewater, whereas biodegradable organics, pathogenic micro-organics, heavy metals, inorganic, organic and toxic containments are usually removed by various advance treatment techniques such as chemical precipitation, ion exchange, carbon adsorption, evaporation, biological degradation, activated sludge and membrane processes [7].

However, all these conventional techniques are having some drawbacks. For instance, toxic loading on adsorbent, adsorbent regeneration and disposal are the major drawbacks of the adsorption process [8]. Biological treatments are time-consuming processes, and they have disposal issues with their nutrient-rich sludge [9]. The electrocoagulation process requires high energy and regular replacement of a sacrificial anode [10]. Similarly, catalytic ozonation has detriment of byproduct's adsorption, catalyst recuperation and presence of interferences in wastewater [11]. The major drawback of liquid–liquid extraction process is the large consumption, toxicity and selection of solvent [12]. The wet air oxidation process requires high capital cost and operating problems, such as scaling up, lack of turbulence and homogenization leading to sedimentation problems and maintenance, among others, that limit its application [13]. On the contrary, the membrane process has become more advantageous due to its selective nature, high efficiency, easy handling and comparatively low capital and maintenance costs [7]. Conventional techniques are comparatively less efficient, and the treatment process is not possible in a single step. Hence, membrane processes have gained more interest over this period of time due to high efficiency and easy handling. It has potential of bridging gap between sustainability and economical gap, eco-friendliness, easy accessibility and so on. Therefore, membrane processes have proved to be a favorable alternative, mostly in water and wastewater treatment recently.

12.3   MEMBRANE SEPARATION PROCESSES

Membrane separation processes involve the separation of chemical species through a membrane interphase by the difference in the transport rate. This transport rate is dependent on the driving force, mobility and concentration of the individual component within the interphase [14]. Solute molecular size, the morphological structure of membrane and chemical affinity are the key factors for the efficient separation of chemical components. The separation efficiency of membranes depends on its types and module. Membranes are usually categorized as isotropic and anisotropic, organic and inorganic, porous and nonporous and composite membranes, as shown in Figure 12.1. Hollow fiber, tubular, flat sheet and spiral wound are some basic configurations of membrane modules.

Significant flux through the membrane is of practical importance, which majorly depends on the driving forces. These driving forces involve concentration gradient, hydrostatic pressure gradient, temperature gradient and electrical potential difference. Driving forces may be interdependent in some membrane processes, for instance, in osmosis phenomena; the concentration gradient is not only responsible for the separation, but hydrostatic pressure also builds up under certain conditions [15]. The mass transport across the membranes can be represented by numerous phenomenological or mathematical models that based on the driving force, such as Fick's law, Ohm's law and Hagen-Poisseulille's law for concentration gradient, electric potential gradient and pressure gradient, respectively (Table 12.1). The main parameter for evaluation of membrane performance is flux, which can be given by a linear relationship between flux and electric charge or volume [13]:

$$J = \Sigma \, D.L,$$

where J, D and L is flux per unit area, generalized driving force and phenomenological constant, respectively.

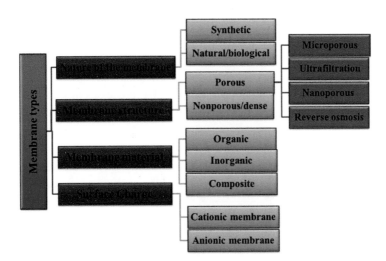

FIGURE 12.1   Schematic of membrane categorization.

TABLE 12.1
Mathematical Relationship between Different Driving Forces and Fluxes

Driving force	Law	Relationship	Flux	Proportionality constant
Pressure gradient	Hagen-Poiseuille's law	$V = h_d.\Delta P$	Volume	Hydrodynamic permeability (h_d)
Concentration gradient	Fick's law	$J = -D.\Delta C$	Mass	Diffusion coefficient (D)
Electric potential	Ohm's law	$I = \Delta U/R$	Electricity	Electric resistant (R)
Thermal gradient	Fourier's law	$Q = k.\Delta T$	Heat	Thermal conductivity (k)

Membrane processes are broadly categorized on the basis of driving forces. Different categories of membrane processes are discussed in this section.

12.3.1 PRESSURE-DRIVEN MEMBRANE PROCESSES

The membrane separation processes that involve a pressure gradient as the driving force include microfiltration (MF), ultrafiltration (UF), nanofiltration (NF), and reverse osmosis (RO) [16]. This classification of pressure driven membrane processes are also according to the pore size and porous structure of membranes. Figure 12.2 represents the main characteristics of pressure-driven membranes with respect to applied pressure, pore size, molecular weight cutoff (MWCO) and permeability [13]. Permeability through the membranes is typically determined by the mobility of certain components through the membrane structure and its concentration in the permeate. A detailed description of pressure-driven membranes and their advantages, disadvantages and applications are covered in this section.

12.3.1.1 MF

MF comprises porous membranes with a pore size ranging from 0.1–10 μm. A hydrostatic pressure gradient of 0.1–2 bar is applicable for the MF process [17]. The mechanism of separation is based on a sieving effect that includes the exclusion of larger particles than the membrane's pores size. The mode of mass transport in MF is by convection [17]. Ceramic or polymeric material can be used as membrane material for a microporous structure. Usually, MF separates macromolecules, suspended particles and colloids from solutions. But it does not have significant removal efficiency for the separation of dissolved solutes [18]. Flat-sheet and spiral-wound membrane modules are available for MF membranes, which can be customized to achieve the required application goal.

MF is widely used in the food and beverages industries for wine, juice and bear clarification, wastewater treatment, pharmaceutical industries, biotechnology and so on.

12.3.1.2 UF

The UF process is operated under hydrostatic pressure of 1–5 bar and is used for separating particles smaller than 3 μm in size. Similar to the MF process, UF processes also exhibit the size exclusion mechanism for the separation of chemical components [18]. Membrane surface adsorption is

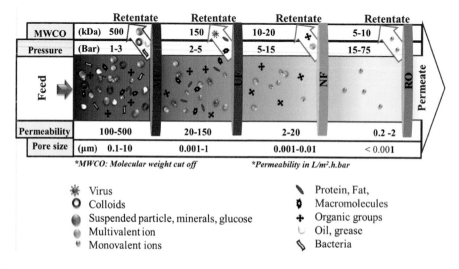

FIGURE 12.2 Illusion for basic characteristics of pressure-driven membranes.

another mechanism associated with the UF process. The mode of transport through a UF membrane is mostly by convection [19]. The pore size for UF membranes ranges from 0.001–1 μm. Low energy consumption, high efficiency, significant flux and long life span are some of the advantages of the UF process. Fouling is a major drawback of the UF process, which is related to the hydrophilicity of membrane. An improvement in hydrophilicity can reduce the chance of fouling to some extent [20].

Applications associated with UF processes are treating industrial wastewater, such as in the oil refinery or petroleum industries, to remove the traces of oils from wastewater; dairy production; cell harvesting; chemical recovery; water reclamation; and more. Moreover, UF is commonly used in concentrating and purifying macromolecules (e.g., protein) and in purification and disinfection processes, such as the removal of bacteria and virus, fouling and paint treatment in metal and textile industries.

12.3.1.3   NF

The NF membrane process was first introduced in late 1980s and used widely in various applications. This process utilizes the hydraulic pressure in between the range of ultrafiltration and reverse osmosis. Typically, the NF process requires hydraulic pressure within a range of 5–15 bars with the pore size ranging from 1 nm–0.001 μm [14]. Tight porous, asymmetric and thin-film composite membranes are used for this type of process [18]. Due to the very small pore size, this process has the capability of rejecting a wide spectrum of solute such as pigments, divalent ions, lactose, sucrose, sulfates, chlorides, multivalent inorganic salts, small organic molecules and so on. The mode of mass transport is partially by diffusion and partially by convection [21]. With the size-exclusion mechanism, adsorption on the membrane surface also plays an important role in the rejection of specific solutes [21]. NF membranes also show some level of charge due to the adsorption of charged molecules or the dissociation of functional group on membrane surface [21]. The presence of ionizable groups, such as carboxylic and sulfonic groups, in the polymeric membrane material is also responsible for the surface charge on membranes. These ionic groups resulted in surface charge when they come in contact with an aqueous solution [22].

NF covers a wide range of applications due to its versatility. Applications of NF are the removal of organic solvent, removal of color, total dissolved solid, chemical oxygen demand and potassium from distillery and other industrial wastewater; desalination; the removal of colors and pigments from textile industries; and the removal of hazardous and toxic contaminants.

12.3.1.4   RO

RO can be defined as the reversal of osmosis phenomena or a process where permeation of the solvent takes place through a semipermeable barrier when subjected to hydrostatic pressure higher than the osmotic pressure [23]. It can also called hyperfiltration. RO is known for its high efficiency among all pressure-driven membrane processes. It can remove very small particles, such as monovalent ions, with almost 99.5% removal efficiency [2]. The driving force for the RO process is pressure gradient and chemical potential gradient, usually 15–75 bar pressure is required to carry out this process. Asymmetric, semiporous and thin-film composite membrane with pore size ranging from 0.01–0.1 nm, in spiral-wound or hollow-fiber configuration is used in RO process [23]. The mode of mass transport in this type of process is by solution diffusion and a preferential sorption mechanism. RO has numerous advantages like high efficiency, eco-friendly and user-friendly. The main drawback is concentration polarization and fouling which can be reduced by proper maintenance, backwash and chemical cleaning.

Application of this process involves removing dissolved, as well as suspended, solids from the feed solution. In recent scenarios, RO process has been used for various applications such as purifying drinking water, desalination, and in the pharmaceutical, bitotechnological, biomedical, power plant, food and beverage, tannery, distillery, textile, and pulp and paper industries, among others. Moreover, it is utilized for recovering valuable components from water and wastewater.

12.3.1.5 Pervaporation

The term *pervoporation* was first reported in 1917, which evolved from a combination of two terms ***permselective*** and ***evaporation*** [24]. The name of this process is based on the operating behaviors, which deals with the permeation of liquid phase feed through selective membrane and the collection of permeate in the form of vapor. Membrane permeation and evaporation both are key to this process. The feed solution is preheated to obtain saturated steam which is known as evaporation. Thereafter, the preheated feed solution is passed through the membrane, where vapor diffuses through the membrane and is collected on the permeate side [24]. Preheating the feed solution enhances the transport rate through the membrane. Pervaporation can be considered to be a pressure-driven process as the main driving force for this process is the partial pressure difference. Vacuum pressure is applied on the permeate side to carry out this process in order to maintain a lower downstream partial pressure than the saturation pressure [21]. Basically, dense or nonporous membranes are applied for the pervaporation process since it is based on the selectivity of the membrane. The mode of mass transfer in this process is by solution–diffusion mechanism and selective sorption [21].

Pervaporation is a complex process due to the maintenance of temperature, although it has numerous applications, such as the separation of azeotropic mixtures; solvent recovery; the separation of water from organic mixtures or dehydration; the removal of volatile organic compounds; wastewater treatment; the separation of organic–organic mixtures such as methanol/methylacetate and ethanol/cyclohexane; the separation of isomers; the separation of transestrification reaction products; the removal of aromatic from gasoline; and so on. The pervaporation process is implemented in distillation industries due to its unique feature of separating azeotropes. Therefore, it is commercialized and is considered an underdeveloped process that could be associated with conventional processes and is known as hybrid pervaporation.

12.3.1.6 Gas Separation

Another class of pressure-driven membrane processes is membrane gas separation (GS). Selectivity is a key factor for GS process. The transport of gaseous molecules through the membrane takes place by the solution–diffusion mechanism [25]. The GS process is also based on the same mechanism as the pervaporation process: Initially, sorption of the feed into the membrane, thereafter a diffusion of permeates through membrane and finally, a desorption of permeate at low-pressure side [25]. Phase change does not take place in GS as happens in the pervaporation process. Generally, a hollow-fiber configuration of the polymeric membrane is used in GS. But the main problem arises with the membrane material when it is applied for high-temperature applications, such as petrochemical and petroleum refineries, natural gas treatment, heavy hydrocarbon separation and the like. For high-temperature applications, carbon and metal oxide membranes and ceramic membranes are proposed.

This process is specifically applicable to the separation of a gaseous mixture and polar vapors using asymmetric, homogeneous or polymeric membranes.

The previously mentioned pressure-driven membrane processes have been used in different combinations or used solely as per the required application. In wastewater treatment plants, these processes serve as pretreatments to other processes. A combination of MF and RO is a common example for the generation of boiler or process water in thermal power plants. Pressure-driven membrane processes are commercially and technically the most relevant processes.

12.3.2 Concentration-Driven

The function of biological membrane system is driven by concentration gradient at isobaric and isothermal conditions. The most common example of synthetic membrane using a concentration-driven membrane process is an artificial kidney. Forward osmosis and dialysis come under this category, in which the concentration gradient becomes the dominant element for separation through a membrane.

12.3.2.1 Forward Osmosis

The forward osmosis (FO) process is similar to the osmosis phenomenon, in which water molecules are transported through the membrane by virtue of a concentration gradient. Dissimilar to the osmosis process, the FO process requires a highly concentrated draw solution to generate the concentration gradient. This concentration gradient is responsible for the osmotic pressure, which, in turn, provides the transport of water molecules from the feed solution to the draw solution until chemical equilibrium is established. The selection of the draw solution for specific applications makes it a more flexible and customized process [26]. Moreover, the regeneration and reuse of draw solution contribute to an economical operation. Diffusion through the membrane is the mode of mass transport. The FO process takes place through a dense, nonporous or selective membrane in numerous membrane modules, such as spiral wound, tubular, hollow fiber and flat sheet. Commercially available asymmetric RO or composite membranes with an ultra-thin selective layer are also reported for application in FO processes [27]. The FO process is energy-efficient compared to the pressure-driven membrane process, but the main drawback associated with FO membranes is the internal and external concentration polarization. This may result in inefficient salt rejection, less water permeation and hydrolysis of the membrane [26].

Applications of the FO process require specific draw solutes to generate the concentration gradient. Numerous applications of FO are reported in the literature, such as treating municipal wastewater, coke-oven wastewater, coal mine wastewater, desalination, sewage and domestic wastewater using $NaCl$, $MgCl_2$, $CaCl_2$ as draw solutions.

12.3.2.2 Dialysis

Another class of concentration-driven membrane processes is dialysis membranes. The device, which is equipped with a dialysis membrane, is called a dialyzes, and the solute-receiving fluid is known as dialysate [14]. This process is operated under optimum conditions in different flow patterns, such as parallel flow, mixed dialysate flow and countercurrent flow. The separation of solutes takes place due to the differences in size of species and diffusion rate. Hence, the mode of transport is by diffusion, with an activity gradient or concentration gradient as the driving force, which is also called diffusive solute transport [14]. A thin polymeric membrane is employed, and a large transmembrane concentration gradient is needed for the dialysis process to be more efficient. Plate-and-frame or hollow-fiber membrane configurations with a membrane pore size of less than 10 nm are used. Usually, the rate of flux through dialysis membrane is less as compared to other pressure-driven membranes as the permeation is based on the concentration gradient.

The application of a dialysis membrane involves hemodialysis for purifying blood outside the body, which is also referred as an artificial kidney; producing less alcoholic beers; removing acids from organic compounds; recovering hydroxide; and more.

12.3.3 THERMAL-DRIVEN

12.3.3.1 Membrane Distillation

The membrane distillation (MD) process is gaining more attention in recent era due to the production of high-quality products. MD is also known as membrane evaporation or a thermal membrane process. The driving force for this process is the transmembrane thermal gradient; therefore, the separation through the membrane is basically due to the difference in the volatilities of substances [2]. A hydrophobic microporous membrane in hollow-fiber or spiral spiral-wound module is employed. Vapor diffusion through pore and vapor–liquid equilibrium is the mass transport mechanism and separation principle, respectively. Thermal polarization phenomena are a drawback of MD [14]. This drawback can be overcome by selection of membrane material with low thermal conductivity, high chemical and thermal stability, stable hydrophobicity and good mechanical properties. Most common applications of the MD process are wastewater treatment and hydrolysis and in the semiconductor, dairy and textile industries, among others [2].

12.3.4 Electric Potential–Driven

12.3.4.1 Electrodialysis

Electrodialysis is a process in which an electrical potential gradient is responsible for the transport of charged components through the ion-exchange membrane. ED membranes are also known as ion-exchange membranes. Nonporous ionic polymeric membranes in a flat-sheet configuration are used for the ED process. Basic principle of separation through ED involves the generation of an electric potential field by means of applied voltage [28]. This potential field is responsible for the migration of anion through the anionic membrane while cations are impermeable. Similarly, cations only pass through the cationic membrane [28]. In this way, a highly concentrated solution and a diluted solution are produced on two different sides of membranes. Therefore, the migration of counter-ions through the ion-exchange membrane is the mode of transport. Low electric resistance, high permselectivity, mechanical and thermal stability and a low degree of swelling are the basic characteristics for the ED or ion-exchange membranes. The ED process exhibits several advantages, such as small space requirement, low cost and power consumption, easy handling and the flexible mode of operation (batch or continuous), provides complete removal of dissolved inorganic components.

ED is very useful in treating wastewater and desalination but is not suitable for high-saline water due to the proportionality between the desalination energy and the removable ions [2]. ED is efficiently applied in wide spectrum of applications such as fuel cell; the treatment of wastewater generated from various industries, agriculture and domestic activity; the purification and separation of organic components; the demineralization and production of baby food, artificial mother's milk and dairy products; desalting of dextran; electrolyte recovery; galvanic bath regeneration; and the like.

12.3.5 Liquid Membrane

In a liquid membrane (LM) process, a thin layer of organic liquid acts as a semipermeable barrier between two aqueous phases of different compositions [28]. Unlike other membrane processes, LM does not require solid membranes. The major drawback associated with LM is the instability at membrane interface that may be due to the difference in pressure and turbulence inside the LM setup. The mode of mass transport through the membrane is diffusion. However, some other mechanisms are also responsible for the separation, which can be defined in stepwise manner. Initially, diffusion in the feed solution across the boundary layer takes place, followed by the sorption on feed–membrane interface. Thereafter, convective transport occurs in the membrane and then diffusion on the receiving side across the boundary layer. Desorption takes place on the interface of the membrane-receiving solution afterward, and finally, diffusion in the receiving solution occurs through the boundary layer. LM possesses attractive features, such as high selectivity, single-stage extraction and stripping and the characteristics of nonequilibrium mass transfer [28]. LM can be categorized as a supported liquid membrane, an emulsion liquid membrane and a bulk liquid membrane. Supported liquid membranes consist of inert microporous support on which the organic phase can be immobilized. In an emulsion liquid membrane, an immiscible liquid layer exists between two miscible liquids. A bulk liquid membrane employs limited diffusion path, distant from the boundary layer [28, 29].

The main application of the LM process includes the separation of metal ions from wastewater, separation, recovery and concentration of acids; bioconversion; GS; and more.

12.3.6 HYBRID MEMBRANE PROCESS

The integration of a membrane process with any conventional process like adsorption, ion exchange, coagulation or another is referred as a hybrid membrane process. It possesses a synergetic effect and high efficiency. The fouling effect can be reduced by using some pretreatment method prior to the membrane process. In water treatment and reclamation, a hybrid membrane process enhances the quality of water for drinking and other applications like irrigation, process water, cooling water and so on. High-purity demineralized boiler water is also produced by a hybrid membrane process. In the treatment of groundwater, a series of MF and UF are combined with adsorption process (activated carbon), which removes particulates and dissolved organic matter, pathogenic bacteria and the like. Several studies reported on the combination of coagulation and UF for potable water production.

Conventional bioreactors are replaced with membrane bioreactors due to high removal efficiency. In contrast to conventional bioreactors, membrane bioreactors are coupled with synthetic membranes with suitable chemical and physical nature to confine the free biocatalyst within the reactor.

12.4 APPLICATIONS

Various applications of different membrane processes are listed in Table 12.2.

TABLE 12.2
Applications of Various Membrane Processes

S.No.	Membrane Process	Applications
1.	Microfiltration	• Urban wastewater treatment process for removal of bacteria, viruses, color, macro- and micropollutants • Sterile filtration • Food and beverage processing • Biotech downstream process • Disinfection and phosphorous removal from municipal wastewater • Oil and petroleum industries (removal of oil traces) • Biomedical therapy and clinical applications • Clarification of fermentation broth • Whey pretreatment • Cheese brine recovery • Ultrapure water processing
2.	Ultrafiltration	• Vegetable oil factory (COD, total suspended solid, total organic carbon, phosphate and chloride ion removal) • Urban wastewater treatment process • Poultry slaughterhouse wastewater treatment • Metal finishing industries • Removal of organic pollutant from paper and pulp industries • Sterile filtration • Hemodialysis • Food and beverage processing • Biotech downstream process • Biomedical diagnostics and therapy
3.	Nanofiltration	• Water recovery from dumpsite leachate • Textile industries • Removal of organic pollutant from paper mill wastewater • Removal of bacteria, viruses, color, macro- and micropollutants from wastewater • Food and beverage processing

(*Continued*)

TABLE 12.2 *(Continued)*

Applications of Various Membrane Processes

S.No.	Membrane Process	Applications
		• Biomedical diagnostics
		• Sugar industry
		• Protein separation
4.	Reverse Osmosis	• Seawater desalination
		• Brackish water desalination
		• Pesticide and pharmaceutical industries
		• Water recovery from dumpsite leachate
		• Potable water
		• Drinking water purification
		• Boiler and process water for thermal power plant
		• Removal of organic compounds from wastewater
		• Concentration of natural color, fruit juice and digested sludge liquid
		• Metal recovery from wastewater
5.	Pervaporation	• Dehydration (water miscible organic solvent)
		• Separation of azeotrope
		• Fermenter (control ethanol concentration)
		• Recovery of alcohol, glycol from water
		• Flavor and fragrance concentration and recovery
6.	Forward Osmosis	• Municipal wastewater treatment
		• Coke-oven industry
		• Coal mine wastewater treatment
		• Desalination
7.	Dialysis	• Metal ion removal and recovery
		• Medical catheter
		• Hemodialysis
8.	Gas separation	• Biogas processing
		• Stack gas purification
		• Removal of CO_2, separation of O_2/N_2, air/hydrocarbon, VOC/air
		• Desiccation
		• Petrochemical industry
		• Facilitated transport
		• Removal of acidic gas
		• Separation of sugar, olefins, etc.
		• Gas sensor
		• Dehumidification of air and gases
9.	Electrodialysis (ion-exchange membrane)	• Brackish water desalination
		• Microelectronic industry for ultrapure water
		• Chloro-alkali electrolysis
		• Food and beverages industries
		• Water electrolysis
		• Pharmaceutical industry
		• Fuel cell
		• Battery (alkali, concentration cell, redox flow)
10.	Membrane distillation	• Hydrolysis
		• Dairy industry
		• Petrochemical industry
		• Dehydrogenation
11.	Liquid membrane	• Enzymatic bioconversion
		• Separation and concentration of amino acids
		• Heavy metal removal
		• Dephenolation
		• Gas separation

12.5 ADVANTAGES AND DISADVANTAGES

The membrane processes have several advantages over conventional processes as well as disadvantages:

Advantages

- Membrane processes can selectively separate a wide range of surfactants, emulsions, organic mixtures and toxicants in a single process.
- They require simple instrumentation and are easy to operate and easy to maintain.
- They are based on a simple, basic concept that is easy to understand.
- Membrane processes require low energy.
- Membrane processes exhibit high efficiency due to the selective nature of membranes.
- Precious and minor components can be recovered from the main stream without any additional energy costs.
- Membrane processes require the use of relatively nonharmful and simple materials, so these processes are potentially environmentally friendly.

Disadvantages

- Polymeric membranes are limited in their use in high-temperature applications.
- Few membranes exhibit chemical incompatibilities such as dissolution, swelling and instability with process solutions.
- Fouling and concentration polarization are some of the major drawbacks of membranes that affect the permeation rate through the membranes.

12.6 RECENT ADVANCES AND CHALLENGES

Recent approaches to membrane development focus on enhancing membranes' versatility, sustainability, reusability and cost-efficiency and on fabricating novel membrane materials, modules and techniques. Membrane applications for energy-conversion processes, biomedical diagnosis and therapy have been investigated intensively and proposed for commercialization. Three-dimensionally printed membranes are replacing conventional methods of membrane synthesis [30]. Moreover, this technique is gaining more attention due to the ease of fabrication and integration with different materials [31]. That, in turn, synthesizes a fully functional membrane with high efficiency. Research on membrane bioreactors with self-healing materials is also under consideration. Dynamic membranes for UF, MF and NF are also being studied widely to convert the fouling effect into an advantage of membrane processes [3]. The stability and industrialization of liquid membranes are also gaining attention as research prospects. Fouling is a major problem specifically for RO membranes. Hence, resistance to fouling still requires some improvements and research.

The selection and modification of membrane used in GS also require some advancement due to the lack of sustainable membrane materials for industrial applications. Perfectly defect-free, biocompatible, multifunctional, selective pore size with rigorous quality control membrane development is needed for biomedical applications. The fabrication of scalable and efficient membranes for real-time approach is in high demand for implementation in industries. Environment protection and waste reduction can also be essential considerations for new membrane materials and membrane processes. Therefore, the future of membrane relies on new innovations, new approaches, high stability, scalability, feasibility and sustainability.

Many challenges still persist in membrane processes, such as zero defects, improved aging, reduced fouling properties, better sealing in modules, high permeability-to-selectivity ratio, robust property for real-time application, zero waste, solvent recovery, membrane recycling, economical ceramic membrane and more [31]. All these challenges associated with different membrane processes needed further improvement and hence require more research and development in this field.

BIBLIOGRAPHY

[1] Quist-Jensen CA, Macedonio F, Drioli E. (2015) Membrane technology for water production in agriculture: Desalination and wastewater reuse. *Desalination*. 364, 17–32.

[2] Ezugbe EO, Rathilal S. (2020) Membrane technologies in wastewater treatment: A review. *Membranes*. 10, 89.

[3] Nunes SP, Culfaz-Emecen PZ, Ramon GZ, Visser T, Koops GH, Jin W, Ulbricht M. (2020) Thinking the future of membranes: Perspectives for advanced and new membrane materials and manufacturing processes. *Journal of Membrane Science*. 598, 117761.

[4] Cao Y, Li B, Zhong G, Li Y, Wang H, Yu H, Peng F. (2018) Catalytic wet air oxidation of phenol over carbon nanotubes: Synergistic effect of carboxyl groups and edge carbons. *Carbon*. 133, 464–473.

[5] Zhang Y, Jin X, Wang Y, Yu Y, Liu G, Zhang Z, Xue W. (2018) Effects of experimental parameters on phenol degradation by cathodic microarc plasma electrolysis. *Separation and Purification Technology*. 201, 179–185.

[6] Gupta V, Mazumdar B, Acharya N. (2017) COD and colour reduction of sugar industry effluent by electrochemical treatment. *International Journal of Energy Technology and Policy*. 13(1/2), 177–187.

[7] Rajasulochana P, Preethy V. (2016) Comparison on efficiency of various techniques in treatment of waste and sewage water- A comprehensive review. *Resource-Efficient Technologies*. 2(4), 175–184.

[8] Water treatment. (2013) Retrieved from www.researchgate.net/post/Water_treatment What are the disadvantages of adsorption process technology Benefits of coupling chemical oxidation with adsorption.

[9] Blue ML. (2017) Retrieved from https://sciencing.com/pros-cons-biological-wastewater-treatments-24030.html.

[10] Siringi DO, Poznyak T, Chairez I. (2012) Is electrocoagulation (ec) a solution to the treatment of wastewater and providing clean water for daily use. *ARPN Journal of Engineering and Applied Sciences*. 7(2), 197–204.

[11] Rodriguez JL, Fuentes I, Aguilar CM, Valenzuela MA. (2018) Chapter 2: Catalytic ozonation as a promising technology for application in water treatment: Advantages and constraints. Editor: Derco J and Koman M. In *Ozone in Nature and Practice*, IntechOpen, London, UK, 17–36. doi:10.5772/intechopen.76228

[12] Dahal R, Moriam K, Seppälä P. (2016) Downstream process: Liquid-liquid extraction (degree dessertation). Aalto university school of chemical technology, Degree Programme in Chemical Technology, Finland.

[13] Debellefontaine H, Foussard JN. (2000) Wet air oxidation for the treatment of industrial wastes: Chemical aspects, reactor design and industrial applications in Europe. *Waste Management*. 20, 15–25.

[14] Strathmann H. (2001) Membrane separation processes: Current relevance and future opportunities. *AIChE Journal*. 47(5), 1077–1087.

[15] Baker RW, Cussler EL, Eykamp W, Koros WJ, Riley RL, Strathmann H. (1991) *Membrane separation systems*, Noyes Data Corp., Park Ridge, NJ.

[16] Chollom M, Rathilal S, Pillay VL. (2014) Treatment and reuse of reactive dye effluent from textile industry using membrane technology. Ph.D. Thesis, Durban University of Technology, Durban, South Africa.

[17] Anis SF, Hashaikeh R, Hilal N. (2019) Microfiltration membrane processes: A review of research trends over the past decade. *Journal of Water Process Engineering*. 32, 100941.

[18] Couto CF, Lange LC, Cristina M, Amaral S. (2018) A critical review on membrane separation processes applied to remove pharmaceutically active compounds from water and wastewater. *Journal of Water Process Engineering*. 26, 156–175.

[19] Aania SA, Mustafa TN, Hilal N. (2020) Ultrafiltration membranes for wastewater and water process engineering: A comprehensive statistical review over the past decade. *Journal of Water Process Engineering*. 35, 101241.

[20] Chao G, Shuili Y, Yufei S, Zhengyang G, Wangzhen Y, Liumo R. (2018) A review of ultrafiltration and forward osmosis: Application and modification. *IOP Conf. Series: Earth and Environmental Science*. 128, 012150.

[21] Oatley-Radcliffe DL, Walters M, Ainscough TJ, Williams PM, Mohammad AW, Hilal N. (2017) Nanofiltration membranes and processes: A review of research trends over the past decade. *Journal of Water Process Engineering*. 19, 164–171.

[22] Tul-Muntha S, Kausar A, Siddiq M. (2017) Advances in polymeric nanofiltration membrane: A review. *Polymer-Plastics Technology and Engineering*. 56(8), 841–856.

[23] Malaeb L, Ayoub GM (2011) Reverse osmosis technology for water treatment: State of the art review. *Desalination*. 267, 1–8.

[24] Luis P. (2018) Chapter 3: Pervaporation. Editor: Luis P In *Fundamental modeling of membrane systems*, Elsevier, Amsterdam, Netherlands, 71–102.

[25] Bernardo P, Drioli E, Golemme G (2009) Membrane gas separation: A review/state of the art. *Industrial & Engineering Chemistry Research*. 48(10), 4638–4663.

[26] Suwaileh W, Pathak NK, Shon H, Hilala N. (2020) Forward osmosis membranes and processes: A comprehensive review of research trends and future outlook. *Desalination*. 485, 114455.

[27] Wang KY, Ong RC, Chung TS. (2010) Double-skinned forward osmosis membranes for reducing internal concentration polarization within the porous sublayer. *Industrial & Engineering Chemistry Research*. 49, 4824–4831.

[28] Campione A, Gurreri L, Ciofalo M, Micale G, Tamburini A, Cipollina A. (2018) Electrodialysis for water desalination: A critical assessment of recent developments on process fundamentals, models and applications. *Desalination*. 434, 121–160.

[29] Parhi PK. (2013) Supported liquid membrane principle and its practices: A short review. *Journal of Chemistry*. 2013, 1–11.

[30] Femmer T, Kuehne AJ, Torres-Rendon J, Walther A, Wessling M. (2015) Print your membrane: Rapid prototyping of complex 3D-PDMS membranes via a sacrificial resist. *Journal of Membrane Science*. 478, 12–18.

[31] Al-Shimmery A, Mazinani S, Ji J, Chew YJ, Mattia D. (2019) 3D printed composite membranes with enhanced anti-fouling behaviour. *Journal of Membrane Science*. 574, 76–85.

13 Application of Membranes in Tannery Wastewater Treatment

Raja. C, J. Anandkumar and B.P. Sahariah

CONTENTS

13.1 INTRODUCTION

The tannery industry plays an essential role in the economic improvement of a country. Argentina, Brazil, Italy, India, and Russia are the foremost countries producing leather. In India, the states of Andhra Pradesh, Maharashtra, Punjab, West Bengal, and Tamil Nadu contribute the most to the tannery industry (Korpe and Rao 2021).

13.1.1 SOURCES AND CHARACTERISTICS OF TANNERY WASTEWATER

In tanneries, leather is processed at different stages, such as beam-house, tanning, and finishing operations (Korpe and Rao 2021). The wastewater generated from parts of beam-house processes such as soaking, unhairing/liming, and deliming/bating. The wastewater coming out from these sections contain a massive amount of ammonia, biochemical oxygen demand (BOD), and salinity due to the many chemicals used in the process. Some tannery industries treat this wastewater

DOI: 10.1201/9781003165019-13

separately to avoid the overload of pollutants (Mpofu et al. 2021). Chrome tanning, vegetable tanning, aldehyde tanning, and synthetic tanning are the typical tanning process. The tanning process produces a considerable volume of wastewater in the overall process (Korpe and Rao 2021). Chrome tanning is the preferable method in the industry. Chrome-tanned leather is superior to vegetable-tanned leather due to its softness, high thermal and water stability, and less time-consuming (Dixit et al. 2015). The skin of sheep, lambs, goats, and pigs are significant sources of leather processing in the chrome-tanning process. The generation of wastewater from soaking unhairing/liming, deliming and bating, chrome tanning, post-tanning, and finishing is 9.0–12.0, 4.0–6.0, 1.5–2.0, 1.0–2.0, 1.0–1.5, and 1.0–2.0 in KL, respectively (Dixit et al. 2015). The tanning process releases wastewater at nearly 15000–45000 L with 0.5 tonnes of sludge for 0.2 tonnes of leather. In chrome tanning, chromium and sulfate are significant pollutants in the tannery wastewater (Mpofu et al. 2021). The schematic presentation of leather processing is shown in Figure 13.1.

In general, leather processing consumes many chemicals and releases various pollutants, such as chromium, sulfide, suspended solids, BOD, and high chemical oxygen demand (COD), into the wastewater. The details of the contaminants present in the different leather-processing operations are given in Table 13.1.

13.1.2 Methods for Contaminants Removal from Tannery Wastewater

The discharge of tannery industry effluent causes a high negative impact on natural water resources and impairs the ecosystem due to its high toxicity. Therefore, the discharge of effluent with zero or

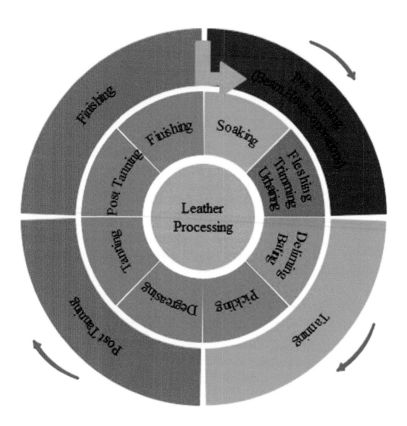

FIGURE 13.1 Schematic representation of leather-processing steps.

TABLE 13.1

Source of Pollutants from Leather Processing

Process	Chemicals	Pollutants
Soaking	Bactericides, sodium chloride	Total dissolved solids
Fleshing	Lime	Total dissolved solids
Trimming & unhairing		Total suspended solids
		Biochemical oxygen demand
		Chemical oxygen demand
Deliming & bating	Enzymes	Biochemical oxygen demand
	Ammonium salts	Chemical oxygen demand,
		Total Kjeldahl nitrogen Ammonia
Pickling	Acid & salts	Total dissolved solids
Degreasing	Solvents & surfactants	Chemical oxygen demand
Tanning	Chromium	Chromium, chemical oxygen demand
Post-tanning	Tanning agents & dyes	Volatile organic compounds
Finishing	Solvents	Chromium, organic compounds and Color

permissible contamination is desirable by using efficient treatment processes. The discharge limits of tannery wastewater in some of the countries are shown in Table 13.2.

Generally, tannery effluent treatment methods are broadly categorized as physical, chemical, and biological treatment methods (Figure 13.2). Physical processes include filtration, sedimentation, and flotation. The chemical treatment methods are coagulation and flocculation, electrocoagulation, precipitation, adsorption, ion exchange, and advanced oxidation. Similarly, in biological methods, aerobic and anaerobic treatments are the most commonly used methods for treating tannery effluent at different stages.

13.1.2.1 Physical Methods

13.1.2.1.1 Sedimentation

Sedimentation facilitates the removal of settled solids, coarser particles, and organic solids from effluent. Plain sedimentation can remove significant amounts of less degradable or non-degradable pollutants. Therefore, this process is mostly suitable as a primary treatment of wastewater. The notable advantages of this process are simplicity in operation and low cost. However, further treatment is a must after sedimentation in tannery wastewater treatment (Song et al. 2000).

13.1.2.1.2 Flotation and Filtration Method

Natural oils, grease, and some fatty substances are released during leather processing. Mixing the suitable solvent in wastewater and allowing it to float all the substance on the top of the effluent and withdrawing the top layer will remove most of the oily, as well as fatty, materials present in the wastewater. A filtration method is efficiently used to handle suspended solids and coarse particles. A sand filter is a simple conventional filtration method that consists of two or more media. The added advantage of this operation is very simple and low cost. However, a sand filter is not suitable for separating many dissolved organic matters from wastewater. Therefore, the advancement in the filtration process by using membrane separation technique in wastewater treatment can be desirable method to handle dissolved organics (Crini and Lichtfouse 2019; Ghumra et al. 2021)

TABLE 13.2

Discharge Limits of Tannery Wastewater in Different Countries

Country	BOD (mg/l)	COD (mg/l)	Total dissolved Solids (mg/l)	Suspended Solids (mg/l)	Chloride (mg/l)	Sulfate (mg/l)	Total chromium (mg/l)	pH
Argentina	50	250	–			3	0.5	5.5–10
Brazil	60	–	–				0.5	5–9
China	150	300	–				1.5	6–9
India	30	250	–	100		1000	2	5.5–9
Italy	40	160	–	40–80	1200	1000	2	5.5–9.5
Nigeria	50	160	2000	30	600	500	–	6–9
Pakistan	80	150	–	200	1000	1000	1	6–9
Saudi Arabia	25	150	–	30–50	–	–	0.1	6–9
Turkey	250	800	–	350	–	–	2	6–9
Thailand	20–60	–	5000	150		–	–	5.5–9

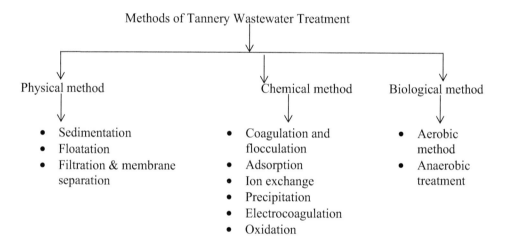

FIGURE 13.2 Different treatment methods used for tannery effluent treatment.

13.1.2.2 Chemical Methods

13.1.2.2.1 Coagulation and Flocculation Method

Coagulation and flocculation using iron/aluminum-based coagulants is a very appropriate method for treating various industrial effluents. Turbidity removal is the prime focus of this process, and further improvements in the process are needed to obtain higher efficiency. Sludge formation is the foremost issue in this process (Ghumra et al. 2021).

13.1.2.2.2 Electrocoagulation

Electrocoagulation is frequently adopted as the alternative method in place of chemical coagulation. In this process, metal species act as coagulating agents, form a complex, and efficiently adsorb the targeted pollutants present in the solution. Generally, an electrode is made by aluminum, iron, and stainless-steel materials. This process has more advantages than conventional chemical coagulation as there is no formation of secondary pollutants after this treatment process. Low cost, less sludge formation, and low energy consumption are some of the added advantages of the electrocoagulation process (Feng et al. 2007).

13.1.2.2.3 Precipitation Method

The precipitation method can proficiently remove heavy metals such as Pb, Cr, As, and others from wastewater. Heavy metals react with precipitating agents to form insoluble precipitates and separate them from water through sedimentation or filtration. However, sludge formation and its disposal problems are substantial disadvantages of this process (Crini and Lichtfouse 2019; Ghumra et al. 2021).

13.1.2.2.4 Adsorption

Adsorption is the process in which the adhesion of molecules on the adsorbent surface occurs via physical or chemical bonding. Activated carbon, zeolite, and ash from agro-waste are a few adsorbents commonly used for tannery effluent treatment in the adsorption step. As a result of this, nondegradable pollutants are removed from the tannery wastewater. The low preparation cost and simple operation and design are a few advantages of the adsorption process (Kumar and Dwivedi 2021).

13.1.2.2.5 Ion Exchange

Ion exchange is the process of exchanging pollutants that are in ionic form between resin and an electrolytic solution (nonhazardous). The ion-exchange process has a high selectivity toward many heavy metals, which facilitates the removal of metals from wastewater. The structure of the resins used in the process do not affect the operation. The noteworthy advantage of the process is cost-effective and straightforward process (Zinicovscaia 2016).

13.1.2.2.6 Advanced Oxidation Processes

Advanced oxidation processes such as Fenton oxidation, photooxidation, ozone-based oxidation, and photocatalysis are currently used when the chief organic pollutants present in the wastewater are oxidized (Lofrano et al. 2013). However, in this process, the oxidizing agents produce some radical ions that further contaminate the wastewater. High removal efficiency, safety in operation, and environmental compatibility are the credits of advanced oxidation processes. However, the major disadvantage of these processes is more sludge formation during the operation.

13.1.2.3 Biological Methods

13.1.2.3.1 Aerobic Treatment Method

In the effluent, microorganisms in presence of molecular oxygen convert the available carbon in the effluent into biomass and carbon dioxide. This method produces a large amount of biomass, and today, a sequencing batch reactor used in tannery effluent treatment is observed to reduce the sludge formation issues. The technique requires a reaction tank with less space and is more flexible than the traditional activated sludge process. As a result, this method can treat a large quantity of tannery wastewater (Zhao and Chen 2019).

13.1.2.3.2 Anaerobic Treatment

In anaerobic treatment process, microorganisms grown in absence of oxygen at various configurations, such as up-flow anaerobic filter, up-flow anaerobic sludge blanket reactor, and down-flow anaerobic filter, among others, are used for tannery effluent treatment. The significant advantages of the anaerobic process are less sludge formation, low energy consumption, and shock loading resistance. Polyurethane foam and polypropylene rings are often used as the filling materials for high microbial presence in the treatment process (Mannucci et al. 2010).

13.2 APPLICATIONS OF MEMBRANE SEPARATION IN TANNERY WASTEWATER TREATMENT

In general, a membrane is a semipermeable barrier that separates a solute from a mixture. The key factors for membrane efficiency are selectivity and flow. Membrane technology has been used in small- and large-scale operations in various modes, including microfiltration, ultrafiltration, nanofiltration, reverse osmosis, membrane distillation, pervaporation, and hemodialysis (Mulder 1996)

13.2.1 Factors Influencing Membrane Performance in the Treatment Process

Different factors that are responsible for efficient membrane treatment of tannery wastewater are membrane materials, the membrane module, and the nature of the feed solution. These principal factors are discussed in the following sections.

13.2.1.1 Membrane Material

Membranes are fabricated from a variety of organic, as well as inorganic, materials and are used for different wastewater treatments. The material selection plays a crucial role in membrane fabrication, which requires the desired packing density of the membrane, stability, the transport mechanism, and the performance of the membrane (Rosman et al. 2018). Tannery wastewater is rich in dissolved inorganics, organic components, colloids, and suspended solids. These components interact with the membrane surface, can reduce the permeate flux after fouling, and thus reduce the performance (Hakami et al. 2020). The significant characteristics of organic membranes include availability, low cost, and ease of fabrication are the reason for choosing as the membrane for wastewater treatment. Organic membranes are made up of polymeric materials. Today, different types of polymeric materials, such as cellulose acetate, polyethersulfone, and polyvinyl alcohol, are used for membrane fabrication. Compared with organic membranes, ceramic membranes have better mechanical and thermal properties and high flux. Ceramic membranes are highly efficient for tannery wastewater treatment due to the regaining the original flux (Du et al. 2020). The selection of membrane material depends on the sources of wastewater. Both organic and inorganic membranes are used for the treatment of beam house effluent and attain better efficiency. The rate of fouling reduced through the hybrid membrane process and cleaning discussed in Section 13.3.

13.2.1.2 Membrane Modules

The membrane module is one of the vital factors for membrane performance for commercial applications. Spiral wound, plate and frame, tubular, and hollow fibers are the crucial modules in membrane fabrication. In addition, the mode of operation, the nature of separation, ease of cleaning and maintenance, and replacement are other key factors considered for the selection of membrane modules.

13.2.1.2.1 Tubular Module

Tubular modules provide a large surface area–to–volume ratio similar to shell and tube heat exchangers. In the tubular module, the feed solution passes the center of the tube and permeates are collected from a wall of the line or vice versa. For tannery wastewater treatment, a single-channel tubular ceramic membrane made up of different materials can remove chromium, tannin, and other pollutants. Kaplan-Bekaroglu and Gode (2016) studied the performance of single-channel tubular ceramic membrane to reduce COD and color in tannery wastewater. Similarly, for chromium removal, Roy Choudhury et al. (2018) experimented CuO/hydroxyethylcellulose ceramic-composite membrane. However, primarily, tubular ceramic membranes are employed for tannery treatment applications. A foam ball is useful to clean a tubular membrane after the treatment process. Less membrane contamination and ease of cleaning are the significant advantages (Ezugbe and Rathilal 2020; Hakami et al. 2020).

13.2.1.2.2   Hollow-Fiber Module

A hollow-fiber module possesses a high packing density. Minimum concentration polarization, low-pressure drop in the permeate side, compact in nature, and high withstand pressure are the significant advantages of the process. Hollow-fiber modules are useful for a relatively clean feed stream, free of very large particulates, as in gas separation, pervaporation, and seawater desalination. Therefore, this module is primarily suitable for all kinds of organic materials (Ezugbe and Rathilal 2020; Hakami et al. 2020).

13.2.1.2.3   Plate-and-Frame Module

The plate-and-frame module is a very old model that consists of membrane and spacers and is similar to plate-and-frame filter press. Easy to clean and concentration polarization reduction are the significant advantages that make the plate-and-frame module highly suitable for treating rich suspended solids wastewater (Ezugbe and Rathilal 2020; Uragami 2017). El-Shafey et al. (2005) used a seven-channel plate and frame filter press for removing sludge from tannery effluent.

13.2.1.2.4   Spiral-Wound Module

Minimum concentration polarization, low-pressure drop in permeate side, compactness in nature, high pressure withstands, easy replacement of module elements, and scale-up are the significant advantages of the spiral-wound module process. Therefore, this module is suitable primarily for reverse osmosis and nanofiltration applications (Ezugbe and Rathilal 2020; Uragami 2017). Stoller et al. (2013) achieved high efficiency with two spiral-wound nanofiltration and reverse osmosis membranes used for removing COD, total suspended solids (TSS), and chromium.

13.2.1.3   Nature of the Influent

Generally, leather industries generate wastewater of different qualities and quantities during various processes, such as soaking, washing, liming, deliming, tanning, and retanning. The soaking and washing process water contains high total dissolved solids, COD, and chloride content with no ammonia. The value of pH is near a neutral condition, that is, a slightly acidic condition. The wastewater from the liming process contains high chloride concentrations, sulfide, and less water volume (Cassano et al. 2001). Following the tanning process, the wastewater becomes highly concentrated with chromium and large total solids, total dissolved solids, suspended solids, COD, chloride, and ammonia. The wastewater from the liming process is rich in COD, and the pH value lies on the alkaline side. Ammonia concentration increases while proceeding from tanning to retaining. Beam-house effluent requires additional attention due to its high COD content (Sawalha et al. 2019). The different stages of the tannery process have different pH values and vary from 3.8 ± 0.2 to 12.5 ± 0.5. The pH value of effluent increases from 8.5 in a soaking process to 12.5 in liming and unhairing process. In deliming and bathing process, the pH value is 6.5, and in the tanning process, the pH is 3.5–4.5. The sulfuric acid reduces the pH value of the effluent after the chrome tanning process. The excessive use of lime and sodium sulfide to increase the effluent's pH, reaching a pH value of about 12.5 ± 0.5. This highly acidic condition affected the polymeric membrane.

In polymeric membrane, acidic group enhance the polymer chain shrinkage and carboxyl group enhance the swelling property of the membrane at low and high pH, respectively. Similarly, basic group change the hydrophilic and hydrophobic character of the membrane at low and high pH (Angelini et al. 2018).

13.2.2 TYPES OF MEMBRANES USED IN THE TREATMENT PROCESS

According to the strength of wastewater and membrane materials, different types of membranes, such as ceramic membranes, organic membranes, and liquid membranes, are utilized for tannery wastewater treatment processes (Samaei et al. 2018). Membranes used to treat effluent from various stages of tannery operation are discussed in this section.

13.2.2.1 Ceramic Membranes

Various types of clay and fly ash–based commercial and low-cost ceramic membranes are used to treat tannery effluent and discussed further in the following sections.

13.2.2.1.1 Kaolin Membrane

Kaolin is used by many researchers as a key material for membrane fabrication due to its refractory properties. Vasanth et al. (2012) prepared a kaolin-based ultrafiltration membrane to remove Cr(VI) from an aqueous solution. High chromium removal efficiency (94%) is reported at pH 1, and a further increase of effluent pH reduces the removal efficiency of the membrane.0

13.2.2.1.2 Clay Membrane

Different types of clays like Moroccan clay and clay materials are used as major membrane precursors due to their abundant availability as well as efficiency. Elomari et al. (2016) used different Moroccan clay ceramic membranes to treat dye tannery effluent and confirmed that Moroccan clays required a low sintering temperature in the membrane fabrication. The porosity of the membranes were 28.1%, 30.8%, and 40%, and the average pore diameters of 1.8, 1.5, and 2.84 μm, respectively, successfully removed higher turbidity. Mouiya et al. (2018) used phosphate as a porosity-making agent in ceramic membrane fabrication and achieved 99.80% of turbidity reduction in beam house effluent. Similarly, a Moroccan perlite-based ceramic microfiltration membrane with a pore size of 1.70 treated industrial tannery effluent and agro-food industrial effluent. The membrane is not suitable for soluble salts present in the effluent, which reflects on the conductivity. The effluent turbidity decreased from 36 to 1.44 NTU (Saja et al. 2018). Mouiya et al. (2019) reported banana peel as a porogen agent incorporated clay membrane to treat tannery and textile effluent. The treatment reduces the considerable level of pH, conductivity, color, and suspended particles in tannery effluent, along with the successful removal of COD with high turbidity recovery. Hatimi et al. (2020) received remarkable turbidity and chromium removal by a clay and pyrrhotite ash microfiltration membrane with the permeability and mechanical strength of 22.88 10^{-7} m³/h m².kPa and 27.42 MPa for industrial tannery effluent treatment. Bhattacharya et al. (2013) proposed a dual-stage membrane system with ceramic microfiltration and polymeric nanofiltration membrane to treat composite tannery wastewater in which the ceramic membrane completely removed the sulfide and total organic carbon (TOC) from the raw sewage at the low operating cost of the microfiltration treatment 3.628 ₹/L. Similarly, Yadav and Bhattacharya (2020) reported tannery effluent treatment using α-Al₂O₃ membrane in a lab and pilot-scale level that reduced total chromium below the detection level along with high removal of COD. The treatment of industrial wastewater significantly reduces the load on freshwater utilization and the discharge of wastewater. Using a single channel, commercial tubular ceramic membranes with different pore sizes such as 200, 50, and 10 are beneficial for the pretreatment of highly polluted industrial tannery wastewater. An ultrafiltration membrane with the pore size of 10 nm can reduce the color to 5 Pt-Co at 2 bar after 10 h operation and 90% COD reduction at 4 bar (Kaplan-Bekaroglu and Gode 2016).

13.2.2.1.3   Ash

Ash is generated from any industrial process like combustion in thermal power plants and natural ash. Beqqour et al. (2019) reported micronized phosphate incorporated pozzolan microfiltration membrane to treat aluminum chloride suspension and tannery industry wastewater microfiltration. For aluminum chloride suspension and tannery wastewater, the membrane removed 99.77% and 97.83% of turbidity, respectively. The efficiency of the ceramic membrane in tannery wastewater treatment is given in Table 13.3.

13.2.2.1.4   Ceramic Composite Membrane

Basumatary et al. (2016) reported FAU-, MCM-41-, MCM-48-coated ceramic composite membranes for Cr(VI) removal from the aqueous solution. MCM 41 has a smaller pore size in comparison with the other two composite membranes. FAU membrane achieved the highest removal efficiency 82% achieved in the Cr removal for a 1000-ppm concentration feed at an applied pressure of 345k Pa.

TABLE 13.3

Efficiency of Ceramic Membranes in Tannery Wastewater Treatment

Type of membrane & material	Porosity %	Pore size	Type of wastewater	Type of pollutant	Initial concentration	Removal %	Reference
Microfiltration kaolin	30	1.32 μm	Model Solution	Cr (VI)	100 mg /l	94	Vasanth et al. (2012)
Microfiltration cordierite	36	–	Beam House Effluent	Cr (VI) Pb	521 mg/l 7 mg/l	99.86 91.4	Bhattacharya et al. (2013)
Ultrafiltration clay–alumina	–	3 nm	Model Wastewater	Cr (VI) Pb	5 mg/l 5 mg/l	91.44 97.14	Choudhury et al. (2018)
Ultrafiltration γ-Al2O3	–	10 nm	Industrial Tannery Effluent	Cr (VI) Color (Pt-Co)	0.89 mg/l 9140	>95	Kaplan-Bekaroglu and Gode (2016)
Microfiltration Moroccan clay	–	2.5 μm	Beam House Effluent	Turbidity	–	99.80%	Mouiya et al. (2018)
Microfiltration clay	40.3	0.45 μm	Tannery effluent	Turbidity	554 NTU	99.99%	Mouiya et al. (2019)
Microfiltration α- alumina	36	0.5 μm	CETP	Turbidity	–	95–98%	Yadav and Bhattacharya et al. (2020)
Microfiltration kaolin	33	0.153 μm	Model Solution	CrVI	1000 mg/l	82%	Basumatary et al. (2015)
Ultrafiltration zirconia	–	30–40 nm	CETP	Turbiditity	1.24 NTU	–	Dey et al. (2018)
Moroccan clay microfiltration	40%	1.5 μm	–	Turbidity	600 NTU	95.17	Elomari et al. (2015)
Nature clay microfiltration	34%	2.5 μm	Tannery Effluent	Turbidity	595 NTU	96%	Hatimi et al. 2020
Microfiltration perlite	52.11%	1.70 μm	Tannery effluent	Turbidity	36 NTU	96%	Saja et al. (2018)

Choudhury et al. (2018) reported hydroxyethyl cellulose and CuO nanoparticles incorporated clay–alumina ceramic composite membrane for wastewater treatment. The maximum rejection percentage of Cr(VI) was obtained at 91.44% at 2-bar transmembrane pressure. CuO nanoparticles contribute to chromium oxidation and reduction during membrane separation. Zirconia-coated multichannel ultrafiltration membranes removed 82 % of COD from the effluent, and turbidity was also reduced to 0.24 NTU. In addition, a considerable range of BOD and TSS were removed from the effluent (Dey et al. 2018).

13.2.2.2 Polymeric Membrane

The different polymeric membranes such as polyvinylidene difluoride, polyethersulfone (PES), polyamide, polyacrylonitrile, and cellulose acetate are available for effluent treatment applications. However, a few polymeric membranes are rarely utilized for tannery wastewater treatment.

13.2.2.2.1 Polyvinylidene Fluoride Membranes

Arif et al. (2020) optimized process parameters using response surface methodology for the treatment of tannery wastewater using titanium dioxide incorporated polyvinylidene fluoride membrane and observed that photocatalytic activity of titanium oxide reduced the hexavalent chromium concentration. It facilitated high chromium removal and favored recycling of the membrane for treatment without compromising efficiency.

13.2.2.2.2 Polyimide Membranes

Polyimide membrane is prepared from polyimide casting solutions. Yadav and Bhattacharya (2020) used polyimide reverse osmosis membrane for tannery effluent treatment and achieved high TOC removal and higher water recycling in leather tanning operations.

13.2.2.2.3 Polysulfone Membranes

Karunanidhi et al. (2020) reported on keratin-incorporated electrospun polysulfone nanofiltration membranes for treatment of simulated post-tanning effluent and the achieved removal of COD, BOD, total dissolved solids, and TSS.

13.2.2.2.4 Polyethersulfone Membranes

Zakmout et al. (2020) have reported on the comparison of commercial and lab-scale membranes used for treating tannery wastewater and observed that commercial nanofiltration and reverse osmosis membranes significantly remove NaCl, $CaCl_2$, and $MgCl_2$. A chitosan-modified polyethersulfone membrane recovered high Cr through an interaction of the amine and hydroxyl groups in chitosan and chromium at a pH of 3.6. A cellulose acetate ultrafiltration membrane with high hydrophilicity provided high chromium removal for 50 and 100 KPa and 95 min at neutral pH (Vinodhini and Sudha 2017). Another investigation conducted on the treatment of tannery effluent using polyethylene glycol and a $CaCl_2$-modified polyethersulfone membrane enlightened that $CaCl_2$ enhances membrane permeability and hydrophilicity whereas a $CaCl_2$ (1%)–blended PES membrane provided high removal of BOD and COD (Rambabu and Velu 2016). The two commercial polysulfone and polyethersulfone ultrafiltration membranes when used for purification of vegetable tanning liquors and synthetic wastewater, the latter showed a higher rejection coefficient for tannins, non-tannins, and total solids. Both membranes showed low permeate with less tannin concentration (Romero-Dondiz et al. 2015).

During different composition of gelatin modified polyethersulfone ultrafiltration membrane investigation (Velu et al. 2015) observed 10% gelatin modified polyethersulfone ultrafiltration membrane provides high removal for BOD, COD, total solids, and chromium (Cr reduced to 1 mg/l) with higher flux rate due to hydrophilicity of the membrane.

Similarly, an aluminosilicate-embedded polyethersulfone ultrafiltration membrane applied for the tannery wastewater treatment process can provide high removal of BOD, COD, and sulfide. Total chromium gets reduced from 5.25 mg/l to 2.2 mg/l at 20% aluminosilicate-blended polyethersulfone ultrafiltration membrane. Alumina silicate particles enhance the adsorption property and improve the separation efficiency (Velu et al. 2021). Fiorentin-Ferrari et al. (2021) reported treating fish-skin tanning effluent using polyethersulfone membrane and obtained high COD removal. The aging of tannery wastewater modifies ion equilibrium and affects the efficiency of the process. The different molecular weight cutoff (MWCO) membranes made up of polyethersulfone and polyamide materials successfully remove suspended solids using a nanofiltration membrane. Kiril Mert and Kestioglu (2014) computed the economic feasibility of the membrane process for polyethersulfone membrane with hydrous ferric oxide membrane used for adsorptive removal of Cr (VI) ion with a flux of 6293.3 l/m^{2}h. The enhancement of permeability occurred due to hydrophilic ferric oxide particles when PES and hydrous ferric oxide ratio were suitable for chromium removal compared with other blending ratios. The removal efficiency of the membrane before and after regeneration remained unaffected for the entire operation. HFO is a hydrophilic group to enhance the permeability of water (Abdullah et al. 2019).

13.2.2.2.5 Cellulose Acetate Membranes

The cellulose acetate nanofiltration membrane can handle high chromium and sulfate recovery from tannery wastewater treatment when commercial nylon and cellulose nitrate are used accompanied by a nylon membrane to retain its high performance after several uses (Religa et al. 2011). The efficiency of polymeric membranes in tannery wastewater treatment is presented in Table 13.4.

13.2.2.3 Liquid Membranes

A liquid membrane is generally used for the separation process with the extraction and stripping for separating molecules using chemical potential. Bulk liquid membranes, emulsion liquid membranes, and supported liquid membranes are the types of liquid membranes used for various separation applications. Here, a few liquid membranes are discussed regarding wastewater treatment. Goyal et al. (2011) achieved high chromium removal using 1-butyl-3-methylimidazolium bis(trifluoromethylsulfonyl) imide as a membrane phase in the emulsion liquid membrane. Hasan et al. (2009) observed that the

TABLE 13.4

Efficiency of Polymeric Membranes in Tannery Wastewater Treatment

Type of membrane	Porosity	Pore size	Type of wastewater	Type of pollutant	Initial concentration	Removal efficiency	Reference
Ultrafiltration PES	–	360 nm	Fish Skin Tanning Effluent	COD	–	88%	Fiorentin-Ferrari et al. (2021)
Ultrafiltration polysulfone	–	1.5725 μm	Synthetic Wastewater	BOD COD	6000 17683	66 53	Karunanidhi et al. (2020)
Nanofiltration chitosan-polyethersulfone	–	–	Industrial Tannery Effluent	Cr	–	>90	Zakmout et al. (2020)
Ultrafiltration gelatin-polyethersulfone	4.77	39.6 A	Common Effluent Treatment Plant	BOD COD Cr	322 1136 5.25	35 233 1.51	Velu et al. (2015)
Gelatin-aluminosilicate	–	6.54 nm	Common Effluent Treatment Plant	BOD COD Sulfate	1020 1136 5.25	26 288 188	Velu et al. (2021)

removal efficiency of the membrane system improved with the combination of tri-octyl phosphine oxide (TOPO), cyclohexane, sodium hydroxide, and sorbitan monooleate when used as a liquid emulsion membrane system for the Cr(VI) removal. Trin-octylamine can also be used as a membrane phase in supported liquid membranes to remove chromium solutions in the range of 2500 ppm to 800 ppm (Chaudry et al. 1998).

13.2.3 NEW STRATEGIES TO IMPROVE THE PERFORMANCE OF MEMBRANE TREATMENT

13.2.3.1 Hybrid Membrane System

Hybrid processes/systems are the processes that extend the efficiency of membranes to achieve and improve the overall performance of the process. A hybrid membrane process/system is the combination of two or more techniques with a membrane system or a combination of multiple membranes for pretreating and treating the wastewater. The different treatment methods, such as coagulation, adsorption, and ion exchange, are attached with the membrane process (Stylianou et al. 2015). A few researchers have implemented the hybrid membrane process to treat tannery wastewater, and those studies are discussed next.

The different physicochemical methods can be integrated with membranes for the treatment of wastewater. Stoller et al. (2013) applied conventional flocculation and flotation for the pretreatment of tannery wastewater before membrane separation. The system failed to bring chromium to bring down to the dischargeable limit from an initial 7.92 mg/l and 102 mg/l of chromium and COD, respectively, in wastewater. Keerthi et al. (2013) achieved complete removal of chromium, as well as high COD, removal from wastewater after coagulation during wastewater treatment using a combination of electrocoagulation and membrane. Pal et al. (2020) treated the tannery wastewater almost completely removed chromium and COD with advanced oxidation before treatment with nanofiltration in similar applications. The possibility of water reuse due to advanced oxidation decomposes the heavy molecules after the combined treatment. Bhattacharya et al. (2013) observed while investigating the role of microfiltration and reverse osmosis membrane for similar applications that dual-stage membrane process performance improves and more possibility to reuse the water after treatment occurs. The hybrid process improves the separation performance and membrane life span in fouling (Rosman et al. 2018).

13.2.3.2 Reusage of Membranes

The performance of membranes in the wastewater treatment process is mainly affected by fouling due to the presence of suspended solids, microbes, and organic materials and reduces the permeate flux. The fouled membrane required high transmembrane pressure for permeation and simultaneously reduced the efficiency of the system. The different types of fouling, such as colloidal, bio, organic, and inorganic fouling, occur on the membrane surface. Generally, molecules are separated primarily on the membrane by the size-exclusion and adsorption mechanism (Madhura et al. 2018). The separation of large molecules like suspended solids in wastewater treatment favors fouling in membranes. Various techniques exist to reduce fouling in membranes. Membrane cleaning is one effective and meaningful method to restore the membrane in terms of flux after fouling. Membrane cleaning can be classified into physical, chemical, and physiochemical cleaning (Jiang et al. 2017).

The cleaning/removal process of pollutants using mechanical energy and different methods, such as periodic backwashing, pneumatic cleaning, ultrasonic cleaning, and sponge ball cleaning, is followed. In addition, the cleaning methods of the membrane are discussed here.

Backwashing is primarily used in industry, and this efficiently regains the flux. It can effectively recover the membrane flux from the fouling problem. The deposition of materials causes fouling on the surfaces of the membranes as a gel or cake layer. Pressure is applied on the permeate side of the membrane, and it creates backward movement of the permeate through the membrane. The pneumatic cleaning process involves the circulation of air through the membrane. The air removes the foulant due to the sheer force of the membrane. In this method, chemicals are not used for cleaning. Ultrasonic cleaning is used to clean the membrane using ultrasound in a liquid medium (Ezugbe and Rathilal 2020). The ultrasound creates energy, forms turbulence on the membrane surface, and weakens the molecules' interaction. Therefore, it is more suitable for membrane surface cleaning (Du et al. 2020). Sponge ball cleaning is the process, cleaning the foulant using a sponge made of polyurethane. This method involves using sponge balls to wipe the surface of membranes. This mechanical cleaning process applies to tubular membranes with large diameters (Ezugbe and Rathilal 2020). Chemical cleaning is the cleaning process to remove the foulant using chemicals, and the interaction between membrane and foulant material is essential. Chemical cleaning agents are generally classified into acid, alkaline, chelating agents, enzymes, and surfactants. Generally, chemical cleaning is carried out through detergent chemicals, and it causes weakened of the foulant materials. Acid-cleaning agents, such as hydrochloric acid, sulfuric acid, nitric acid, and phosphoric acid, are used (Du et al. 2020). Alkaline cleaning agents like sodium hydroxide are suitable mainly for organic materials. Biological/biochemical cleaning enzymes are used for cleaning purposes and are more suitable for biological membranes. The membrane-based bioprocess, cleaning of the membrane using chemical reagents, affects the membrane and the process. Combining the physical and chemical methods involves removing the foulant from the membrane. Adding chemical agents to physical cleaning methods enhances the cleaning process's effectiveness (Ezugbe and Rathilal 2020).

13.3 FUTURE SCOPE

Recently, numerous studies are conducted in wastewater treatment using membranes, and still, the process needs some improvement and modification required. A revision of membrane fabrication and utilization in the tannery wastewater treatment process is necessary. Generally, fouling is the major issue in the membrane separation process; however, tannery wastewater contains a vast volume of suspended solids, soluble organic, and inorganic molecules. Continuous research is needed to find a suitable membrane process for pretreatment and posttreatment. In a hybrid membrane system, the treatments of industrial feed solution with lab-scale membrane continuous studies are needed to develop for industrial needs adequately. Future research should look at the possibility of a membrane system for high concentrated tannery effluent. Further studies in membrane stability and fouling should focus more on membrane development and efficiency.

13.4 CONCLUSION

Leather industries generate wastewater from various steps of leather processing which contains numerous pollutants. To attain the permissible limits of effluent quality and protect the ecosystem, tannery effluent must be properly treated by a suitable technique to remove the highly toxic substances from wastewater before discharging it into the environment. The membrane is the key candidate for efficient treatment of tannery wastewater. The most significant benefits of ceramic membranes are high chemical, thermal, and mechanical stability and regeneration capacity in compared with polymeric membranes. Ceramic membranes with different modules could be used for tannery effluent treatment along with other treatment processes to improve the overall efficiency of the process. Electrocoagulation, membrane bioreactors, and adsorption are some notable methods that can be used with membrane separation in the system. Other than process efficiency, the combination of membranes with other treatment techniques also reduces the membrane's fouling. Periodic backwashing, pneumatic, ultrasonic, and sponge ball cleaning can be used as significant membrane cleaning methods during membrane usage. Therefore, the proper selection and implementation of the membrane treatment techniques, pretreatments, and maintaining of optimum conditions during the treatment process would improve the treatment efficiency and increase the life of membranes with long life.

BIBLIOGRAPHY

Abdullah, Norfadhilatuladha, Norhaniza Yusof, Muhammad Hafiz Abu Shah, Syarifah Nazirah Wan Ikhsan, Zhi Chien Ng, Subrata Maji, Woei Jye Lau, Juhana Jaafar, Ahmad Fauzi Ismail, and Katsuhiko Ariga. 2019. "Hydrous Ferric Oxide Nanoparticles Hosted Porous Polyethersulfone Adsorptive Membrane: Chromium (VI) Adsorptive Studies and Its Applicability for Water/Wastewater Treatment." *Environmental Science and Pollution Research* 26 (20). Environmental Science and Pollution Research: 20386–99. doi:10.1007/s11356-019-05208-9.

Angelini, Alessandro, Csaba Fodor, Wilfredo Yave, Luigi Leva, Anja Car, and Wolfgang Meier. 2018. "PH-Triggered Membrane in Pervaporation Process." *ACS Omega* 3 (12): 18950–18957. doi:10.1021/acsomega.8b03155.

Arif, Z., N. K. Sethy, P. K. Mishra, and B. Verma. 2020. "Green Approach for the Synthesis of Ultrafiltration Photocatalytic Membrane for Tannery Wastewater: Modeling and Optimization." *International Journal of Environmental Science and Technology* 17 (7). Springer Berlin Heidelberg: 3397–3410. doi:10.1007/s13762-020-02719-8.

Basumatary, Ashim Kumar, R. Vinoth Kumar, Aloke Kumar Ghoshal, and G. Pugazhenthi. 2015. "Synthesis and Characterization of MCM-41-Ceramic Composite Membrane for the Separation of Chromic Acid from Aqueous Solution." *Journal of Membrane Science* 475: 521–532.

Basumatary, Ashim Kumar, R. Vinoth Kumar, Aloke Kumar Ghoshal, and G. Pugazhenthi. 2016. "Cross Flow Ultrafiltration of Cr (VI) Using MCM-41, MCM-48 and Faujasite (FAU) Zeolite-Ceramic Composite Membranes." *Chemosphere* 153: 436–446. doi:10.1016/j.chemosphere.2016.03.077.

Beqqour, D., B. Achiou, A. Bouazizi, H. Ouaddari, H. Elomari, M. Ouammou, J. Bennazha, and S. Alami Younssi. 2019. "Enhancement of Microfiltration Performances of Pozzolan Membrane by Incorporation of Micronized Phosphate and Its Application for Industrial Wastewater Treatment." *Journal of Environmental Chemical Engineering* 7 (2). Elsevier: 102981. doi:10.1016/j.jece.2019.102981.

Bhattacharya, Priyankari, Arpan Roy, Subhendu Sarkar, Sourja Ghosh, Swachchha Majumdar, Sanjay Chakraborty, Samir Mandal, Aniruddha Mukhopadhyay, and Sibdas Bandyopadhyay. 2013. "Combination Technology of Ceramic Microfiltration and Reverse Osmosis for Tannery Wastewater Recovery." *Water Resources and Industry* 3: 48–62. doi:10.1016/j.wri.2013.09.002.

Cassano, A., R. Molinari, M. Romano, and E. Drioli. 2001. "Treatment of Aqueous Effluents of the Leather Industry by Membrane Processes: A Review." *Journal of Membrane Science* 181 (1): 111–126. doi:10.1016/S0376-7388(00)00399-9.

Chaudry, M. Ashraf, S. Ahmad, and M. T. Malik. 1998. "Supported Liquid Membrane Technique Applicability for Removal of Chromium from Tannery Wastes." *Waste Management* 17 (4): 211–218. doi:10.1016/S0956-053X(97)10007-1.

Choudhury, Roy Piyali, Swachchha Majumdar, Ganesh C. Sahoo, Sudeshna Saha, and Priyanka Mondal. 2018. "High Pressure Ultrafiltration CuO/Hydroxyethyl Cellulose Composite Ceramic Membrane for Separation of Cr (VI) and Pb (II) from Contaminated Water." *Chemical Engineering Journal* 336 (Vi): 570–578. doi:10.1016/j.cej.2017.12.062.

Crini, Grégorio, and Eric Lichtfouse. 2019. "Advantages and Disadvantages of Techniques Used for Wastewater Treatment." *Environmental Chemistry Letters* 17 (1). Springer International Publishing: 145–155. doi:10.1007/s10311-018-0785-9.

Dey, Surajit, Priyankari Bhattacharya, Sibdas Bandyopadhyay, Somendra N. Roy, Swachchha Majumdar, and Ganesh C. Sahoo. 2018. "Single Step Preparation of Zirconia Ultrafiltration Membrane over Clay-Alumina Based Multichannel Ceramic Support for Wastewater Treatment." *Journal of Membrane Science and Research* 4 (1): 28–33. doi:10.22079/jmsr.2017.58311.1126.

Dixit, Sumita, Ashish Yadav, Premendra D. Dwivedi, and Mukul Das. 2015. "Toxic Hazards of Leather Industry and Technologies to Combat Threat: A Review." *Journal of Cleaner Production* 87 (C): 39–49. doi:10.1016/j.jclepro.2014.10.017.

Du, Xianjun, Yaoke Shi, Veeriah Jegatheesan, and Izaz Ul Haq. 2020. "A review on the mechanism, impacts and control methods of membrane fouling in MBR system." *Membranes* 10. doi:10.3390/membranes10020024.

Elomari, H., B. Achiou, M. Ouammou, A. Albizane, J. Bennazha, S. Alami Younssi, and I. Elamrani. 2016. "Elaboration and Characterization of Flat Membrane Supports from Moroccan Clays: Application for the Treatment of Wastewater." *Desalination and Water Treatment* 57 (43): 20298–2306. doi:10.1080/19443994.2015.1110722.

El-Shafey, E. I., P. F. M. Correia, and J. M. R. De Carvalho. 2005. "Tannery Waste Treatment: Leaching, Filtration and Cake Dewatering Using a Membrane Filter Press (a Pilot Plant Study)." *Separation Science and Technology* 40 (11): 2297–2323. doi:10.1080/01496390500201474.

Ezugbe, Elorm Obotey, and Sudesh Rathilal. 2020. "Membrane Technologies in Wastewater Treatment: A Review." *Membranes* 10 (5). doi:10.3390/membranes10050089.

Feng, Jing wei, Ya Bing Sun, Zheng Zheng, Ji Biao Zhang, Shu Li, and Yuan chun Tian. 2007. "Treatment of Tannery Wastewater by Electrocoagulation." *Journal of Environmental Sciences* 19 (12): 1409–1415. doi:10.1016/S1001-0742(07)60230-7.

Fiorentin-Ferrari, L. D., K. M. Celant, B. C. Gonçalves, S. M. Teixeira, V. Slusarski-Santana, and A. N. Módenes. 2021. "Fabrication and Characterization of Polysulfone and Polyethersulfone Membranes Applied in the Treatment of Fish Skin Tanning Effluent." *Journal of Cleaner Production* 294: 126127.

Ghumra, Dishit P., Chandrodai Agarkoti, and Parag R. Gogate. 2021. "Improvements in Effluent Treatment Technologies in Common Effluent Treatment Plants (CETPs): Review and Recent Advances." *Process Safety and Environmental Protection* 147: 1018–1051. doi:10.1016/j.psep.2021.01.021.

Goyal, Rahul Kumar, N. S. Jayakumar, and M. A. Hashim. 2011. "Chromium Removal by Emulsion Liquid Membrane Using [BMIM]+[NTf2]- as Stabilizer and TOMAC as Extractant." *Desalination* 278 (1–3): 50–56. doi:10.1016/j.desal.2011.05.001.

Hakami, Mohammed Wali, Abdullah Alkhudhiri, Sirhan Al-Batty, Myrto Panagiota Zacharof, Jon Maddy, and Nidal Hilal. 2020. "Ceramic Microfiltration Membranes in Wastewater Treatment: Filtration Behavior, Fouling and Prevention." *Membranes* 10 (9): 1–34. doi:10.3390/membranes10090248.

Hasan, M. A., Y. T. Selim, and K. M. Mohamed. 2009. "Removal of Chromium from Aqueous Waste Solution Using Liquid Emulsion Membrane." *Journal of Hazardous Materials* 168 (2–3): 1537–1541. doi:10.1016/j.jhazmat.2009.03.030.

Hatimi, B., J. Mouldar, A. Loudiki, H. Hafdi, M. Joudi, E. M. Daoudi, H. Nasrellah, I. T. Lançar, M. A. El Mhammedi, and M. Bakasse. 2020. "Low Cost Pyrrhotite Ash/Clay-Based Inorganic Membrane for Industrial Wastewaters Treatment." *Journal of Environmental Chemical Engineering* 8 (1). doi:10.1016/j.jece.2019.103646.

Jiang, S., Y. Li, and B. P. Ladewig. 2017. "A Review of Reverse Osmosis Membrane Fouling and Control Strategies." *Science of the Total Environment* 595: 567–583.

Kaplan-Bekaroglu, Sehnaz S., and Serife Gode. 2016. "Investigation of Ceramic Membranes Performance for Tannery Wastewater Treatment." *Desalination and Water Treatment* 57 (37): 17300–17307. doi:10.1080/19443994.2015.1084595.

Karunanidhi, Arthi, Padma Sheeba David, and Nishter Nishad Fathima. 2020. "Electrospun Keratin-Polysulfone Blend Membranes for Treatment of Tannery Effluents." *Water, Air, and Soil Pollution* 231 (6): 1–11. doi:10.1007/s11270-020-04682-z.

Keerthi, V. Vinduja, and N. Balasubramanian. 2013. "Electrocoagulation-Integrated Hybrid Membrane Processes for the Treatment of Tannery Wastewater." *Environmental Science and Pollution Research* 20 (10): 7441–7449. doi:10.1007/s11356-013-1766-y.

Kiril Mert, Berna, and Kadir Kestioglu. 2014. "Recovery of Cr(III) from Tanning Process Using Membrane Separation Processes." *Clean Technologies and Environmental Policy* 16 (8): 1615–1624. doi:10.1007/s10098-014-0737-4.

Korpe, Sneha, and P. Venkateswara Rao. 2021. "Application of Advanced Oxidation Processes and Cavitation Techniques for Treatment of Tannery Wastewater: A Review." *Journal of Environmental Chemical Engineering* 9 (3). Elsevier Ltd: 105234. doi:10.1016/j.jece.2021.105234.

Kumar, Vinay, and S. K. Dwivedi. 2021. "A Review on Accessible Techniques for Removal of Hexavalent Chromium and Divalent Nickel from Industrial Wastewater: Recent Research and Future Outlook." *Journal of Cleaner Production* 295. doi:10.1016/j.jclepro.2021.126229.

Lofrano, Giusy, Sureyya Meriç, Gülsüm Emel Zengin, and Derin Orhon. 2013. "Chemical and Biological Treatment Technologies for Leather Tannery Chemicals and Wastewaters: A Review." *Science of the Total Environment* 461–462: 265–281. doi:10.1016/j.scitotenv.2013.05.004.

Madhura, Lavanya, Suvardhan Kanchi, Myalowenkosi I. Sabela, Shalini Singh, Krishna Bisetty, and Inamuddin. 2018. "Membrane Technology for Water Purification." *Environmental Chemistry Letters* 16 (2). Springer International Publishing: 343–65. doi:10.1007/s10311-017-0699-y.

Mannucci, Alberto, Giulio Munz, Gualtiero Mori, and Claudio Lubello. 2010. "Anaerobic Treatment of Vegetable Tannery Wastewaters: A Review." *Desalination* 264 (1–2): 1–8. doi:10.1016/j.desal.2010.07.021.

Mouiya, M., A. Abourriche, A. Bouazizi, A. Benhammou, Y. El Hafiane, Y. Abouliatim, L. Nibou, et al. 2018. "Flat Ceramic Microfiltration Membrane Based on Natural Clay and Moroccan Phosphate for Desalination and Industrial Wastewater Treatment." *Desalination* 427 (November 2017): 42–50. doi:10.1016/j.desal.2017.11.005.

Mouiya, Mossaab, Abdelmjid Bouazizi, Abdelkrim Abourriche, Abdelaziz Benhammou, Youssef El Hafiane, Mohamed Ouammou, Younes Abouliatim, Saad Alami Younssi, Agnès Smith, and Hassan Hannache. 2019. "Fabrication and Characterization of a Ceramic Membrane from Clay and Banana Peel Powder: Application to Industrial Wastewater Treatment." *Materials Chemistry and Physics* 227 (January): 291–301. doi:10.1016/j.matchemphys.2019.02.011.

Mpofu, A.B., O.O. Oyekola, and P.J. Welz. 2021. "Anaerobic Treatment of Tannery Wastewater in the Context of a Circular Bioeconomy for Developing Countries." *Journal of Cleaner Production* 296. doi:10.1016/j.jclepro.2021.126490.

Mulder, M. 1996. "Basic Principles of Membrane Technology, 2E.Pdf."

Pal, M., M. Malhotra, M. K. Mandal, T. K. Paine, and P. Pal. 2020. "Recycling of Wastewater from Tannery Industry Through Membrane-Integrated Hybrid Treatment Using a Novel Graphene Oxide Nanocomposite." *Journal of Water Process Engineering* 36: 101324.

Rambabu, Krishnamoorthy, and Sagadevan Velu. 2016. "Modified Polyethersulfone Ultrafiltration Membrane for the Treatment of Tannery Wastewater." *International Journal of Environmental Studies* 73 (5): 819–826. doi:10.1080/00207233.2016.1153900.

Religa, P., A. Kowalik, and P. Gierycz. 2011. "Application of Nanofiltration for Chromium Concentration in the Tannery Wastewater." *Journal of Hazardous Materials* 186 (1). Elsevier B.V.: 288–292. doi:10.1016/j.jhazmat.2010.10.112.

Romero-Dondiz, Estela María, Jorge Emilio Almazán, Verónica Beatriz Rajal, and Elza Fani Castro-Vidaurre. 2015. "Removal of Vegetable Tannins to Recover Water in the Leather Industry by Ultrafiltration Polymeric Membranes." *Chemical Engineering Research and Design* 93 (June): 727–735. doi:10.1016/j.cherd.2014.06.022.

Rosman, Nurafiqah, W. N. W. Salleh, Mohamad Azuwa Mohamed, J. Jaafar, A. F. Ismail, and Z. Harun. 2018. "Hybrid Membrane Filtration-Advanced Oxidation Processes for Removal of Pharmaceutical Residue." *Journal of Colloid and Interface Science* 532. Elsevier Inc.: 236–260. doi:10.1016/j.jcis.2018.07.118.

Saja, Souad, Abdelmjid Bouazizi, Brahim Achiou, Mohamed Ouammou, Abderrahman Albizane, Jamal Bennazha, and Saad Alami Younssi. 2018. "Elaboration and Characterization of Low-Cost Ceramic Membrane Made from Natural Moroccan Perlite for Treatment of Industrial Wastewater." *Journal of Environmental Chemical Engineering* 6 (1). Elsevier: 451–458. doi:10.1016/j.jece.2017.12.004.

Samaei, Seyed Mohsen, Shirley Gato-Trinidad, and Ali Altaee. 2018. "The Application of Pressure-Driven Ceramic Membrane Technology for the Treatment of Industrial Wastewaters: A Review." *Separation and Purification Technology* 200 (February). Elsevier: 198–220. doi:10.1016/j.seppur.2018.02.041.

Sawalha, Hassan, Razan Alsharabaty, Sawsan Sarsour, and Maher Al-Jabari. 2019. "Wastewater from Leather Tanning and Processing in Palestine: Characterization and Management Aspects." *Journal of Environmental Management* 251 (September). Elsevier: 109596. doi:10.1016/j.jenvman.2019.109596.

Song, Z., C. J. Williams, and R. G. J. Edyvean. 2000. "Sedimentation of Tannery Wastewater." *Water Research* 34 (7): 2171–2176. doi:10.1016/S0043-1354(99)00358-9.

Stoller, Marco, Olga Sacco, Diana Sannino, and Angelo Chianese. 2013. "Successful Integration of Membrane Technologies in a Conventional Purification Process of Tannery Wastewater Streams." *Membranes* 3 (3): 126–135. doi:10.3390/membranes3030126.

Stylianou, Stylianos K., Katarzyna Szymanska, Ioannis A. Katsoyiannis, and Anastasios I. Zouboulis. 2015. "Novel Water Treatment Processes Based on Hybrid Membrane-Ozonation Systems: A Novel Ceramic Membrane Contactor for Bubbleless Ozonation of Emerging Micropollutants." *Journal of Chemistry* 2015. doi:10.1155/2015/214927.

Uragami, Tadashi. 2017. "Membrane Shapes and Modules." *Science and Technology of Separation Membranes*, 87–103. doi:10.1002/9781118932551.ch4.

Vasanth, D., G. Pugazhenthi, and R. Uppaluri. 2012. "Biomass Assisted Microfiltration of Chromium(VI) Using Baker's Yeast by Ceramic Membrane Prepared from Low Cost Raw Materials." *Desalination* 285. Elsevier B.V.: 239–244. doi:10.1016/j.desal.2011.09.055.

Velu, M., B. Balasubramanian, P. Velmurugan, H. Kamyab, A. V. Ravi, S. Chelliapan, . . . & J. Palaniyappan. 2021. "Fabrication of Nanocomposites Mediated from Aluminium Nanoparticles/Moringa oleifera Gum Activated Carbon for Effective Photocatalytic Removal of Nitrate and Phosphate in Aqueous Solution." *Journal of Cleaner Production* 281: 124553.

Velu, S., L. Muruganandam, and G. Arthanareeswaran. 2015. "Preparation and Performance Studies on Polyethersulfone Ultrafiltration Membranes Modified with Gelatin for Treatment of Tannery and Distillery Wastewater." *Brazilian Journal of Chemical Engineering* 32 (1): 179–189. doi:10.1590/0104-6632.20150321 s00002965.

Vinodhini, P. Angelin, and P. N. Sudha. 2017. "Removal of Heavy Metal Chromium from Tannery Effluent Using Ultrafiltration Membrane." *Textiles and Clothing Sustainability* 2 (1). Textiles and Clothing Sustainability. doi:10.1186/s40689-016-0016-3.

Yadav, A. K., and S. Bhattacharyya. 2020. "Preparation of Porous Alumina Adsorbent from Kaolin Using Acid Leach Method: Studies on Removal of Fluoride Toxic Ions from an Aqueous System." *Adsorption* 26 (7): 1073–1082.

Zakmout, Asmaa, Fatma Sadi, Carla A. M. Portugal, João G. Crespo, and Svetlozar Velizarov. 2020. "Tannery Effluent Treatment by Nanofiltration, Reverse Osmosis and Chitosan Modified Membranes." *Membranes* 10 (12): 1–20. doi:10.3390/membranes10120378.

Zhao, Changqing, and Wuyong Chen. 2019. "A Review for Tannery Wastewater Treatment: Some Thoughts under Stricter Discharge Requirements." *Environmental Science and Pollution Research* 26 (25): 26102–26111. doi:10.1007/s11356-019-05699-6.

Zinicovscaia, I. 2016. "Conventional Methods of Wastewater Treatment." In: Zinicovscaia I., Cepoi L. (eds) Cyanobacteria for Bioremediation of Wastewaters. Springer, Cham. https://doi.org/10.1007/978-3-319-26751-7_3

14 Removal of Cr (VI) and Pb from Electroplating Effluent Using Ceramic Membrane

Deepak Sharma, Abhinesh Prajapati,
Raghwendra Singh Thakur, Ghoshna Jyoti,
and Parmesh Kumar Chaudhari

CONTENTS

14.1 INTRODUCTION

14.1.1 ELECTROPLATING INDUSTRY

Electroplating industries (EPIs) are one of the important segments of the economy in many countries and are the highest-polluting industries. It discharges toxic materials and heavy metals through wastewater (effluents), air emissions and solid wastes in the environment; hence, it is mandatory to treat the effluent before discharging it in such a manner that it can be regenerated and reused. Several types of metals are used for the metal plating, including zinc, lead (Pb), chromium (Cr) and others, that appear in the EPI effluent. Electroplating takes four steps, namely, surface preparation,

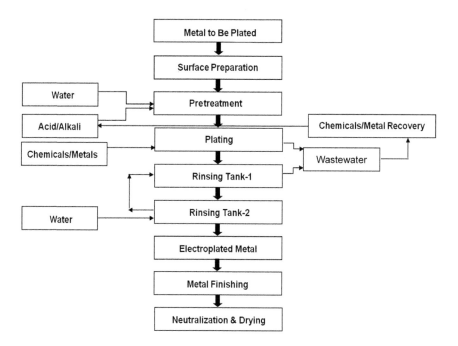

FIGURE 14.1   Flow sheet showing steps of the process in EPI.

pretreatment, electroplating and post treatment. An overview of the metal plating industry is shown in Figure 14.1.

Surface preparation: Surface preparation is the initial and important part of the electroplating process. It is mandatory to ensure a uniform adhesion of the coating on the object (metal). This step involves smoothing the object surface (item to be coated) before the plating operation. It is only a physical process, and no other material is required. During the surface preparation, the buffing is done by a scrapper and may be done by manually or mechanically.

Pretreatment of object: The object material contains several types of contaminants, such as oil, grease, dirt, mineral oils, organic soils (i.e., paints, fingerprints) and miscellaneous solid particles (i.e., dust, abrasive grits, chips), The purpose of this process is to remove such undesired matter, because it is responsible for promoting an unstable coating.

Electroplating: This technique involves the deposition of a fine layer of one metal on the object through an electrolytic process to save the original material from corrosion and enhance surface hardness, luster, color, aesthetics, value addition and so on. During the process, the anode and the cathode in the electroplating cell are both connected to direct current (DC) from external supply. The anode (coating material, i.e., Cr and Pb) is connected to the positive terminal of the supply, while the cathode (object) is connected to the negative terminal. When the direct current is passed, the electrolyte dissociates to produce positively and negatively charged ions. The positively charged ions (cations) move toward the cathode, whereas negatively charged ions (anions) move toward the anode. On reaching their respective electrodes, ions lose their charges and become neutral particles. The cations accept electrons from the cathode, become neutral and get deposited in the form of metal on the cathode, whereas anions give electrons to the anode to become neutral, thus forming electrolytes. Both the cathode (the item to be coated) and the anode (the coating substance) are immersed in the bath solution. However, if an inert electrode is used, the coating substance would be the metal salts in liquid form added to the solution. The metal salts subsequently dissociate into anions and cations, which then are deposited onto the items to be plated.

Post treatment: After the deposition of the coating material on the object, most of the plated objects require post treatment operations. Post treatment includes sealing, dying and conversion coating. Post treatment operations improve the physical appearance of the object and enhance the corrosion resistance and the aesthetic values.

14.1.1.1 Environmental Pollution from EPIs

Out of the total water used in the electroplating process only 40% water is consumed during the process and remaining comes out as effluent, known as the electroplating effluent (EPE). Electroplating wastewaters contain various kinds of toxic substances, such as cyanides, alkaline cleaning agents, degreasing solvents, oils, fats and metals. Most of the metals such as chromium, lead, copper, nickel, chromium, silver and zinc in the EPE are harmful if discharged without treatment.

The EPE characteristics depend on the metal object and the coating material. For example, effluent from the chromium- and lead-based industry has a very high content of these materials. Similarly, other metals, such as copper, nickel, silver and zinc, have high concentrations as they belong to the same metal plating industry. Since chromium and lead provide resistivity against the atmospheric reaction, consequently, they are more stable as compared to the other metals and, hence, are commonly used in surface coating practice.

During the plating, a large amount of EPE is generated in each run. A large-scale EPI generates about 500,000 L of EPE per day and contributes to the increase pollution load, which is more than the domestic sewage of a city with a population of 0.5 million and is capable of polluting any big water body. The most significant sources of wastewater in an EPI are surface preparation, pretreatment, electroplating and post treatment. The typical composition of EPE as reported by various researchers is presented in Table 14.1.

TABLE 14.1

Chemical Composition of Electroplating Industry Effluent

S. No.	Authors Parameters	Akbal et al. (2011) (metal plating)	Golder et al. (2009) (copper plating)	Kobya et al. (2010) (electroplating rinse water)
1	Copper	45	74.4	–
2	Chromium	44.5	–	–
3	Nickel	394	–	175
4	Zinc	–	6.8	–
5	Total iron	–	2.2	–
6	Cyanide	–	–	120
7	Arsenic	–	–	–
8	Lead	–	–	–
9	Cadmium	–	–	102
10	Chemical oxygen demand	–	22	180
11	pH	3	4.8	8.6
12	Conductivity (mS/cm)	2	.48	1
13	Sulphate	–	187.3	–
14	Total disolved solid	–	500	–
15	Total solid	–	570	–
16	Total suspended solid	–	70	175

14.1.1.2 Harmful Effects of Heavy Metals

A toxic heavy metal is a relatively dense metal or metalloid that is noted for its potential toxicity, especially in environmental contexts. The term has particular application to cadmium, mercury, lead, chromium and arsenic, all of which appear in the World Health Organization's list of 10 chemicals of major public concern. Under acute exposure (for a day or less) to these metals, lung inflammation (cadmium), diarrhea (mercury), brain dysfunction (lead), acute renal failure (chromium) and nausea (arsenic) are observed in people. Under chronic exposure (months or years), they cause lung cancer (cadmium), inflammation in gums and mouth (mercury), anemia (lead), lung scarring (chromium) and cancer (arsenic).

As the administration of reusing water is strict, most EPIs send wastewater to the treatment plant to be reutilized and reduce the demand for water. Partially treated/untreated types of water cannot be directly discharged in any pure water stream because it can change the quality of the original water; hence, treating EPE is mandatory.

14.1.1.3 Membrane Technology for Separation of Heavy Metals

The removal of heavy metals from EPE is difficult by any simple physical, chemical or biological method. A number of methods are available to treat the EPE including precipitation, adsorption, biosorption, ionexchange, electrodialysis, and membrane separation, among others. This section focuses on membrane technology for removing heavy metals from wastewater.

A membrane separation process has a very important role in the separation industry. This process differs from others based on separation mechanisms and the size of the separated particles. The membrane acts as a semipermeable barrier, and separation occurs by controlling the rate of movement of various molecules between two liquid phases, two gas phases, or a liquid and a gas phase. The two fluid phases are usually miscible, and the membrane barrier prevents actual, ordinary hydrodynamic flow. In the due course of separation, the primary species that are rejected and retained are termed retentate solutes, and the species that pass through the membrane are termed permeate solutes. In general, the driving force to accomplish the desired separation is brought forward by the application of pressure, concentration or voltage difference across the membrane. The widely used membrane processes include microfiltration (MF), ultrafiltration (UF), nanofiltration (NF), reverse osmosis (RO), electrolysis, dialysis, electrodialysis, gas separation, vapor permeation, evaporation, membrane distillation and membrane contactors. All processes except for pervaporation involve no phase change. Also, except for electrodialysis, other processes are pressure-driven. MF and UF are widely used in food and beverage processing (beer MF, apple juice UF), biotechnological applications, the pharmaceutical industry (antibiotic production, protein purification) and water purification and wastewater treatment. NF and RO membranes are mainly used for water purification purposes.

Based on the membrane transport mechanism, various membrane separation processes can be classified as follows:

1. Pressure-driven processes: MF, UF, NF and RO
2. Concentration-driven processes: Gas separation through dense membranes, pervaporation (PV), dialysis, membrane extraction, supported liquid membrane and emulsion liquid membrane
3. Temperature-driven processes: Membrane distillation and thermo-osmosis
4. Electrically driven processes: Electrodialysis, electrofiltration and electrochemical ion exchange

14.2 ADVANCEMENT IN MEMBRANE APPLICATIONS

In the past two decades, significant advances in membrane technology research have been reported. Numerous applications have been proposed, of which MF and UF are more common. Today, the membrane separation process has become economically competitive due to the availability of membranes with higher flux and lower process costs. Some of the applications of ceramic membranes used for metal removal from water are presented in the following works reported by various authors:

Choudhury et al. (2018) have fabricated a clay–alumina ceramic composite membrane composed of hydroxyethyl cellulose and CuO nanoparticles for the separation of Cr (VI) and Pb (II) from contaminated water. For the improvement of ceramic composite membranes, CuO nanoparticles, in combination with a biopolymer, have been added, which causes are duction in the pore size of the ceramic substrate from 0.5–1.5 μm to 3 nm. The reduced pore size has improved the heavy metal rejection rate, with a permeability of 34.99 L/(m^2 h.1 bar). The investigators selected the operational pressure range of 0–5 bar. The maximum percentage of rejection achieved to 97.14% for Pb (II) and 91.44% for Cr (VI) at 2 bar transmembrane pressure.

The performance of ceramic monolith in MF for treating a solid–liquid wastewater has been evaluated by Arzani et al. (2018). They used kaolin to prepare the membrane. The optimized values of the process variables were estimated as0.5 wt.%, 1150 °C and 5 h for poly (vinyl alcohol) concentration, sintering temperature and sintering time, respectively. The membrane exhibited optimum 89.8% turbidity rejection. The prepared membrane gave high practical separation potential, and it could be used for treating solid–liquid wastewaters.

Muthumareeswaran et al. (2017) analyzed UF membrane separation for removing chromium ions from potable water. A rejection of ≥90% was achieved at pH ≥ 7 and a low chromate concentration (≤25 ppm) in feed. The rejection mechanism of chromium ions followed to Donnan exclusion principle. They have found that the pH of the solution had a vital role in changing the porosity of the membrane and on the retention behavior of chromate ions. They also found that, at higher feed concentrations (≥400 ppm), the concentration polarization became prominent, and it reduced chromate rejection. In another study, Kaplan-Bekaroglu and Gode (2016) treated tannery wastewater using ceramic MF and UF membranes. They focused on the impact of membrane pore size and pressure on permeate flux, chemical oxygen demand (COD), and color reduction. Three different single-channel tubular ceramic membrane modules with average pore sizes of 10, 50, and 200 nm were used. More than 95% color removal was consistently achieved with both UF membranes (10 and 50 nm). COD reductions ranged between 58 and 90% at all pressures for UF membranes.

Piedra et al. (2014) studied the removal of hexavalent chromium from metal plating industry wastewater using NF and RO. The results revealed chromium rejection above 97%. The NF90 membrane showed the best performance (highest flux and excellent selectivity, typically above 99%). Among the various membranes, the highest flux was obtained with NF90, followed by BW30 and MPS-34, respectively. They ascertained that NF90 was, overall, the best-performing membrane with the highest flux and observed rejection higher than 99.5%.

Vasanth et al. (2012) have reported removal of Cr(VI) from an aqueous solution using a ceramic membrane prepared from low-cost raw materials. The obtained result reveals that the removal of Cr (VI) strongly depended on the pH of the solution and the highest removal of 94% was at pH 1. The removal of Cr (VI) increased with an increase in the biomass concentration and decreased with an increase in Cr (VI) ion concentration. The prepared membrane gave good mechanical strength (34 MPa flexural strength) and chemical stability along with pore size of 1.32 μm.

Jana et al. (2011) used a chitosan-based ceramic UF membrane and found it to be highly effective for removing Hg(II) and As(III) from an aqueous solution. They have found almost 100% removal of mercury and arsenic at its low concentration. The average pore size of 1093 nm and a porosity of 0.37 were reported. The chitosan-impregnated ceramic membranes have form to applicable for both MF and UF applications.

Murthy et al. (2008) have worked on rejection of nickel ions from aqueous solutions through ceramic membrane. They examined the effect of various parameters like feed concentration (5–250 ppm), applied pressure (4–20 atm), feed flowrate (5–15 dm^3/min) and pH (2–8). In the process, the rejection of nickel ions were found to increase with an increase in the feed pressure and decreased with an increase in feed concentration. Up to 98% and 92% nickel removal obtained for an initial feed concentration of 5 and 250 ppm, respectively.

14.3 PREPARATION OF CERAMIC MEMBRANE AND EXPERIMENTAL PROCEDURE

The ceramic membrane is placed in the middle part of the reactor (i.e., between the upper and bottom section). Effluent is fed into the top section of the reactor, and permeate comes out the bottom. During the process, the pressure inside the reactor has been maintained at a certain level by an air compressor. Thus, the process could be carried out for batch and continuous studies. For the removal of Cr and Pb, a batch reactor data are presented.

14.3.1 PREPARATION OF THE CERAMIC SUPPORT MEMBRANES

Ceramic membranes have been prepared in a disc shape with a 50-mm diameter and a 5-mm thickness by uniaxial followed by sintering at 900 °C. For the preparation of the membrane, first, the top layer of the ceramic support is coated with chitosan using a spin-coating technique. A solution of chitosan (1–2 wt.%) prepared by dissolving chitosan flakes in a 2 wt.% aqueous acetic acid solution is used. Then the solution (chitosan and acetic acid) is mixed with 0.12%(v/v) glutaraldehyde solution in a 3:2 ratio with stirring for 2 minutes. It promotes a crosslink reaction. Glutaraldehyde releases aldehyde groups, which include the amino groups of chitosan that consequently form covalent amine bonds due to the resonance established with the adjacent double ethylenic bonds via a Schiff reaction. To stabilize the pore penetration of chitosan in the coating process, the ceramic membranes are dipped in water for 5 h before spin coating. During this, the air present in the porous structure of the membrane is displaced by water. Then the membranes are taken out from water and covered by aluminum foil to prevent chitosan deposition. The prepared membrane is then coated by using a spin-coating machine. The operating speed of the machine is set to 3500 rpm. The coating time may be 1–3 minutes. Less coating time consequently means less coating, which gives a higher pore size, and high coating produces a small pore size. After the coating process, the membrane is recovered from the solution and dried at 105 °C for 5 h in a hot-air oven to remove water from the membrane. For experimental purposes, the ceramic membrane which has been coated is called as MD-2, while without coating is referred as MD-1.

14.3.2 FILTRATION STUDIES

A dead-end filtration setup was used for the ceramic membrane filtration to carry out pure water flux, and MF experiments of EPE are presented in Figure 14.2. This has also been used to treat EPE. This setup had two parts made of stainless steel with circular base plate having circular host for membrane. The work reported by Sharma et al. (2020) used a 50-mm-diameter, 5-mm-deep membrane. The top part had a cylindrical compartment attached with circular flanges. The upper part had two inlets, one for the liquid feed and the other for the compressed gas to maintain the pressure and one outlet for retentate stream. The bottom part had one outlet for permeate stream. The top inlet was connected to compressor. Permeate flux, J dm^3 (h.m^2) was calculated at different applied pressures by collecting volume of permeate at specified time interval for every applied pressure using Equation 14.1:

$$J = \frac{Q}{A \times t},$$

(14.1)

Where Q (dm^3) is the volume of permeate collected in time t (hr) and A (m^2) is the effective membrane area for permeation.

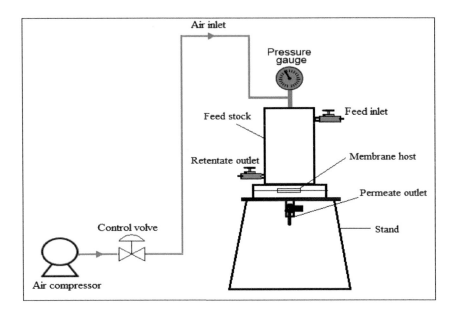

FIGURE 14.2 Filtration setup diagram.

14.4 REMOVAL OF CR AND PB USING CERAMIC MEMBRANES

Locally available clay material can be used to prepare the ceramic membrane in the laboratory via a uniaxial compaction method followed by sintering. A spin-coating technique is employed for the deposition of chitosan on the ceramic membrane. Sharma et al. (2020) have used two types of membrane, namely, ceramic membrane MD-1 (ceramic membrane without a chitosan coating) and MD-2 (ceramic membrane with a chitosancoating), to remove Cr (VI) and Pb from EPE. The pollutants parameters after removal obtained in our studies are presented in Table 14.2. The characterization of prepared material has been performed using X-ray diffraction, scanning electron microscopy (SEM), Fourier transform infrared spectroscopy (FTIR), thermal gravity analyser, surface pore density, pore size and mechanical strength. The detailed composition of both membranes, MD-1 and MD-2, are presented in Table 14.2.

14.4.1 Physical Properties of Membrane

Some physical properties of ceramic membranes that widely affect filtration follow.

14.4.1.1 Surface Pore Density

The average number of pores present per unit area of the membrane surface is known as surface pore density. The pore density can be calculated from SEM images as described by Jana et al. (2011). The membrane reported by Sharma et al. (2020)hada surface pore density 1.34×10^{10} for MD-1 and 1.64×10^{10} has for MD-2. Jana et al. (2011) have reported a pore density in the range of $2.09–13.3 \times 10^{10}/m^2$ for ceramic membrane prepared from muddy clay.

14.4.1.2 Pore Size Distribution and Average Pore Diameter

For the calculation of pore diameter, the SEM images of the membrane can also be considered. For MD1 and MD2 membranes, about 600 pore diameters have been determined by the Image J software (developed at the National Institute of Health and the Laboratory for Optical

TABLE 14.2

Typical Composition of Electroplating Effluent before and after Treatment by Membrane Separation at pH = 3.5, Pressure = 300 kPa

Characteristics	EPE	Treated EPE under optimum condition (MD-1)	Treated EPE under optimum condition (MD-2)
Chromium (VI) (mg/dm³)	55.3	19.35	10.50
Lead (mg/dm³)	3.5	1.12	0.245
pH	5.4	3.5	3.5
Conductivity (mS/cm)	3	2.8	2.5
TDS (mg/dm³)	5563	550	100
TSS (mg/dm³)	4350	450	150
TS (mg/dm³)	9913	1000	250
Turbidity (NTU)	5350	300	100
color	Dark brown	Light brown	Light brown

Computational Instrumentation). The average pore diameter (d_s) can be calculated with the help of Equation 14.2:

$$d_s = \left(\frac{\sum_{i=1}^{n} n_i d_i^2}{\sum_{i=1}^{n} n_i} \right)^{0.5}$$

(14.2)

Where n is the number of pores and d_i is the pore diameter (μm) of the ith pore. The average pore size of membrane MD-2 (1.41 μm) and that of membrane MD-1 (2.56 μm) were evaluated. This happened due to the chitosan covering the pores of the ceramic membrane. A pore size of 2.16–4.73 μm was evaluated by Jana et al. (2011) for clay-based ceramic membranes. The pore size of the membranes, determined by permeability experiments, were 1.46 μm and 1.10 μm for MD-1 and MD-2, respectively, which is smaller than pore size obtained from the SEM images (Sharma et al., 2020). This happens due to the presence of dead-end pores that are not incorporated in water permeability, while in the case of SEM, image analysis was integrated.

14.4.1.3 Porosity

The total porosity of the membrane can be determined using Archimedes's principle by measuring the weight of the membrane at different conditions using Equation 14.3:

$$\varepsilon = \frac{M_w - M_d}{M_w - M_a} \times 100$$

(14.3)

Where M_w(g) is the weight of the membrane in the wet saturation condition, M_d(g) is the weight of the membranes in the dry condition and M_a(g) is the suspended weight of the membrane in water. Sharma et al. (2020) observed the porosity of membrane MD-1 (0.423) to be greater than the porosity of membrane MD-2(0.264). The chitosan reduces the porosity of membrane. During sintering, the gaseous products that make the surface porous are formed, and the void spaces generated are filled by the other materials through structural densification. Furthermore, the densification increases with an increase in sintering temperatures followed by the transformation of phase from amorphous to crystalline of the clay material (Jana et al., 2010). The low porosity favors the retention of impurities over the membrane (Ghosh et al., 2013). Some of the clay-based membranes have found porosity in the range of 0.43 to 0.85 (Ghosh et al., 2013).

14.4.1.4 Mechanical Strength

The membrane is fixed in a module for filtration, and pressure is applied on the liquid by air or nitrogen that is transferred to the membrane; thus, having the proper strength of membrane is essential. The flexural strengths of MD-1 and MD-2, calculated by a three-point bending strength method, were reported to 2.25 MPa (Sharma et al., 2020). This strength is sufficiently high.

14.4.2 Water Permeation Experiment and Removal of Cr (VI) and Pb

For a good membrane, the permeation flux and metal retention should be high. The effect of pressure on water flux and metals removal is discussed in the following sections.

14.4.2.1 Effect of Pressure on Water Flux

It is seen that during MF, the pure water flux collection increases with an increase in pressure over the liquid that transfers to the membrane. Furthermore, the permeate flux varies almost linearly with increasing applied pressure, which is because there is no significant contribution of additional transport resistance from concentration polarization and adsorption (Jana et al., 2011). The permeate flux decreases with time due to the deposition of metal on the membrane surface. The value of permeate flux has been reported to be between $1-7 \times 10^5$ m³/(m²s¹) for MD-2 and $7-58 \times 10^5$ m³/(m².s) for MD-1. Sharma et al. (2020) also reported that the permeate flux of the MD-1 and MD-2 is to be slightly lower for EPE than for the pure water. This is due to the osmotic pressure generated by the retained ions, which resultsin reducing the effective pressure across the membrane. In addition, the flux collection rate in MD-1 was higher than MD-2 due to the larger pore size of MD-1 compared to MD-2 (Sharma et al., 2020). Ghosh et al. (2013) reported a permeate flux in the range of $5.4-313.4 \times 10^5$ m³/(m².s).

14.4.2.2 Effect of pH on Cr (VI) and Pb Removal

The pH of the effluent plays an important role in rejection of metals during the filtration process. Work reported in Sharma et al. (2020) is shown in Figures 14.3a and 14.3b. It can be seen that the percentage rejection of Cr (VI) and Pb decreases with an increase in the pH of the effluent for both membranes (MD-1 and MD-2). At pH 3.5, a maximum 63% Cr (VI) and 67% (Pb) removal can be seen over membrane MD-1, while membrane MD-2 had a maximum 81% Cr (VI) and 93% (Pb) removal. The coating material (chitosan) reduces the pores of the membrane; due to this, the metal removal is greater. ApH of 3.5 is the optimum for the pH study between 2 to 9. The metal rejection is almost the same at pH 2 and pH 3.5 for MD-2, while for MD-1, the metal rejection is about 2% less at pH 2 compared to pH 3.5. The Cr (VI) existed in solution in different ionic forms ($HCrO_4^-$, CrO_4^{2-}, $Cr_2O_7^{2-}$), which depend on the solution's pH and the concentration of Cr (VI) (Piedra et al., 2014). At a low pH, $HCrO_4^-$ is the dominant species. The $HCrO_4^-$ has the properties to exchange easily with OH- ions onan active surface under acidic conditions. Furthermore, at a low pH, hydronium ions are present, which increases the Cr (VI) interaction, causing a retention of Cr over the membrane (Piedra et al., 2014). At near-to-neutral pH, the Pb remains in four oxidation stages, which changes to two oxidation stages at an acidic pH. The formation of lead oxides is expected because of the reaction of lead with dissolved oxygen at a high acidic pH. Lead sulfate ($PbSO_4$) could also form at a low pH in the presence of sulfate anions, which is quite insoluble (Ghosh et al., 2013). The size of ions and compounds formed are also different at different pH values; thus, its rejection varies due to its size. As the pH increases, the overall surface charge of the cell becomes negative, and hence, binding capacity decreases (Piedra et al., 2014). All these affect the removal of Cr (VI) and Pb.

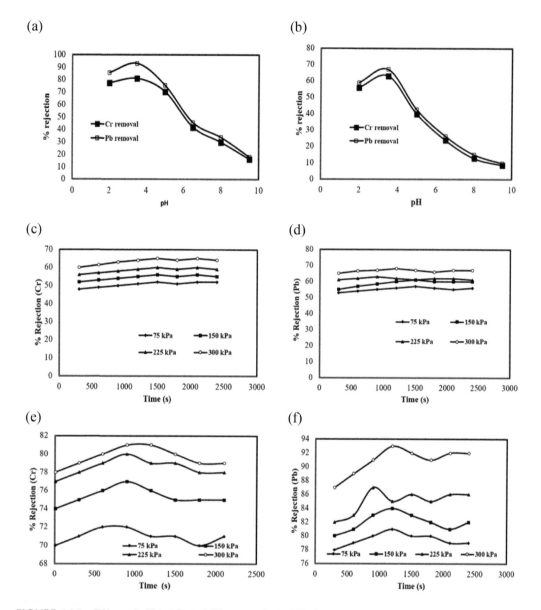

FIGURE 14.3 Effect of pH on Cr and Pb removal: (a) MD-2; (b) MD-1 at applied pressure = 300 kpa; (c) effect of pressure on Cr removal; (d) Pb removal over MD-1; (e) effect of pressure on Cr; (f) Pb removal over MD-2 (Sharma et al., 2020).

14.4.2.3 Effect of Applied Pressure on Cr (VI) and Pb Removal

The applied pressure on effluent highly affects the permeate water flux and the removal efficiency of the membrane. The pressure applied to the water is transferred to the membrane, thus affecting metal removal. Figures 14.3c and 14.3d present the Cr(VI) and Pb removal rate at different pressure as reported by Sharma et al. (2020). With membrane type MD-1, at the optimum pH of 3.5 and with 300 KPa of applied pressure, 65% Cr (VI) and 68% Pb removal can be seen. Figures 14.3e and 14.3f also present the metal removal rate at different applied pressures for membrane MD-2. At the optimum condition (pH 3.5 and pressure 300 kPa), 81% Cr(VI) and 93% Pb were removed. These figures show that the metal removal rate increases with an increase in the applied pressure.

TABLE 14.3

Cost Analysis of the MD-2 Membranes from the Unit Cost of Raw Materials (Sharma et al., 2020)

	Raw materials	Material required (kg/m²)	Unit price ($/kg or $/dm³)	Total cost ($/m²)
Support	Clay	9	–	–
	Kaolin	2.1	8.1	17.0
	Sodiumcarbonate	0.65	8.31	5.401
	Sodium meta silicate	0.395	17.48	4.78
	Boric acid	0.395	10.7	4.22
Coating material	Water	11.1	–	–
	Acetic acid	0.21	8.9	1.86
	Glutaraldehyde solution (25%)	5.34	93.89	418.37
	Chitosan	0.21	660.39	138.7
	Total			590.68

14.5   MEMBRANE COST CALCULATION

A variety of membranes are available in the market for industrial purposes in a cost range of 600–2100 $/m²(Jana et al., 2011). Sharma et al. (2020) reported the cost estimation based on 5-mm-thick, 50-mm-diameter membrane. A total 8.5 dm³/m² solution was needed for successful spin coating. The details of the chemicals used and their prices for a 1-m² membrane are presented in Table 14.3. The total price is calculated to be 590.68 $/m². Apart from this, additional costs, including manufacturing and shipment, are required; thus, the total cost may reach 649 $/m².

14.6   CONCLUSION

The two types of ceramic membranes–MD-1 (a ceramic membrane without a chitosan coating) and MD-2 (a ceramic membrane with a chitosan coating) – can be prepared to remove Cr (VI) and Pb from EPE. The optimum condition with MD-1 is noted at pH 3.5 and 300 KPa applied pressure where 65% Cr (VI) and 68% Pb removal is feasible with MD-1. On the other hand, with the MD-2 type, at the optimum condition (pH 3.5 and pressure 300 kPa), 81% Cr (VI) and 93% Pb could be removed. A ceramic membrane with a reduction in pore size due to a coating of chitosan has been found effective for removing metal ions Cr (VI) and Pb from EPE. The pH and the applied pressure have been found to have an effect on removing these metals' cations. After a coating of chitosan, the Pb removal increased to 93% from 68% and Cr (VI) removal to 81% from 65%.

BIBLIOGRAPHY

Akbal F, Camci S. Copper, chromium and nickel removal from metal plating wastewater by electrocoagulation. Desalination. 2011; 269(1–3):214–222.

Arzani M, Mahdavi HR, Sheikhi M, Mohammadi T, Bakhtiari O. Ceramic monolith as microfiltration membrane: Preparation, characterization and performance evaluation. Applied Clay Science. 2018;161:456–463.

Choudhury PR, Majumdar S, Sahoo GC, Saha S, Mondal P. High pressure ultrafiltration CuO/hydroxyethyl cellulose composite ceramic membrane for separation of Cr (VI) and Pb (II) from contaminated water. Chemical Engineering Journal. 2018;336:570–578.

Ghosh D, Sinha MK, Purkait MK. A comparative analysis of low-cost ceramic membrane preparation for effective fluoride removal using hybrid technique. Desalination. 2013;327:2–13.

Golder AK, Dhaneesh VS, Samanta AN, Ray S. Electrotreatment of industrial copper plating rinse effluent using mild steel and aluminum electrodes. Journal of Chemical Technology & Biotechnology. 2009;84(12):1803–1810.

Jana S, Purkait MK, Mohanty K. Removal of crystal violet by advanced oxidation and microfiltration. Applied Clay Science. 2010;50(3):337–341.

Jana S, Saikia A, Purkait MK, Mohanty K. Chitosan based ceramic ultrafiltration membrane: preparation, characterization and application to remove Hg (II) and As (III) using polymer enhanced ultrafiltration. Chemical Engineering Journal. 2011;170(1):209–219.

Kaplan-Bekaroglu Sehnaz S, Gode S. Investigation of ceramic membranes performance for tannery wastewater treatment. Desalination and Water Treatment. 2016;37:17300–17307.

Kobya M, Demirbas E, Dedeli A, Sensoy MT. Treatment of rinse water from zinc phosphate coating by batch and continuous electrocoagulation processes. Journal of Hazardous Materials. 2010;173 (1–3):326–334.

Murthy ZV, Chaudhari LB. Application of nanofiltration for the rejection of nickel ions from aqueous solutions and estimation of membrane transport parameters. Journal of Hazardous Materials. 2008;160(1):70–77.

Muthumareeswaran MR, Alhoshan M, Agarwal GP. Ultrafiltration membrane for effective removal of chromium ions from potable water. Scientific Reports. 2017;7:41423.

Piedra E, Álvarez JR, Luque S. Hexavalent chromium removal from chromium plating rinsing water with membrane technology. Desalination and Water Treatment. 2014;53(6):1431–1439.

Sharma D, Chaudhari Parmesh Kumar, Pawar N, Prajapati AK. Preparation and characterization of ceramic microfiltration membranes for removal of Cr (VI) and Pb from electroplating effluent. Indian Journal of Chemical Technology. 2020;27:294–302.

Vasanth D, Pugazhenthi G, Uppaluri R. Biomass assisted microfiltration of chromium (VI) using Baker's yeast by ceramic membrane prepared from low cost raw materials. Desalination. 2012;285:239–244.

15 A Combined Coagulation and Membrane Filtration Approach for Fluoride Removal

*Nitin Pawar, Sandeep Dharmadhikari, Vijyendra Kumar,
Vijay Kumar, and Parmesh Kumar Chaudhari*

CONTENTS

15.1 INTRODUCTION

Membrane separation plays a crucial role in wastewater treatment. Before 1980, technically membranes were not considered for separation purposes. Since then, their application in wastewater treatment has increased, which has created an opportunity for researchers to study and investigate membrane separation as a potential technique in domestic and industrial wastewater treatment. There are various types of membrane separation processes present, which include microfiltration, ultrafiltration, nanofiltration, reverse osmosis, gas separation, electrolysis, dialysis, electrodialysis and pervaporation [1]. None of the processes involve a phase change except for pervaporation, which involves the vaporization of the liquid mixture through a membrane surface [2]. Most membrane

DOI: 10.1201/9781003165019-15

processes require a pressure difference as the driving force for separation; some processes utilize a concentration gradient and an electrical potential for separation [3].

15.1.1 APPLICATIONS OF MEMBRANE FILTRATION

Membrane filtration finds a very wide application in the chemical, food, dairy, and pharmaceutical industries.

Chemical industry: The chemical industry produces a lot of waste, which includes organic and inorganic waste, metal ions, dyes, phenolic compounds, and cyanides. They used membrane filtration to remove these pollutants in order to meet standard water acceptable limit before discharging effluents into the water bodies. Another common application is desalination of water and production of clean and potable water from industrial wastewater.

Food industry: The diverse application of membrane filtration in the food industry covers a very wide area. The most common applications are the concentration of fruit juices, egg white, ashes of porcine and bovine or bone gelatin. The other applications are the clarification of meat brine to exclude the left-out bacteria and reuse of brine; clarification of vegetables and plants, for example, soy, oats and canola; and alcohol separation in wine and beer. Membrane filtration can also find uses in the sugar industry for clarifying and concentrating sugar syrup.

Dairy industry: Membrane filtration is an integral part of the separation and manufacturing industry, exceptionally in the production of dairy ingredients. Its important usage is application to milk, whey and clarification of cheese brine.

Pharmaceutical industry: Membrane filtration is applied for collecting cells from a culture or recovering biomass produced during fermentation, especially in the manufacturing of antibiotics. Membranes also play an important role in the production of enzymes and their concentration before other processes. Membrane filtration also helps improve productivity as well as reduce the human workload and manpower costs.

15.1.2 CLASSIFICATION OF MEMBRANE FILTRATION AND ITS TYPES

Membrane filtration has garnered significant attention in the treatment of wastewater. On the basis of the size of the particles to be separated, membrane filtration is classified as microfiltration, ultrafiltration, nanofiltration and reverse osmosis. Microfiltration has a pore size in the range of 0.1–10 μm and is capable of separating suspended solids, bacteria, sugars, proteins, salts, and low-molecular-weight molecules. The driving force required for microfiltration is the pressure difference in the range of 1–5 bar. Ultrafiltration has a pore size in the range of 5–20 nm and is used to separate heavy metals, macromolecules and suspended solids from inorganic solutions. The driving force required is 1–10 bar. Both nanofiltration and reverse osmosis require a pressure difference of 5–200 bar and are capable of separating small molecules, divalent and monovalent ions and water–solvent mixtures and have proved to be the best and reliable methods in the wastewater treatment. The disadvantage associated with reverse osmosis and nanofiltration is that they require very high pressures [4].

The different membranes used in wastewater treatment are polymeric and ceramic.

15.1.2.1 Polymeric Membranes

Polymeric membranes are thin films of thicknesses 10–100 μm. The various types of polymers that are widely used to fabricate polymeric membranes are polysulfone, cellulose acetate, polyamide, polyethersulfone, polyvinylidene fluoride, polyacrylonitrile, polytetrafluoroethylene, polyetherimide and polypropylene.

Advantages of Polymeric Membranes

1. Available in a wide range of pore sizes varying from microfiltration to reverse osmosis
2. Both hydrophobic and hydrophilic membranes are available
3. Cheaper than ceramic membrane
4. Easy to fabricate and use
5. Ease to scale up

Disadvantages of Polymeric Membranes

1. Low solvent resistance
2. Lower applicable range of pH and hence low corrosion resistance
3. Low temperature ranges
4. Short life span

Recently, modified polymeric membranes have been developed that can be applied to wider pH ranges and that have good corrosion resistance. Also, they are resistant to organic solvents with better industrial applications [5].

15.1.2.2   Ceramic Membranes

Typically ceramic membranes are made of various inorganic materials such as α-alumina, γ-alumina, zirconia, silica, titania, kaolin and others. Compared to polymeric membranes, ceramic membranes possess superior chemical, thermal and mechanical stability. The thickness of ceramic membranes is in the range of 5–8 mm and sometimes higher, depending on the specific applications.

Advantages of Ceramic Membranes

1. Very high corrosion resistance. There are a few chemicals, such as strong acids, for which the ceramic membranes do not have high corrosion resistance. The ceramic membrane also has a strong ability to tolerate high doses of chlorine.
2. Applicable to wider pH ranges (0.5–14)
3. Applicability to wider temperature ranges (350–500 °C). As a result, they are used in industrial-scale separations without any feed preconditioning steps.
4. Longer life span (5–10 years)
5. Less fouling tendency
6. Inertness to common solvents and chemicals
7. Higher mechanical strength

Since ceramic membranes do not get damaged by the nature and frequency of cleaning, they can be subjected to intensive cleaning regimes and agents, which is very predominant in industrial manufacturing units.

Disadvantages of Ceramic Membranes

1. Frequently, ceramic membranes are available in the micro-and ultrafiltration range of pore diameter 0.010–10 mm
2. Comparatively higher cost. Although the price of ceramic membranes may have reduced in due course of time, these costs have not been very competitive with polymeric membranes
3. They are brittle in nature. If dropped or subjected to undue vibrations, they may be damaged [6].

Ceramic membranes find wide application in wastewater treatment. The literature reports a lot of studies on the use of ceramic membranes for dye removal [7, 8], treating oil emulsion [9–10] and removing toxic metals [11, 12]. In its early stages, the research on ceramic membrane fabrication was mainly focused on utilizing an expensive precursor to fabricate the membrane [13–17]. The cost of these inorganic precursors is higher, and therefore, it contributes significantly to the operating cost of membrane for industrial applications. This drives the need of utilizing cheaper materials such as apatite powder [18], fly ash [19], natural raw clay [20], and kaolin [21] for membrane fabrication. Belouatek et al. investigated the optimum membrane formulation combination using inorganic precursors such as clay (21 wt.%), feldspar (20 wt.%), kaolin (35 wt.%) and sand (24 wt.%) (all on a dry basis) for fabricating membrane supports capable of wastewater treatment [22]. Of these precursors, quartz, feldspar and pyrophyllite could be expensive compared to kaolin, ball clay and calcium carbonate. Few studies have reported the synthesis of membrane supports using a combination of different clays [23, 24]. Some researchers also examined coal fly ash (a by-product of coal combustion in thermal power plants), a good candidate for preparing low-cost ceramic membranes due to its high alumina and silica content. Bose and Das suggested the use of sawdust as a pore former instead of the conventional calcium carbonate for the manufacturing ceramic membranes. Although calcium carbonate is commonly used as a pore former in laboratory-scale fabrication, its use in industrial-scale manufacturing can enhance the fabrication cost of membranes [25]. To overcome this, several researchers have investigated alternative raw materials, like sawdust, as pore formers [26]. Economically, sawdust is advantageous over conventional pore formers as it can provide the highly porous structure required for efficient separation. In addition, sawdust can provide good performance with the desired workability and strength for the membrane as it has been used as an alternative for cement in concrete mixes [27, 28]. Bose and Das examined that the sawdust has been used in various applications such as filters, ceramic bricks and membrane support, among others, under a controlled fabrication cost.

This chapter focuses on the application of the ceramic membrane for fluoride removal. Fluoride is an essential element occurring in minerals, geochemical deposits, and natural water systems. It enters food chains through either plants or cereals that are eaten or drinking water [29]. Generally, a small quantity of fluoride is added to drinking water to prevent dental caries [30]. Minerals rocks, volcanic activities and phosphate fertilizers used in agricultural and industrial activities, such as clays in ceramics, are responsible for the presence of fluorides in ground- and surface water. The presence of fluoride in drinking water has many health benefits unless it exceeds a certain limiting concentration. According to the World Health Organization (WHO), the maximum limit of fluoride in potable water is below 1.5 mg/L [31]. For concentrations higher than1.5 mg/L, it may cause the mottling of the teeth, thyroid and liver, whereas concentrations from 3 to 6 mg/L cause skeletal fluorosis [31]. High concentrations of fluoride in water may also cause neurologic manifestation, depression, male sterility and painful skin rashes. It also affects the intelligence of kids [32, 33].

Various physicochemical techniques have been extensively studied for removing fluoride from water, such as coagulation, electrocoagulation, filtration, flocculation, chemical precipitation, ion exchange, and adsorption and membrane technology. Among these, coagulation is an effective and simple technique for removal of fluoride from drinking water and industrial wastewater [34]. Other techniques widely studied in literature are electrocoagulation and adsorption. The adsorption has also potential applications in the treatment of toxic and volatile organic compounds, natural organic matter and inorganic pollutants [35].

In this chapter, the application of a combination of coagulation and membrane filtration for fluoride removal has been presented. Coagulation alone is not efficient because the flocs formed by coagulation enhance water turbidity; therefore, membrane filtration can be used as a complementary technique for reducing turbidity within the limits. The coagulation of prepared membrane and the cost estimation of the synthesized membrane have been also carried out based on the raw materials utilized for membrane preparation.

15.2 PREPARATION OF THE MEMBRANE

15.2.1 RAW MATERIALS

The inorganic precursors, such as kaolin ($Al_2Si_2O_5(OH)_4$), sodium metasilicate ($Na_2SiO_3 \cdot 9H_2O$) and boric acid (H_3BO_3), used for preparing the membranes were made by Loba Chemie Pvt. Ltd, Mumbai. The clay used for membrane synthesis were collected from the NIT Raipur campus. Rice husk was obtained from a local rice mill located in Raipur, India. Sodium fluoride was procured from Loba Chemie Pvt. Ltd, Mumbai. Alum rock was used as a coagulant.

15.2.2 THERMAL MODIFICATION OF RICE HUSK

Rice husk is used as a pore former. It is a lignocellulosic material containing cellulose, hemicellulose and lignin. It can be used as a pore former in membrane preparation as it easily forms pore by removing cellulose, hemicellulose and lignin compounds during the sintering of membranes and it is cheaply available. First, the raw rice husk was washed with deionized water to remove any clay, sand and rock impurities and then dried at room temperature for 24 h. It was then dehydrated at 120°C for 12 h in a silica crucible placed in a muffle furnace followed by heating at 250°C for 24 h [25]. The burnt rice husk was then crushed in a ball mill to get a fine powder. Finally, the powder was dried and sieved with a 200-BSS sieve. This fine powder was used as a pore former in a ceramic membrane.

15.2.3 MEMBRANE SYNTHESIS

Four membranes having different compositions were prepared by varying the quantities of inorganic precursors. These precursors used are clay, kaolin, rice husk, sodium metasilicate and boric acid, and the synthesized membranes are named as MD1, MD2, MD3 and MD4. The detailed composition of membranes is given in Table 15.1.

The various precursors used in the preparation of ceramic membranes provide different functional attributes. Kaolin is responsible for low plasticity and high refractory properties to the membrane. Rice husk acts as a pore former. It dissociates into smaller compounds during sintering of membranes and releases carbon dioxide (CO_2) gas. The path followed by this released CO_2 is responsible for the porous texture in the ceramic membrane and contributes to the membrane porosity. Boric acid enhances the mechanical strength by the formation of metallic metaborates at sintering temperatures and also provides dispersion properties. Sodium metasilicate binds all the membrane's elements by creating silicate bonds among them and acts as a binder. It also provides high mechanical strength in the membrane [36].

TABLE 15.1
Composition of Membranes on a Dry Basis

Materials	Name of the membrane			
	MD1	MD2	MD3	MD4
	Weight % (dry basis)			
Clay	80	70	60	20
Kaolin	10	15	15	20
Rice husk	5	10	15	20
Boric acid	2.5	2.5	5	5
Sodium meta silicate	2.5	2.5	5	5

To prepare a membrane disc, a uniaxial compaction method was applied, and the desired amount of dry finely ground raw material mixture was placed in a pestle mortar and mixed thoroughly. The mixture was then placed in a circular stainless steel mold, and the powder was uniformly layered inside the mold. The mold was then placed in a hydraulic press operated manually under a required pressure for 1 minute. As a result, a membrane disc was formed. Finally, membrane disc was removed from the mold and sent to the muffle furnace for sintering. In a lab scale, 30 g of finally ground material was taken to make a 50-mm-diameter, 7.5-mm-thickness membrane.

15.3 PREPARATIONS OF CERAMIC MEMBRANE

The ceramic membrane had the following properties.

15.3.1 POROSITY

Archimedes's principle was used to evaluate the porosity and structural density of the membranes, where the volume of the wetting liquid that displaces air in a dry membrane is measured. Total porosity (ε) was evaluated using Equation 15.1 [37].

$$\varepsilon = \frac{M_w - M_d}{M_w - M_a} \times 100, \tag{15.1}$$

Where M_w is the weight of the membrane in the wet condition, M_d is the dry weight of the membrane and M_a is the suspended weight of the membrane in water. Variation in the porosity of the membranes with increasing in composition of rice husk (5–20%) and decreasing composition of clay from 80% to 50% was obtained from experimental work. The porosity of membranes sintered at various temperatures depend on three factors, such as composition of raw materials, decomposition of rice husk and sintering temperature. The porosity of MD1 was found to vary in the range of 33.62% to 41.29% when the temperature was changed from 550 to 750°C and 41.29% to 48.12% for membranes MD1 to MD4 sintered at 750°C. The increase in porosity with increase in temperature is due to the decomposition of rice husk. The higher the amount of rice husk, the higher is the porosity, as particles are loosely packed, that is, less densification of membranes.

15.3.2 PURE WATER PERMEATION

Permeation is a one of the important properties of membranes. The pure water flux in the batch process was measured experimentally for all the membranes. The results are reported for different applied air pressures, that is, 50, 100, 200 and 300 kPa (Figure 15.1). It was observed that the water flux increased with applied air pressure for all membranes. The flux also increased due to an increase in pore size and porosity of membranes, as seen in the present case. The pure water flux increased to 99.2 from 3.84 L h^{-1} m^{-2} for MD1, 16 to 144 L h^{-1} m^{-2} for MD2, 28.8 to 256 L h^{-1} m^{-2} for MD3 and 48 to 352 L h^{-1} m^{-2} for MD4 when the air pressure was increased from 50 kPa to 300 kPa. A similar trend for increased in flux with applied pressure has been also reported by Ghosh et al. [37].

15.3.3 HYDRAULIC PERMEABILITY AND AVERAGE PORE SIZE

Hydraulic permeability (L_p), and average pore radius (r_h) of the membrane discs can be estimated from the pure water flux data, assuming the pores are cylindrical and parallel. The Hagen-Poiseuille expression can be used for this as shown by Equation 15.2 [37].

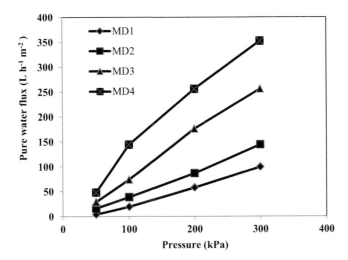

FIGURE 15.1 Variation of the membrane permeate flux with applied pressure at room temperature.

$$J = \frac{n\Pi r^4 \Delta P}{8\mu l} = L_p \times \Delta P, \tag{15.2}$$

where J (L h^{-1} m^{-2}) is the liquid flux through the membrane, L_p (m Pa^{-1} s^{-1}) is the hydraulic permeability, r_h is the hydraulic radius, μ (kg m^{-1} s^{-1}) is the viscosity of the liquid, l (mm) is the pore length, ΔP (kPa) is the applied air pressure and ε (nπr^2)is the porosity of the membranes.

$$r_h = \sqrt{\frac{8\mu l/Lp}{\varepsilon}} \tag{15.3}$$

The hydraulic permeability of membranes are varied with the composition of membranes and with the variation in applied air pressure. The permeability has increased from 0.344 to 1.468 L h^{-1} m^{-2} kPa^{-1} for membranes MD1 to MD4 sintered at 550°C, and 0.172 to 1.176 L h^{-1} m^{-2} kPa^{-1} for membranes MD1 to MD4 sintered at 750°C.

The average pore sizes of membranes change with the composition of the inorganic precursors and the sintering temperature. The membranes MD1, MD2, MD3 and MD4 had an average pore size of 0.228, 0.364, 0.488 and 0.506 µm, respectively, sintered at 550°C, whereas at 750°C, an average pore size of 0.016, 0.232, 0.436 and 0.468 µm, respectively, was noted. This change in average pore sizes was due to the less densification of membranes as the composition of inorganic precursors, binding material and pore-former changed for MD1 to MD4.

15.3.4 CHEMICAL STABILITY AND MECHANICAL STRENGTH

The membrane should have nonreactive. The chemical stability is analyzed using HCl (pH 1) and NaOH (pH 13). For this, the membranes are kept in different pH solutions for 15 consecutive days at atmospheric conditions. First, the weight of the membranes before keeping them in contact with acid and base solutions was measured. Then the membranes were left in contact with the acid and base solutions for 15 days under atmospheric conditions. Thereafter, the wet membranes were dried, and the weights of dried membranes were measured. The difference in the weights of membranes before and after the acid and base treatment gives the weight loss. In laboratory experiments, the weight loss for all membranes for both acid and base solutions was found to be less than 4%. The

TABLE 15.2
Chemical Stability Test Results for Membranes Sintered at 750°C

a. 0.1 N HCl solution pH 1

Membrane disk	Sintering temperature	Initial weight (W1) gm	Final weight (W2) gm	Weight loss = (W1 − W2) = ΔW	Weight loss (%) = (ΔW/ W1) *100
MD1	750°C	24.31	23.59	0.72	2.96
MD2	750°C	24.55	23.8	0.75	3.05
MD3	750°C	25.11	24.36	0.75	2.98
MD4	750°C	24.12	23.28	0.9	3.74

b. 0.1 N NaOH solution pH 13

Membrane disk	Sintering temperature	Initial weight (W1) gm	Final weight (W2) gm	Weight loss = (W1 − W2) = ΔW	Weight loss (%) =(ΔW/ W1) *100
MD1	750°C	24.31	24.13	0.18	0.74
MD2	750°C	24.55	24.36	0.19	0.77
MD3	750°C	25.11	24.81	0.3	1.19
MD4	750°C	24.12	23.81	0.31	1.28

weight loss of the membranes is presented in Table 15.2a and 15.2b. Membranes MD1, MD2, MD3 and MD4 show a weight loss of up to 4% in the acidic media and 2% in the basic media.

Similarly the membrane should have proper mechanical strength. To know their strength, the membranes were subjected to compressive stresses in a tensile machine until cracks appeared to know the maximum applicable pressure. In the laboratory-prepared membranes, MD1, MD2, MD3 and MD4 were found to a flexural strength of 60.48, 54.43, 45.36 and 36.28 MPa, respectively.

The membrane formulation pressure affects both porosity and the average pore size of the membrane. In theMD1 membrane, when different formulation pressures 60, 110, 160, 210 and 260 kN has applied, the porosities obtained at these pressures are 45.78, 44.61, 43.78, 42.43 and 41.29, and average pore sizes obtained are 1.235, 0.785, 0.641, 0.452 and 0.167, respectively. Both the porosity and average pore size were reduced to 41.29% and 0.167 μm from 45.78% and 1.253 μm when applied pressure changed to 260 kN from 60 kN. These changes are due to the strong binding of particles with increasing membrane formulation pressure. In the laboratory-prepared membrane, MD1 had a smaller pore size compared to other membranes. There was no significant change in the porosity and permeability for membranes MD1, MD2, MD3 and MD4. Also, they had good chemical and mechanical resistance. Membranes were also characterized for surface texture, elemental composition, and phase analysis.

15.3.5 MEMBRANE CHARACTERIZATION

15.3.5.1 Scanning Electron Microscopy and Energy-Dispersive X-ray Spectroscopy

Scanning electron microscopy (SEM) analysis was performed to analyze the morphology and change in surface texture with the variation in the composition of the membrane material. The SEM analysis for membranes MD1 and MD2 is shown in Figure15.2a. A surface with a rough morphological structure and no cracks or surface defects can be seen in MD1, whereas a more defected surface

(a)

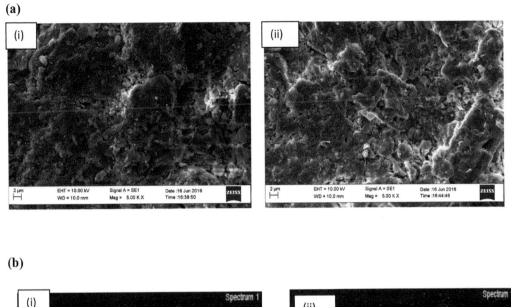

(b)

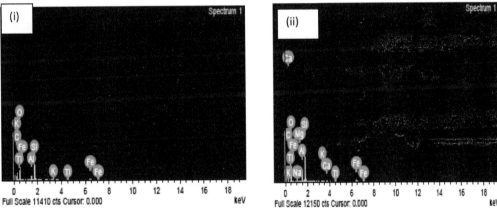

(c)

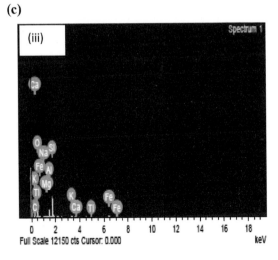

FIGURE 15.2   Analysis of membranes: (a) SEM of (i) MD1 and (ii) MD2; (b) EDX of (i) clay (ii) MD1 (iii) MD2.

is seen in MD2, which is due to less densification of particles. The overall observation reveals that there was a significant change in the surface texture of both membranes. This is due to the change in the composition of the raw materials used for preparing the membrane.

The composition of ceramic membranes can be determined using a dispersive X-ray spectroscopy (EDX) machine. EDX analysis of clay and sintered membranes MD1 and MD2 are presented in Figure15.2b, which shows the elemental composition. The major elements in clay were found to be C, O, Al, Si and Fe, while K, and Ti were present in traces. An elemental analysis of membranes MD1 and MD2 confirms the presence of C, O, Al and Si as the major elements and Na, Ca, Fe, K, Ti and Mg as traces. The Si was the dominant element present in MD1 and MD2, with a maximum percentage of 29.93% and 24.81%, respectively. There was no significant variation in the elemental analysis of membranesMD1 and MD2. C% decreased to 8.16% and 6.29% from 36.92% in MD1 and MD2, respectively, compared to that of clay. This was due to the presence of inorganic precursors along with clay for preparing the membranes.

The preceding analysis shows that membrane MD1 had better, superior properties compared to the other membranes. Membrane MD1 had a small pore size, good porosity, permeation rates, high mechanical strength and better surface texture.

15.4 FLUORIDE REMOVAL

As membrane MD1 was found to have better characteristics compared to other membranes, its utilization for the removal of fluoride from water is presented. The effect of various operating parameters such as pH, coagulant dose and initial concentration of F− for its removal is presented.

15.4.1 Effect of pH

The F− removal can be increased by adding a coagulant on F− being water. The purpose of adding alum is to create flocs. The size of fluoride ions (F−) is less than μm; thus, it is not retained by a ceramic membrane, but when a coagulant is added, it creates flocs in which F− are entrapped. F− also neutralize by cation of alum, and the pH effect to flock formation. A study was performed to evaluate the impact of pH in presence of 2 g/L alum coagulant. Figure 15.3a reflects the removal of fluoride at different pH. The minimum concentration of fluoride reached was 0.78 mg/L at pH 6 from an initial fluoride concentration of 10 mg/L; thus, it is the optimum pH. Fluoride removal was decreased with an increase in pH from 6 to 10. The figure also shows that fluoride removal strongly depended on the pH level. The reason for such a large variation in pH is due to the nature of floc formation at different pH levels. The maximum fluoride removal obtained is 92.2 % at a pH of 6.

15.4.2 Effect of Coagulant Dose

Figure15.3b illustrates the impact of coagulant dose on the removal of fluoride from water at optimum pH of 6. The fluoride removal increased with an increase in the dose and attained a maximum of 92.2% at 2 g/L of alum; then there was no change in removing fluoride by increasing the coagulant dose. After an optimum dose of coagulant, its efficiency decreased due to the charge reduction. After a certain dose of coagulant, pollutant removal remained constant or decreased, which has been also reported by many authors.

15.4.3 Effect of Initial Concentration

The impact of initial fluoride concentration on its removal has been studied. The removal of fluoride at different initial concentrations, that is, 5, 10, 15, 20 and 30 mg/L was performed at a fixed pH of 6 and a coagulant dose of 2 g/L. The percentage of fluoride removed decreased as the initial concentration increased. Removal of fluoride has been found to be 93, 92.1, 85.65, 82.06 and 77.1% for

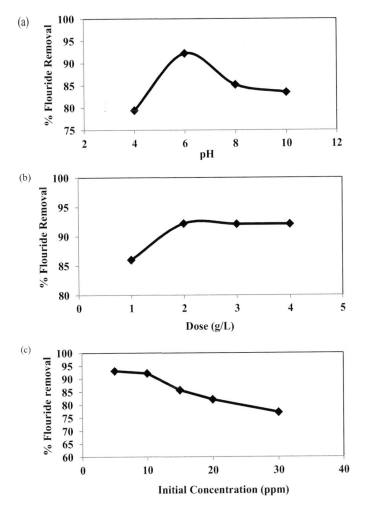

FIGURE 15.3 Removal of fluoride using alum: (a) Effect of pH; (b) effect of mass loading; (c) effect of initial concentration of F−.

initial concentrations of 5, 10, 20, 30 and 40 mg/L, respectively. As high concentration of fluoride it passes through the membrane and resulted in less percentage removal. The industrial effluent contains much fluoride but less in ground water; thus, the fluoride removal data at different doses of initial concentration is necessary.

15.4.4 STATISTICAL ANALYSIS AND MODELING FOR FLOURIDE REMOVAL

The process optimization for fluoride removal at different pH levels, initial fluoride concentrations and coagulant dose are presented. These are the variables that considerably affect the removal of species in the membrane separation process. Table 15.3a gives the chosen variable and its level. The encoded values, along with a set of data used for statistical analysis and the corresponding percentages of fluoride removal, are given in Table 15.3b.

The Box-Behnken design (BBD) has been used to model the experimental data using response surface methodology (RSM). The experimental runs are conducted for design and statistical analysis and the final equation obtained in terms of coded factors for % fluoride removal is expressed

TABLE 15.3

a. Process Parameters and Their Levels for the Removal of Fluoride

Variables	−1	0	1
pH, A	4	6	8
Initial conc (mg/L), B	5	10	15
Coagulant Conc. (mg), C	1	2	3

b. Design of RSM and Its Actual and Predicted Values for Fluoride Removal

Standard order	pH	Initial conc (mg/L)	Coagulant Conc. (2 g/L)	% F- removal actual	%F- removal predicted
1	8	15	2	85.2	86.85
2	6	15	1	88.8	88.90
3	4	15	2	74.05	75.044
4	4	10	1	77.15	76.05
5	4	10	3	72.16	73.91
6	8	10	2	81.32	84.28
7	6	5	3	90	89.89
8	6	15	3	83.15	80.39
9	8	10	1	81.86	80.10
10	6	5	1	84.21	86.96
11	8	5	2	83.21	82.21
12	6	10	2	92.2	92.2004
13	4	5	2	88.89	87.23
14	6	15	2	85.65	90.55
15	6	5	2	93	94.33

c. Analysis of Variance for Percentage of Fluoride Removal Quadratic Model

Source	DF	Sum of square	Mean square	F	P
Regression	9	618.159	68.68	11	0.008
Linear	3	67.131	113.99	18.26	0.004
A	1	23.052	224.194	35.91	0.002
B	1	28.539	1.412	1.99	0.218
C	1	15.54	108.642	17.40	0.009
Square	3	447.068	149.023	23.87	0.002
A^2	1	316.382	340.992	54.61	0.001
B^2	1	1.83	0.226	0.04	0.857
C^2	1	128.856	128.856	20.64	0.006
Interaction	3	103.96	34.653	5.55	0.048
AB	1	70.812	70.812	11.34	0.020
AC	1	0.429	0.429	0.07	0.804
BC	1	32.718	32.718	5.24	0.071

% fluoride removal = $-3.6175 + 25.7988A - 1.95625B + 28.9388C - 2.4025A^2 + 0.0099B^2 - 5.9075C^2 + 0.42075\ AB - 0.16375\ AC - 0.572\ BC$ (4).

TABLE 15.4
Membrane Raw Material Cost

Material	Usage (g)/ membrane area	Cost/ 500 g (Rs)	Cost/g (Rs)	Cost/ membrane Disc (Rs)
Clay	24	0	0	0
Kaolin	3	220	0.44	1.32
Rice husk	0.75	0	0	0
Boric acid	0.75	250	0.5	0.375
Sodium metasilicate	2.5	320	0.64	1.6
Total				3.295

by Equation 15.4. The predicted values mentioned in Table 15.3b, which were determined from Equation15.4, are very close to the experimental values, which also confers validity to the model.

The results of the experimental BBD at different variable levels and the percentage of fluoride removal predicted for individual experiments are presented in Table 15.3b. Both graphical and statistical tests were used to predict the validity of the model using p value, determination coefficients, and a lack-of-fit test. The analysis of variance showed that the developed model is statistically significant (p value <0.05; Table 15.3c).

15.4.5 ESTIMATION OF MEMBRANE COST

The material cost of the prepared membrane MD1 has been estimated in terms of Indian currency (rupees) per square meter of membrane. Thirty grams of raw materials, including clay and other materials, in their respective compositions, were taken as a base for preparing a 50-mm-diameter, 7.5-mm-thick membrane. The diameter and thickness of the membrane were measured using a vernier caliper. The total surface area of the membrane was calculated assuming a cylindrical shape. The cost of all chemicals per gram was known, which gives the total material cost required for membrane fabrication. The cost of materials and the total surface area of the membrane lead to the material cost for preparing a ceramic membrane per square meter. The raw material cost for preparation of per square meter of membrane area is given in Table 15.4.

This 3.295 INR was for 30 g of material that was used to prepare a 5-cm disc (19.6 cm^2) membrane. The cost to prepare 1-m^2 sheet membrane was $\frac{3.295}{19.6} \times 100 \times 100 = 1681\, INR$.

This is for laboratory-grade chemicals, for commercial, chemicals that less costly are required.

15.5 CONCLUSION

A low-cost ceramic membrane can be successfully prepared. Characteristics like permeation properties, with chemical stability, mechanical strength, surface texture and elemental composition of the membrane, were presented. The best membrane (MD1) was used to remove fluoride from water. Upto 92.2% F-removal was achieved with a coagulant dose of 2 g/L. The total material cost of the synthesized membrane was about INR 1681 per m^2 of membrane.

BIBLIOGRAPHY

[1] I. Pinnau, B.D. Freeman, Formation and modification of polymeric membranes: Overview, ACS Symp. Ser. 744 (1999) 1–22.

[2] G. Jyoti, A. Keshav, J. Anandkumar, Review on pervaporation: Theory, membrane performance, and application to intensification of esterification reaction, J. Eng. (United Kingdom). 2015 (2015) 1–24.

[3] M.A.A. El-Ghaffar, H.A. Tieama, A review of membranes classifications, configurations, surface modifications, characteristics and its applications in water purification, Chem. Biomol. Eng. 2 (2017) 57.

[4] K. Nath, Overview of membrane separation processes, in: Membr. Sep. Process. (2011) 5–9.

[5] H. Lin, Y. Ding, Polymeric membranes: Chemistry, physics, and applications, J. Polym. Sci. 58 (2020) 2433–2434.

[6] F. Lin, S. Zhang, G. Ma, L. Qiu, H. Sun, Application of ceramic membrane in water and wastewater treatment, E3S Web Conf. 53 (2018) 4–7.

[7] E. Zuriaga-Agustí, E. Alventosa-deLara, S. Barredo-Damas, M.I. Alcaina-Miranda, M.I. Iborra-Clar, J.A. Mendoza-Roca, Performance of ceramic ultrafiltration membranes and fouling behavior of a dye-polysaccharide binary system, Water Res. 54 (2014) 199–210.

[8] K.M. Majewska-Nowak, Application of ceramic membranes for the separation of dye particles, Desalination. 254 (2010) 185–191.

[9] J. Fang, G. Qin, W. Wei, X. Zhao, L. Jiang, Elaboration of new ceramic membrane from spherical fly ash for microfiltration of rigid particle suspension and oil-in-water emulsion, Desalination. 311 (2013) 113–126.

[10] B.K. Nandi, A. Moparthi, R. Uppaluri, M.K. Purkait, Treatment of oily wastewater using low cost ceramic membrane: Comparative assessment of pore blocking and artificial neural network models, Chem. Eng. Res. Des. 88 (2010) 881–892.

[11] S. Jana, A. Saikia, M.K. Purkait, K. Mohanty, Chitosan based ceramic ultrafiltration membrane: Preparation, characterization and application to remove Hg (II) and As (III) using polymer enhanced ultrafiltration, Chem. Eng. J. 170 (2011) 209–219.

[12] A.E. Pagana, S.D. Sklari, E.S. Kikkinides, V.T. Zaspalis, Microporous ceramic membrane technology for the removal of arsenic and chromium ions from contaminated water, Microporous Mesoporous Mater. 110 (2008) 150–156.

[13] K.A. DeFriend, M.R. Wiesner, A.R. Barron, Alumina and aluminate ultrafiltration membranes derived from alumina nanoparticles, J. Memb. Sci. 224 (2003) 11–28.

[14] Y. Yoshino, T. Suzuki, B.N. Nair, H. Taguchi, N. Itoh, Development of tubular substrates, silica based membranes and membrane modules for hydrogen separation at high temperature, J. Memb. Sci. 267 (2005) 8–17.

[15] T. Tsuru, Inorganic porous membranes for liquid phase separation, Sep. Purif. Rev. 30 (2001) 191–220.

[16] C. Falamaki, M.S. Afarani, A. Aghaie, Initial sintering stage pore growth mechanism applied to the manufacture of ceramic membrane supports, J. Eur. Ceram. Soc. 24 (2004) 2285–2292.

[17] Y.H. Wang, T.F. Tian, X.Q. Liu, G.Y. Meng, Titania membrane preparation with chemical stability for very hash environments applications, J. Memb. Sci. 280 (2006) 261–269.

[18] S. Masmoudi, A. Larbot, H. El Feki, R. Ben Amar, Elaboration and characterisation of apatite based mineral supports for microfiltration and ultrafiltration membranes, Ceram. Int. 33 (2007) 337–344.

[19] R. Ben, Elaboration and characterisation of fly ash based mineral supports for microfiltration and ultrafiltration membranes, 35 (2009) 2747–2753.

[20] N. Saffaj, M. Persin, S.A. Younssi, A. Albizane, M. Bouhria, H. Loukili, H. Dach, A. Larbot, Removal of salts and dyes by low $ZnAl_2O_4$-TiO_2 ultrafiltration membrane deposited on support made from raw clay, Sep. Purif. Technol. 47 (2005) 36–42.

[21] M.C. Almandoz, J. Marchese, P. Prádanos, L. Palacio, A. Hernández, Preparation and characterization of non-supported microfiltration membranes from aluminosilicates, J. Memb. Sci. 241 (2004) 95–103.

[22] A. Belouatek, A. Ouagued, M. Belhakem, A. Addou, Filtration performance of microporous ceramic supports, J. Biochem. Biophys. Methods. 70 (2008) 1174–1179.

[23] F. Bouzerara, A. Harabi, B. Ghouil, N. Medjemem, B. Boudaira, S. Condom, Elaboration and properties of zirconia microfiltration membranes, Procedia Eng. 33 (2012) 278–284.

[24] S. Jana, M.K. Purkait, K. Mohanty, Preparation and characterization of low-cost ceramic micro filtration membranes for the removal of chromate from aqueous solutions, Appl. Clay Sci. 47 (2010) 317–324.

[25] S. Bose, C. Das, Preparation and characterization of low cost tubular ceramic support membranes using sawdust as a pore-former, Mater. Lett. 110 (2013) 152–155.

[26] P. Turgut, H. Murat Algin, Limestone dust and wood sawdust as brick material, Build. Environ. 42 (2007) 3399–3403.

[27] A.U. Elinwa, Y.A. Mahmood, Ash from timber waste as cement replacement material, Cem. Concr. Compos. 24 (2002) 219–222.

[28] F.F. Udoeyo, P.U. Dashibil, Sawdust ash as concrete material, J. Mater. Civ. Eng. 14 (2002) 173–176.

[29] S.S. Tripathy, J.L. Bersillon, K. Gopal, Removal of fluoride from drinking water by adsorption onto alum-impregnated activated alumina, Sep. Purif. Technol. 50 (2006) 310–317.

[30] Meenakshi, R.C. Maheshwari, Fluoride in drinking water and its removal, J. Hazard. Mater. 137 (2006) 456–463.

[31] International Standards for Drinking Water, Occup. Environ. Med. 29 (1972) 349–349.

[32] I.M. Umlong, B. Das, R.R. Devi, K. Borah, L.B. Saikia, P.K. Raul, S. Banerjee, L. Singh, Defluoridation from aqueous solution using stone dust and activated alumina at a fixed ratio, Appl. Water Sci. 2 (2012) 29–36.

[33] N.J. Miller-Ihli, P.R. Pehrsson, R.L. Cutrifelli, J.M. Holden, Fluoride content of municipal water in the United States: What percentage is fluoridated?, J. Food Compos. Anal. 16 (2003) 621–628.

[34] W.X. Gong, J.H. Qu, R.P. Liu, H.C. Lan, Effect of aluminum fluoride complexation on fluoride removal by coagulation, Colloids Surfaces a Physicochem. Eng. Asp. 395 (2012) 88–93.

[35] A. Addoun, W. Bencheikh, L. Temdrara, M. Belhachemi, A. Khelifi, Adsorption behavior of phenol on activated carbons prepared from Algerian coals, Desalin. Water Treat. 52 (2014) 1674–1682.

[36] B.K. Nandi, R. Uppaluri, M.K. Purkait, Preparation and characterization of low cost ceramic membranes for micro-filtration applications, Appl. Clay Sci. 42 (2008) 102–110.

[37] D. Ghosh, M.K. Sinha, M.K. Purkait, A comparative analysis of low-cost ceramic membrane preparation for effective fluoride removal using hybrid technique, Desalination. 327 (2013) 2–13.

For Product Safety Concerns and Information please contact our
EU representative GPSR@taylorandfrancis.com Taylor & Francis
Verlag GmbH, Kaufingerstraße 24, 80331 München, Germany